Oil in the 21st Century

Oil in the 21ˢᵗ Century:

Issues, Challenges and Opportunities

Edited by
Robert Mabro

Published by the Oxford University Press
For the Organization of the Petroleum Exporting Countries
2006

OXFORD

UNIVERSITY PRESS

Great Clarendon Street, Oxford OX2 6DP

Oxford University Press is a department of the University of Oxford.
It furthers the University's objective of excellence in research, scholarship
and education by publishing worldwide in

Oxford New York

Auckland Cape Town Dar es Salaam Hong Kong Karachi
Kuala Lumpur Madrid Melbourne Mexico City Nairobi
New Delhi Shanghai Taipei Toronto

with offices in

Argentina Austria Brazil Chile Czech Republic France Greece
Guatemala Hungary Italy Japan Poland Portugal Singapore
South Korea Switzerland Thailand Turkey Ukraine Vietnam

Oxford is a registered trade mark of Oxford University Press
in the UK and in certain other countries

Published in the United States
by Oxford University Press Inc., New York

British Library Cataloguing in Publication Data
Data available

Library of Congress Cataloguing in Publication Data
Data available

Cover design by Alaa Al-Saigh
Typeset by Philip Armstrong, Sheffield, United Kingdom
Printed by Ueberreuter Print and Digimedia, Vienna, Austria

ISBN 0-19-920738-0   978-0-19-920738-1

1 3 5 7 9 10 8 6 4 2

CONTENTS

LIST OF FIGURES AND TABLES

Chapter 4

Figures

Chapter 5

Figures

Tables

Chapter 6

Figures

Chapter 7

Figures

Tables

Chapter 8

Figures

Chapter 9

Figures

Tables

ix

CONTRIBUTORS

Thomas S. Ahlbrandt is World Energy Project Chief at U.S. Geological Survey

Olivier Appert is President of Institut Français du Pétrole

Bassam Fattouh is Reader in Finance and Management, School of Oriental and African Studies, University of London

Andrew Gould is Chairman & CEO, Schlumberger Limited

Karl S. Lackner is Ewing-Worzel Professor of Geophysics at Columbia University, New York

Robert Mabro is a Fellow of St Antony's College, Oxford University

Benito Müller is Senior Research Fellow at the Oxford Institute for Energy Studies and Managing Director of Oxford Climate Policy

Philippe Pinchon is Director, Moteur Energie, Institut Français du Pétrole

Adnan Shihab-Eldin is Director of Research, OPEC Secretariat and Acting Secretary General in 2005

FOREWORD

It is a well established fact that oil played a critical role in transforming the economies of today's industrialized countries during the last century. It powered the engines of economic growth, hastened technological innovations, expanded production possibilities and increased productivity in those economies. In doing so, it significantly increased the wealth and raised the living standards of the peoples of those societies. There is a consensus, both in the oil industry and among energy experts and leading specialized energy agencies and organizations, that hydrocarbons, especially oil, will make similar contributions in the twenty-first century, in particular to the emerging economies of Asia, Latin America and Africa.

Indeed, this transformation has already started, and we believe that it should be encouraged. The successful transformation of these economies will help promote a more equitable global economic system, reduce poverty and other injustices affecting developing countries and establish a fairer foundation for interdependence between the nations of the North and those of the South, as was envisaged in the first Solemn Declaration of OPEC Sovereigns and Heads of State drawn up in Algiers some thirty years ago.

To sustain this laudable process, and indeed accelerate it, for the benefit of the world economy, in particular developing countries' economies, energy availability is critical. It is, therefore, essential that all stakeholders address the imminent challenges currently facing the oil industry, as well as potential future challenges, to ensure that adequate supplies to fuel growth are available. This is a collective responsibility, to which all parties, producers as well as consumers, developed as well as developing economies, must contribute, if it is to be achieved.

The challenges facing the oil industry in the twenty-first century are a lot more daunting than those of the previous century. Concern has been expressed regarding the investment challenges facing the oil industry. The availability of future supplies has also given cause for concern, thereby raising questions as to the wisdom of continued dependence on oil as the dominant energy source driving world economic growth in the present century. Global environmental concerns have become considerably more pronounced, and there are many more. Yet, these challenges are not insurmountable, as technological advances in the last few decades have proved with regard to the development of cleaner and more environmentally friendly fuels from oil.

As to the fear expressed regarding future availability of oil, the results of work carried out by serious, specialized organizations, such as the US Geological Survey, do not justify that fear. The record shows that there are sufficient hydrocarbon resources under the ground, in terms of proven reserves, reserve growth and new discoveries, to ensure that the world is adequately supplied with its oil requirements well into the first half of the twenty-first century, and perhaps for even longer. There is enough oil to ensure that the process of transforming the developing economies is sustained. The challenge facing stakeholders is how best to harness this existing resource, and make it readily available to those who need it. There are also other challenges facing the industry today – and more may arise in the future. What is required is determination on the part of all stakeholders to take whatever steps are necessary to bring this oil to the market. Knowing and agreeing on what this means will require some frank dialogue among all stakeholders, producers as well as consumers, investors as well as governments.

OPEC strongly supports this process of dialogue and exchanging ideas and thoughts which is already taking place under the auspices of various organizations, such as the International Energy Forum (IEF), with headquarters in Riyadh, and the recently initiated joint OPEC-EU Dialogue. A similar forum for dialogue is being established between OPEC and China. In addition, collaboration on technical issues as well as on joint efforts to improve oil and energy data are ongoing on the part of the OPEC Secretariat, the International Energy Agency (IEA) and other specialized international and regional organizations. Cooperation among producers – OPEC and non-OPEC – has long existed and has been enhanced over the years, with many leading non-OPEC producers attending OPEC Ministerial Conferences as observers, testimony to the realization that oil market stability is a responsibility to be shared by all players in the market.

Ample fora clearly exist to facilitate dialogue among all stakeholders. It is also the case that interest in dialogue among stakeholders has increased in recent years, evidenced by the regular increase in the number of participants at the biennial International Energy Forum. What remains is the need to ensure that this dialogue is not only sustained, but also addresses, in frank and open discussions, the legitimate concerns of all parties and the solutions they envisage, critically evaluating all possible options, in order to identify solutions to the problems facing the industry.

It is in this spirit, benefiting from dialogue to address the current and future challenges facing our industry and seek timely solutions, that OPEC sponsored the production of this seminal work *Oil in the 21st Century Issues, Challenges and Opportunities*.

The editor of the collection of articles in this book is the renowned authority on oil economics, Robert Mabro, CBE, who headed the Oxford Institute for Energy Studies for over twenty years, until his retirement in 2003, and who was also the first recipient of the biennial OPEC Award, which honours distinguished individuals who have made an outstanding contribution to knowledge of the petroleum industry and oil-related issues. The essays provide critical analyses and new insights into such contemporary, important and sometimes contentious subjects as, inter alia, Global Petroleum Reserves, the Oil Price Regime, Technological Developments in Enhanced Oil Recovery, Technological Progress in Motor Engines, the Environment and Climate Change, Renewable Energy, and the Investment Challenge. Their authors come from both consumer and producer nations, from academia and from the industry, including international oil companies, which wide variety of backgrounds is reflected in their essays.

As noted by the editor in his introduction to the book, although produced by OPEC, this book is not about OPEC and the Organization does not necessarily subscribe to the views of its authors. In fact, on some topics, OPEC holds views that differ strongly from those of the authors. In keeping with the spirit of dialogue, however, which the Organization believes is critical in addressing the problems currently facing the industry and with which it will most likely be confronted in the future OPEC has provided this forum for the expression of divergent views.

At this point, I should like to restate OPEC's statutory objectives, whose origins date from the establishment of our Organization, some 45 years ago. These are a) the coordination and unification of the petroleum policies of our Member Countries and the determination of the best means of safeguarding their interests, individually and collectively; b) devising ways and means of ensuring the stability of prices in the international oil markets with a view to eliminating harmful and unnecessary fluctuations, and c) giving due regard to the interests of the producing nations and to the necessity of securing a steady income for them, an efficient, economic and regular supply of petroleum to consuming nations, and a fair return on their capital to those investing in the petroleum industry.

It is, therefore, clear from these objectives that OPEC can have no interest in either very high or very low oil prices. OPEC would rather see oil prices sustained at levels that are reasonable, that consumers can afford and that are conducive to robust, global economic growth, attractive for investors, and securing reasonable income for the development needs of the producing nations.

OPEC's commitment to supporting measures that will promote market stability was again emphasized in the OPEC Long-term Strategy, adopted by the OPEC Conference at its 137th Meeting in September 2005.

The strategy explicitly recognizes the important role oil plays in the world economy, recalls and further delineates the Organization's objectives, identifies the key challenges it faces currently and in the future, and explores scenarios for the energy scene. In so doing, it covers such important elements as the oil price, upstream and downstream investment, technology, the role of OPEC members' national oil companies, multilateral agreements and negotiations related to energy and relationships with both producers and consumers, as well as with international organizations and institutions.

Within OPEC, the strategy calls for Member Countries to strengthen cooperation in upstream and downstream scientific research and technological development, both among themselves and with international institutions. It advocates supporting research in the production and use of cleaner petroleum-based fuels and taking an active role in the development of technologies that address climate change concerns, while improving and expanding the role of oil in meeting future world energy demand, such as carbon dioxide sequestration.

Regarding oil prices, the strategy builds upon the fundamental recognition that extreme price levels, either too high or too low, are damaging for both producers and consumers, and points to the need to be proactive under all market conditions. It also re-emphasizes OPEC's commitment to supporting market stability and stresses the role of other producers and consuming countries in achieving this, especially in the downstream sector. Hence, dialogue among producers, and between producers and consumers, constitutes a crucial element of the strategy, which recommends that such dialogue be widened and deepened to cover more issues of mutual concern, such as security of demand and supply, market stability, investment, technology and the downstream sector, as well as policy issues.

I should like to express my sincere appreciation to Robert Mabro, the editor of this book, and all the eminent contributors for producing what is inarguably OPEC's major contribution to date to the literature on oil. I should also like to thank the staff of the OPEC Secretariat, led by the Director of the Research Division and Acting Secretary General,

Dr Adnan Shihab-Eldin, for coordinating the production of the book. Finally I should like to express my gratitude to Kuwait Petroleum Corporation for making available the funds for this project.

Sheikh Ahmad Fahad Al-Ahmad Al-Sabah
President of the OPEC Conference,
Minister of Energy of the State of Kuwait, and
Chairman, Kuwait Petroleum Corporation.

October, 2005

CHAPTER 1

INTRODUCTION

Robert Mabro

1.   The Significance of Oil

Oil involves a double dependence. All countries depend on oil as the fuel of choice in the transport sector. Cars, trucks, aircrafts and ships use almost exclusively oil products to fuel their engines. The exceptions lie in vehicles, usually taxis or buses, which have been adapted for natural gas or LPG as a substitute; and in cars with engines designed to burn alcohol as in Brazil. Nevertheless oil continues to dominate in a sector which is rightly considered as the network of blood vessels of the economic organism. Any problem that affects the transport sector severely damages the economy and in extreme cases can bring it to a complete standstill.

In this sense, oil is a strategic commodity.

In turn, the oil-exporting countries are themselves vitally dependent on oil revenues for their economic development and the welfare of their population. All oil-exporting countries with the exception of Norway, Canada, Russia and very few others[1] are part of the third world. Many have large populations and are therefore poor in terms of per capita incomes. The few with high average incomes face a long-term dependence on hydrocarbon export revenues because of the paucity of other resources and the time needed to diversify the economy. To establish the basis of self-sustained economic growth independently of oil is often a very long-term challenge. For these countries oil is indeed a strategic commodity.

This double dependence is asymmetrical however. Countries that depend on oil imports for a large part or the whole of their require-ments worry about the security of supply. They seek ways, often but not always successfully, to reduce the demand for oil through policy

---

1   The UK was a net oil exporter for more than two decades but is at present on the point of becoming a net importer.

Table 1:   Gross National Income per Capita, US Dollars. 2004

OPEC Member Countries:

Algeria	2,623
Indonesia	1,187
IR Iran	2,480
Iraq	891
Kuwait	19,587
SP Libyan AJ	5,013
Nigeria	549
Qatar	45,953
Saudi Arabia	10,677
United Arab Emirates	32,235
Venezuela	4,050
OPEC average	2,196

Non-OPEC Developing Countries.Oil Exporters:

Oman	7,890
Mexico	6,770
Malaysia	4,650
Gabon	3,940
(Russia)	3,410
Kazakhstan	2,260
Ecuador	2,180
Colombia	2,000
Egypt	1,310
Syria	1,190
Angola	1,030
Yemen	570

World Average	6,280
Range	56,230−90

Sources: For OPEC Member Countries, *OPEC Annual Statistical Bulletin 2004.* For non-OPEC countries, World Bank, *World Development Indicators data base,* 15 July 2005.

measures that enhance energy efficiency or substitute other fuels for oil. Countries whose economies are highly dependent on oil export revenues, on the contrary, worry about the security of demand, or more precisely of their exports. Hence, their desire to stabilize the oil market by holding a certain volume of surplus capacity which can be used to meet an unexpected surge in demand or to compensate for a supply disruption, and thus prevent prices from exploding with detrimental effects on the world economy and on the long-term demand for oil. This is not always possible of course as much depends on the volume of surplus capacity available and the sentiments and behaviour of traders in the futures markets where the marker prices for oil in international trade (WTI and Brent) are formed today.

Hence, the critical importance of oil that calls for continuing research and debate to enhance understanding of fundamental issues and challenges.

2.   The Current Oil Situation

The current oil situation may be succinctly described as follows. On the *demand* side growth in the period beginning in the mid-1970s until 2004 has been lacklustre. Thus oil consumption increased from 55.990 million barrels per day (mb/d) in 1974 to 80.76 mb/d in 2004,[2] at an average annual compound rate of 1.19 percent. In the same period world primary energy consumption grew from 6.07 billion tonnes of oil equivalent (toe) in 1974 to 10.2 billion toe in 2004, that is at an average annual compound rate of 1.8 percent. The growth rate differential implies that the share of oil in the energy slate has been declining during this period. The change was from an oil share of about 45 percent in 1974 to 36 percent in 2004. The shift in shares was largely due to the substitution of fuel oil by coal, gas, nuclear and hydro in power generation and steam-raising in industry; and of middle distillates by natural gas in other applications. Oil retained however its dominance in the transport sector, the area where technological developments – their nature, rate and the timing of their introduction – are most critical to its future.

The determinants of oil demand are income (the rate of economic growth), price, the technology of oil-using vehicles and appliances, changes in consumers' tastes which are partly correlated to income (the proxy for standards of living), government energy policies and the availability of competitive substitutes for uses where substitution is technically feasible.

Clearly, income growth will raise the demand for oil in *ceteris paribus* conditions. To give one example, people become able to travel more when they are better off as is evident with the significant growth of air travel in recent decades. Changes in real prices paid by final consumers, the resultant of changes in international prices, indirect taxes and inflation rates, will influence demand depending on the direction, extent and duration of the price movements and the price elasticity (small in

---

2   The data are from BP *Statistical Review of World Energy*.

the case of automotive fuels but high in uses where oil has substitutes). Technical progress in vehicles, aircrafts, ships, oil-using industrial plants and so on has been energy saving on the whole, but the potential oil demand reduction is being partly compensated for by the increased attraction of bigger, more powered cars and SUVs. Policies that subsidize substitutes, encourage the use of more efficient energy equipment and appliances, introduce specifications in their manufacturing, or heavily tax petroleum products are among those which have an impact on oil demand. Finally, as mentioned earlier, petroleum products have ready substitutes in most sectors other than transport, and this may have been the most significant determinant of declines in demand.

Demand for oil is the highest in OECD countries both in absolute terms and per capita.[3] However, broadly speaking, the growth in oil demand has been relatively low in OECD countries and much higher in the developing world, particularly in the new industrialized countries of Asia. The latter is due to the economic development process itself which goes through stages and involves structural changes. The advanced countries have reached a development stage where the shift away from non-commercial energy has already taken place in all of them. In many, a shift from solid to liquid fuels has also occurred. As importantly, per capita incomes in industrial countries have reached long ago the level that leads to the massive penetration of private cars in the economy. Finally, an important structural change has occurred in advanced countries – a relative shift away from industry toward services. The latter are less energy intensive than the former. In developing countries an income threshold is attained sooner or later which leads to the introduction on a large scale of modern means of transportation and other energy using appliances. The migration of industry from advanced to developing countries causes demand for energy, including oil, to grow at a faster rate in parts of the third world. And, as mentioned earlier, a shift in the composition of the energy slate away from traditional to modern fuels occurs, slowly in certain instances and very suddenly in others. Yet, the relatively high rates of growth of both energy and oil in developing countries should not divert attention from the fact that a large proportion of the world population still lacks access to modern energy.[4]

On the *supply* side, the changes in the OPEC share of global oil production went through a cycle in the period 1974−2003. Significant

---

3   In 2004, the OECD share of world oil consumption was 59.8 percent

4   To give one example, according to the IEA, *World Energy Outlook 2004*, 1.6 billion persons in the world have no access to electricity.

increases in non-OPEC production at a time when world demand was either stagnant or growing very slightly were accommodated by an equivalent reduction in OPEC's output. The reason is that non-OPEC producers are volume maximizers while OPEC, having assumed the role of defending an administered price (between 1973 and 1985), became the residual oil supplier to the world. Thus OPEC's production share which stood at about 52 percent in 1974[5] was as low as 30.7 percent in 1986, but then recovered to 41 percent in 2004.

Important changes also occurred in the regional structure of non-OPEC production in the past 30 years. The second half of the 1970s saw the arrival of new producers, the most significant being the UK and Norway in the North Sea, Mexico and Alaska. Oil production in the USSR increased until the early 1980s reaching a peak of 12.520 mb/d in 1983. A huge decline set in after the break-up of the Soviet Union, with oil production of the FSU falling to 7.035 mb/d in 1996.

This fall in production was matched by a similar decline in consumption however, and thus turned out to be close to neutral in terms of the world supply/demand balance. This declining trend was reversed in the year 2000; and in the five years to 2004 oil output in the Russian Federation increased by a mammoth 3.1 million barrels per day. This trend was not sustained in 2005 however, and it is unlikely that Russia will be able to add annually 500,000–700,000 barrels per day in the coming years as it did in 2000–2004. One reason is that, more than in the recent past, growth now depends on work to be done on new and difficult fields.[6]

Non-OPEC production will undoubtedly continue to grow for a number of years particularly in the deep offshore, the Caspian, Angola, other West Coast African countries, and at a reduced rate in Russia. Yet, in some countries production decline has already set in. This is the case for the UK, Oman, Syria, Egypt and US onshore for example. The moot question is about the share of the demand increment that will be supplied by additional production from the non-OPEC region. This parameter is of some importance for OPEC as it determines the

---

5   To enable comparisons with the current situation OPEC here means the OPEC 11. Data are taken from BP *Statistical Review of World Energy*. These data initially covered only oil except for the USA for which NGLs were included. Later the BP production data included NGLs (where it is recovered from natural gas separately). This distorts the comparisons with earlier years.

6   To be sure there are still technology-driven prospects for increasing production in the mature fields responsible for the achievements of 2000–2004. But they are unlikely to add on their own 600,000–700,000 barrels per day in the coming years.

size of the future call on OPEC and has a bearing therefore on its Member Countries' investment plans.

During the period covered by this discussion, OPEC production moved from some 30 million barrels per day in 1974 to 27 mb/d in 1980 after which significant production falls occurred year after year until 1985. Thus, in 1981, OPEC's oil output had declined to 22.8 mb/d, by more than 4 mb/d compared with the previous year. In 1985, a bottom was hit at 16.8 million barrels per day. This huge reduction in the production volume was in the nature of a shock leading to the 1986 events. This shock had a more significant adverse economic impact on the OPEC nations, and indeed on all the oil-exporters of the third world, than the much talked about 1973 shock on the advanced countries of the North. OPEC's production did recover after 1985 however. In 1992, production (including NGLs) reached 26 mb/d. It peaked at 31 mb/d in 2000 and again at almost 33mb/d in 2004. Let us note that during this period production in some OPEC Member Countries suffered from disruption or stagnation because of wars or oil sanctions imposed by the UN or the USA. The country distribution of OPEC production has thus varied from time to time, and on occasions in a significant way.

Two important issues fall under the supply heading. The first is the *surplus capacity* issue and the second the *peak oil theory*.

There is no doubt that a production system that churns out more than 80 million barrels per day of a commodity of varying qualities that is transported round the globe between production points and markets, and that in addition has to be processed in heavy plants to yield final products must involve a certain amount of surplus capacities in all its stages. This is necessary for the smooth operation of such a system given the inevitable fluctuations in supply and demand that arise in the short term.[7] A host of factors cause these fluctuations from changes in weather conditions which have an impact, albeit in different ways, on both supply and demand to industrial relation problems causing strikes that shut down production for a while, technical accidents and geo-political crises. Every engineer or production manager knows that.

Ideally one should have a surplus capacity equivalent to 5 percent of average world production in the upstream and 7.5 percent in refineries. In the latter case the greater provision is needed because of maintenance shutdowns which are more significant than in the upstream.

---

7   Fluctuations in the medium and long term have to be coped with in a different way, mainly by varying the levels of planned investment/disinvestment.

The difficult question, given that holding surplus capacity is costly, is: Who should bear these costs? The issue is rarely, if at all, discussed. As regards the *upstream*, the world has become accustomed to the idea that OPEC has an interest in holding surplus capacity as this enables it to stabilize the market in certain circumstances. The fact, however, is that a large volume of surplus capacity – perhaps 12mb/d or more – emerged (a large part of it involuntarily) in the 1980s as a result of the increase in non-OPEC production and falling demand. Over the past 20 years this volume has been slowly declining.

By 2004, the market began to worry about the adequacy of the volume available given concerns, whether justified or not, about supply disruptions. When traders worry the discussion or analysis of the issue that raises concerns becomes very confused. Estimates of the volume alleged to be available are mentioned without reference to the particular capacity concepts to which they are supposed to refer. Every estimate is considered by its proponent as representing the ultimate truth even when the proponent is an outsider who has no way of knowing the truth. An important difficulty relates to the fact that there are several capacity concepts – the rated capacity of a plant as posted by its manufacturer, optimum capacity, surge capacity, capacity with or without an operational tolerance and so on. It is also important to specify a time dimension. The available surplus capacity cannot be fully activated at the drop of a hat. Some of it would have been mothballed and some time is required to bring it back on stream; some will need to undergo some tests before the re-start of operations, and so on. It is thus critical to know about the time profile of capacity re-activation when estimates are produced, discussed or challenged by different authorities and commentators.

We are left with the fundamental questions: Who should carry the costs? Is it possible to distribute the burden equitably between producers and importing countries since both sides have an equal interest in stable markets? And how can we be sure that the required volume will be available? The view of the industrialized countries and their agencies, which seem to have become an implicit part of the conventional wisdom of the past 20 years, is that the role of non-OPEC is to meet as big a part of the world oil demand increment as possible and the role of OPEC is to hold in the upstream the surplus capacity buffer required to cope with small and big emergencies.[8] That was accepted, also implicitly, so long as surplus capacity happened to be there.

8   The strong reluctance of IEA Member Countries to release oil from strategic stocks in previous emergencies supports the point. The exception, of course, was the IEA

Things have changed today. The questions stated here above deserve to be placed on the agenda of discussions that take place within the producer/consumer dialogue.

Privately owned oil companies do not hold surplus capacity upstream as their objective is to maximize the returns to their investments and the flow of commercial profits. Their shareholders demand the maximization of shareholder value and the return to them of monies that cannot be very profitably invested in good projects. This is an imperative that management can only ignore at their peril. This means that they will resist, strong and fast, any suggestion that they should keep some capacity idle in the interest of the public good without proper remuneration or compensation. There are cases, however, where the terms of a concession or contract with an OPEC country in which they have equity oil requires them to agree to a production cut in accordance with an OPEC decision on output ceilings and Member's quotas. These clauses were accepted in the 1970s but they give rise to tough bargaining in negotiating more recent contracts.

The private oil companies have more downstream than upstream capacity. Given that the reference price of oil is determined in futures markets (NYMEX and IPE) and the significant role of the NYMEX, based in New York, the oil situation in the USA, particularly in PADD 1, 2 and 3, has a considerable influence on oil price movements. And it is in the USA that the refinery system is presently constrained. As the winter season approaches the market begins to worry about a possible shortage of heating oil in the cold season. And as the summer approaches the worries shift to the supply of gasoline during the holidays and driving season. That refining capacity is too tight is now widely acknowledged, and the hurricanes, Katrina and Rita, which are not the cause of the fundamental refining problem graphically revealed the vulnerability of the system. Whether the private oil companies have the intention to invest in order to solve the refining problem is not evident. In any case, the investment lead-in times are so long that relief in the short term cannot be expected.

As the refining problem will beset the world petroleum market either continually or cyclically for many years to come a question arises about the role that Western governments can play through policy and regulations to ensure that companies hold the required surplus capacity in their systems. There is no sign at present that the key IEA member countries are ready or willing to address these problems. So far their

quick response to the dislocation caused by the hurricane Katrina. In that case, however, the disruption took place in the USA, the most important IEA member.

automatic reaction when oil prices rise is to request OPEC immediately to increase its crude oil production. Such requests are at best irrelevant and at worst offensive when the problem lies in refineries and not in the upstream.

The second issue related to supplies is the peak oil theory. Worries about oil depletion are not new phenomena in the history of oil. These worries have surfaced from time to time. They have recently come back to the fore in the works of Campbell and Laherrère. That oil production will reach a peak at some point in the future does not mean, of course, that the world will run out of oil at that point. It only tells us that it will not be possible to supply demand increments if *ex ante* oil consumption wants to grow at prevailing prices. Prices will then have to give way and rise to the levels that choke off demand.

The relevant question is not *whether* oil production will peak at some future date but *when* it will reach this maximum point. Predictions of the date of the peak were initially put by some authors at the year 2000, then at 2005. But as these dates have now passed and oil production is still rising new predictions from the pessimists have ranged from 2010 and 2015.

The assumptions, definitions and arguments used by the proponents of extreme versions of the peak oil theory give rise to criticisms.[9] The oil concept used is narrow as it does not take sufficient account of unconventional oil. In fact, the distinctions between different types of oil resources are becoming increasingly blurred, and attention is focusing on the concept of liquid fuels rather than on the conventional narrow definition of oil. The role of technology which enables increases of the recovery factor from say 30 percent of the oil in place to perhaps more than 50 percent is not given its due. Ultimate recoverable resources have continually increased in the past and are likely to continue to do so.[10] It is true that as we are moving away from easy to difficult oil costs of production rise. But is this a problem? Oil prices have most of the time tended to be higher than both average and marginal costs.[11]

The *structure of the oil industry* has been changing significantly in the past 30 years. The de-integration of the system that prevailed until the 1970s in the world outside North America and the communist countries, the result of the nationalization of oil concessions,[12] means

Structure of the industry

---

9   See my discussion of the peak oil theory in Chapter 10.

10 On this point see Chapter 5.

11 Marginal costs here refer to the costs of fields required to meet demand at the time, not the costs of fields that are not needed at that stage.

that the upstream became disconnected from the downstream. Under the old system most crude oil in international trade was moving within the vertical channels of the majors, or between them under long-term contracts. An arm's-length market did exist but was not very liquid. De-integration increased the size of the outside market, caused it to grow initially at a high rate and to develop new types of transactions such as the 15-day Brent forward physical contract, oil futures and a range of other derivatives – different types of options, swaps and Contracts for Differences for example. This complex market structure has implications for the determination of oil prices, an issue discussed later.

The private companies emerged in the late 1970s with a heavy baggage of downstream assets (including tankers and refineries) the outcome of earlier investment decisions when world oil demand was rising annually by 6 or 7 percent year after year. They were entering a new world in the 1980s, the 1990s and until the early 2000s where the rates of oil demand growth stood at between 1.0 and in the best years 2.0 percent. Their responses were to close refineries wherever that was possible and not prohibitively expensive, to seek cost cutting and synergies through mergers, to savagely cut employment and other costs through restructuring and to focus emphatically on upstream projects whose profitability is tested at unrealistically low oil prices, rather than on downstream where historically returns have been low.

To prefer upstream to downstream investments is perfectly consistent with the logic of a system that privileges shareholders at the expense of other objectives or considerations. The relevant consideration as mentioned at the beginning of this chapter is that oil is a strategic commodity not so much today as a means of war, but as the fuel that keeps the economy's wheels turning given its dominance in the transport sector. In times of crisis oil is a public utility. A shortage of petroleum products elicits strong reactions from those who depend on the supply of diesel or gasoline for their trips to work or to the super-market. Similarly all those in industry and commerce who depend on the smooth supply of goods and commodities will blame both the oil industry and their government for the emergency and its consequences. On these occasions governments have to intervene. They could release strategic stocks of petroleum products. European countries and Japan hold such stocks, but paradoxically the USA does not. Private oil companies cannot be relied upon to release oil from their inventories

---

12 These happened over time at different dates in different countries. In some cases the foreign private oil corporations retained under new agreements some equity oil.

in a crisis unless forcibly mandated to do so. The reason, obviously, is that it does not make commercial sense to part with stocks if prices are expected to rise as the crisis unfolds.

As regards crude oil the 'public utility' aspect is *continually* covered by the existence of surplus capacity in OPEC and particularly Saudi Arabia. In extreme emergencies the strategic stocks held by IEA countries are brought to bear as in the aftermath of the devastating hurricane Katrina that disabled both upstream and downstream facilities in the US Gulf of Mexico region. Yet, OPEC cannot do very much when the problems, as happened recently, arise in the downstream.

Private companies abhor regulation. Their failure to recognize effectively the public utility dimension of the oil issue can be damaging, even to their own interests, in certain circumstances.

The world petroleum market has developed to a degree of complexity that defies attempts to seek transparency and appropriate understanding. Marker prices are now taken directly or indirectly from futures exchanges where the transactions are for financial instruments denominated in oil. The determinants of price movements have much to do with traders' perceptions about the future state of the supply/demand balance but they are also influenced by the need to optimize returns of a portfolio of diverse financial instruments. Positions held with other oil derivatives also have an influence. More significantly, a large proportion of trades in paper markets are about price differentials which raises questions about the determination of flat prices.

This market structure and the behaviour of the so-called 'non-commercial' entities which often seem to lead prices in the upswing or the downswing, depending on whether they hold long or short positions, complicates OPEC attempts to stabilize prices through signals about its production policies.

They may not be at this time a feasible alternative to the current oil pricing system in international trade. But this should not distract us from the need to continually assess this system and compare it with alternatives.

3. Challenges

The challenges that the world oil industry – producers and consumers – will be facing in the twenty-first century are not new. They will not suddenly emerge in the years to come. They are already with us, some of them since many years. The previous section mentioned or hinted at a few. In this section further points will be developed.

One set of challenges has a bearing on two most important economic variables – price and investments. As indicated earlier, the price regime for oil in international trade which has been in place since the second half of the 1980s involves much volatility. This is partly because it seeks marker prices in narrow and fairly illiquid spot markets and in futures markets notorious for their tendency to under- and over-shoot in response to news and a variety of small or major shocks. OPEC's attempts to stabilize prices, though sometimes successful, often meet with complex difficulties.

Significant price volatility affects adversely the economic interests of both producers and consumers. The role of prices is to signal relative scarcity or abundance which in turn causes adjustments to the allocation of resources. High volatility confuses the signals and leads to inefficiencies. I have never been convinced by views widely held by traders and finance economists that volatility does not matter because hedging instruments are available. Hedging which seeks to manage given risks by taking on another set of risks may make sense at the micro-level when it applies to a cargo, a project or, as is often the case, price differentials or spreads. But how can those who invest in upstream and downstream projects with a long lead-in time cope with swings that, in the past seven years, moved the WTI price from $10 per barrel in 1998–99 to almost $70 per barrel in September 2005?

A number of uncertainties have an impact on the investment decisions of both national and international companies. Uncertainties affect the determinants of the future course of oil demand as this depends on the rate of growth of the world economy, technological developments in the manufacturing of oil-using vehicles, aircrafts, plants and appliances, on the oil market encroachment of substitutes such as natural gas or nuclear, and on the energy and environmental policies of oil-importing countries. Uncertainties also affect the supply side. Questions such as, 'What is the growth potential of non-OPEC production?' are relevant to the investment decisions of OPEC countries.

There are risks of either under- or over-investing in the upstream. Both have negative consequences albeit of a different type. The former may cause prices to rise to levels that are damaging to the world economy and to the subsequent course of oil demand. The latter could push prices to levels so low as to cause serious hardships to oil-exporting countries of the developing world and Russia through a significant reduction in oil revenues. The private oil companies will also suffer from a reduction in cash flows causing them to slash their capital expenditures budget as they did in 1986 and on some occasions in the 1990s.

Investment in the entire supply chain is necessary. Until very recently, the international oil discourse has focused almost exclusively on the upstream and OPEC's behaviour and policies. Yet, the upstream is only one stage in the oil supply chain, and OPEC is but one party of the world petroleum scene which includes non-OPEC crude oil producers, private oil companies and the governments of oil-importing countries, both industrialized and developing.

The realization that problems lie in the midstream and downstream (tankers and refineries) is just beginning to emerge. It is still weak however. It has not yet induced any significant responses from Western governments and international oil companies. Western governments lost interest in oil in the mid-1980s partly because of the persistence of a glut, and partly because the prevailing economic ideology holds that commodities (and in this view what is oil but a commodity like any other) are for markets to deal with free from government intervention. Some went as far as abolishing their energy ministry (for example, the UK) or radically reducing the size and scope of the department in charge of hydrocarbons (for example France where the famous and much feared DHYCA is now a shadow of its past self). Today most Western governments are not in a position to address oil issues with sufficient knowledge and competence. Things may change in the future but will inevitably take much time. Yet, the solution of the refining problem calls for government policies regarding planning rules, petroleum product specifications and mandatory holdings of spare capacity.

Most private oil companies, although aware of the problem, are not inclined to invest significantly in refining. Their reasons are as follows. First, the historical rates of return on these investments have been low compared with the upstream. Secondly, they do not believe that high refining margins, even somewhat lower than those enjoyed recently, will continue to obtain. Thirdly, they do not want to carry the costs of holding the amount of spare capacity that ensures the smooth operation of the system. In short, refining poses a challenge that requires the co-operation of all the parties involved. But is this forthcoming?

The climate change issue involves a number of difficult challenges. At the core is a set of uncertainties. I am not referring here to uncertainties relating to the scientific verdict. They do exist, of course, because of the complexity of the climate and the factors that determine its changing behaviour. There is however a majority view among the scientific community that the risks of damaging climate change phenomena are too high to be ignored.

The uncertainties I am referring to relate to policy. Which policies are going to be implemented, when will they be implemented and with what short- and long-term effects? Will new international commitments be agreed upon in the future and will they be implemented as agreed? As for all issues that require international co-operation on a wide scale countries find it difficult to design equitable formulae for the distribution of commitments, as these involve costs that everybody seeks to minimize, or to accept in an effective way the principle of 'common but differentiated responsibilities and respective capabilities' on which OPEC and developing countries insist. This principle is stated in the UN Framework Convention on Climate Change that most countries have ratified.

These uncertainties about agreements and policies matter for a large number of issues, not least their impact on climate change itself, but in our context here they clearly have a bearing on the two economic variables of concern – price and investments.

It is well beyond the scope of this introduction to detail the positions of the different parties involved, and to speculate on what they may do or not do. Readers may however be interested in the points of view of developing and OPEC's member countries which are not as widely known as those of the USA, the EU and other industrialized countries.

As part of the international community and stakeholders in a global problem most developing countries, including OPEC, are willing to co-operate and participate in efforts to reduce risks associated with the climate change phenomenon according to a fair share. They remind us however, that economic development is their first and most urgent objective, and point to the fact that they do not have the resources to tackle in a significant way the climate change problem. Furthermore, they cannot accept more than a share of the responsibility for a problem that has largely arisen from past accumulation of emissions from industrialized countries.

OPEC, among others, recognizes however that environmental issues can provide an opportunity for developing counties to leapfrog some of the technological stages that industrialized countries have gone through during the long developmental process. There are also opportunities for co-operation with other international parties in seeking a more efficient use of available energy resources and developing appropriate technologies that serve the climate objective and meet their development needs.

One of the significant uncertainties mentioned earlier on relates to the policies that industrialized countries are likely to implement. The position of some countries has varied from one negotiation round to the next one. Their views about the solution of the climate change problem differ and their sense of urgency is difficult to assess.

An important area of disagreement between industrialized countries on the one hand and developing countries, including OPEC, on the other is that of technology transfer. The industrialized countries argue that the transfer of technology will enable most countries to make significant contributions to the climate change problem. Investment and transfer of technology are construed as private sector activities, and developing countries are urged to use new technologies that the private sectors of the North may develop in order to reduce their emissions.

The developing countries have strong qualifications about this approach. First, they note that the industrialized countries which have access to these technologies have not yet shown much success in limiting emissions as per their commitments. Secondly, there are doubts as to whether technologies developed in the North will be appropriate to the circumstances of the South. Thirdly, there are fears that costs arising from teething problems in the adoption of new technologies will be disproportionately borne by the developing countries.

In short OPEC and developing countries in general want industrialized countries to honour their international commitments, put in place incentives to the provider and developer of new technologies in their countries and finance much needed R&D.

Will the industrialized countries respond in a way that will enable the climate change agenda to move forward?

There are another two most demanding challenges.

The first concerns oil-exporting countries. It is the challenge of economic development. The objective of any oil country is to establish the basis of sustainable economic growth in the long run. To many, oil will continue to provide much needed foreign exchange revenues over years to come. Irrespective, however, of the future evolution of oil prices and export volumes, it is likely that the level or growth of revenues will not be continually sufficient to cope with demographic pressures and increasing development needs. Broadly speaking, oil-exporting countries of the developing world fall in two categories: (a) countries with relatively small but rapidly growing populations but without significant agricultural or mineral resources. Oil and/or gas is their only significant resource, and (b) countries with large, and in some cases very large populations, and agricultural resources. The first group generally enjoy high per capita incomes at present. The diversification of their economies is very challenging however. The obstacles are the limited resource base outside the oil sector and an insufficient endowment of human capital relative to the economic tasks to be undertaken. In the second group, per capita incomes are in some cases very low and oil revenues although they provide some much needed relief are not big enough to act as a powerful engine of growth.

I do not believe in the 'oil curse' theory. All oil-exporting countries, although in different ways and degrees, would be worse off today if they never had oil. It is sufficient to illustrate the point to recall the underdevelopment and dire poverty conditions of many of these countries only 30 or 40 years ago. To say that oil revenues were not always well spent, that some expenditures were very wasteful, is a different point. It should not distract from the fact that what was wisely spent improved health, nutrition, education, housing, public utilities and more generally the standard of living of at least part of the population. I believe, however, that oil reduces the incentive to introduce painful policy reform, but we observe this phenomenon also in non-oil developing countries.

The challenge remains. The economies of oil-exporting countries need to be transformed to ensure a continually improving standard of living for their citizens in years to come and irrespective of the vagaries of oil. The key to success is investment in human capital, a process that takes a long lead time to yield its fruit.

The second challenge is to develop a dialogue between the various parties of the energy scene conducive to meaningful co-operation on fundamental oil issues. Enormous progress has been achieved already on this front with the holding on a bi-annual basis of a conference between oil-exporting and oil-importing countries and the establishment of an International Energy Forum Secretariat in Riyadh. The difficult issues about oil market stabilization and the maintenance of adequate spare capacity along the oil supply chain have not yet been addressed in a way that leads to an agreement on co-operation in policy.

4.   The Book

This book is not about OPEC, its history, policies, or behaviour. It only includes one chapter by the current Acting Secretary General of the Organization, and the topic is an outlook of likely oil developments in the next fifteen years. It involves scenarios and the author's views about recent oil developments. All other chapters have been written by independent authorities.

The objective was to address in a single book a number of issues relating to oil today and to the future of this important fuel. We did not attempt to be comprehensive as this would have resulted in the publication of a small encyclopaedia. The choice of topics was by necessity selective, and I am aware that different editors would each have made a slightly or fundamentally different selection. With more time at our disposal and more resources I would have liked to add a chapter on demand, one on the relationships between gas, nuclear and coal on the one hand and oil on the other, and possibly one on the economic development of oil-exporting countries. And I can hear some people saying: 'and what about non-conventional oil?' We would then be well on our way to produce an encyclopaedia.

The ten chapters of this book fall logically in three parts. This chapter is an Introduction which describes the current oil situation and defines the challenges that producers and consumers already face and will continue to face, with the hope that some of them will be overcome, in the twenty-first century. Chapter Two, by Dr Adnan Shihab-Eldin, looks at world oil as things stand today and at possible developments between now and 2020.

The second part includes two chapters. The first (Chapter 3) on the evolution of the oil price regime over the past fifty years, the operations of the world oil market and the implications for oil price volatility among other things. The author is Dr Bassam Fattouh of the University of London. The investment challenge is the subject of Chapter 4. The issue is of critical importance because of the destabilizing impacts of investment cycles and the difficulties of carrying along the whole supply chain the volumes of surplus capacity required for the smooth operation of the complex oil system. This chapter is the product of a co-operative effort in which Dr Fattouh took the lead.

The third part is comprised of issues that have a bearing on the future of oil. These are (a) the geological resource and reserve base and the technologies that enhance recovery and discovery rates; (b) the environmental problem and the possibilities of solutions provided by carbon sequestration; (c) the likely technical progress in car engines and powertrains; (d) the possible role and impact of renewable sources of energy, the subject of much interest to environmentalists and to those concerned with what will happen 'beyond oil'. Chapter 5 written by Dr Thomas Ahlbrandt of the US Geological Survey emphasizes the significance of reserve growth and presents a wealth of data on resources, reserves, reserves growth and oil to be still discovered. I find in his paper some reassurance that we are not on the point of hitting troublesome problems caused by paucity of resources. Chapter 6 by Andrew Gould, the chairman of Schlumberger, is a useful complement as it details a number of technologies now in use which, among other things, enhance recovery. Naturally, many of the technologies described are the company's but they illustrate what is in use under different patents by other companies in the world oil services industry.

Chapter 7 is on the environment issue which, as argued in this introduction, is not only contentious but represents one of the most significant challenges facing fossil fuels in this century. It was written by Dr Benito Müller of Oxford who has focused selectively on three aspects – the science, the politics and the impact on competitiveness in energy-intensive industrial sectors. Chapter 8 by Professor Klaus Lackner of Columbia University is a fascinating survey of the carbon sequestration issue, technologies and possible policies. Both Müller's and Lackner's views on policy are not identical on certain points to those of OPEC, and this adds to their interest.

Oil dominates in the transport sector. The long-term future of oil depends critically on what technology will do in the future to the engines of cars, trucks, coaches and the likes. Chapter 9 by Olivier Appert, the President of the Institut Français du Pétrole, and his colleague Philippe Pinchon covers the issue in interesting detail. The small Chapter 10 which I contributed on renewables implicitly raises the question: If renewable sources of energy are not the successor to oil in the foreseeable future, which fuel is the heir to the throne? Is it natural gas, nuclear or hydrogen? I have no answer to this question which may have to await another book.

Let me now turn to the pleasant task of thanking the authors for their patience and willingness to contribute within a very tight time schedule to this book. Dr Adnan Shihab-Eldin and Mr Mohamed Hamel gave invaluable support by drawing my attention to errors of facts or loose arguments and supplying references. Dr Omar Ibrahim and his colleagues in the OPEC Secretariat helped with the organization of production. Mr Ramiro Ramirez of the OPEC Secretariat enlightened me on some complex environmental issues. Many others provided useful comments to myself and the authors of various chapters. Let them be sure that we are very grateful to them even though it is impossible to name them all here.

Finally, special thanks are due to HE Sheikh Ahmad Fahad Al-Ahmad Al-Sabah who initiated the project and to Kuwait Petroleum Corporation for providing the grant without which this book would not have materialized.

CHAPTER 2

THE OUTLOOK FOR OIL TO 2020

Adnan Shihab-Eldin

1. Introduction

Since its inception in the middle of the nineteenth century, the modern petroleum industry has had to adapt to many changes in a constantly evolving global landscape, and has had to face repeatedly the risks posed by global challenges and the benefits that, ultimately, can be derived from successfully meeting them. However, the accelerating pace, strength and scope of recent globalization forces and processes are more pervasive, with far-reaching impacts, and are presenting many new, formidable challenges to the way the industry functions. At the same time, they are creating many exciting opportunities for the industry.

Oil has been the leading source of energy since early in the twentieth century, due to its perceived benefits vis-à-vis other types of energy. These relate to accessibility, transportability, versatility and low costs, and, as a consequence of all these, its established, elaborate infrastructure right across the supply-chain. The development of the industry has gone hand-in-hand with that of the global economy, to the benefit, principally, of what are now seen as the advanced consumer societies.

A conscious effort has been made in recent years to set in motion actions that will help spread the benefits of oil across the world community at large, with a view to supporting, in particular, the economic and social development of countries of the South and the eradication of poverty, as well as enhancing the state of the global environment − in line with the plan of implementation from the World Summit on Sustainable Development in Johannesburg in 2002.

These are among the significant developments that have presented new challenges to the way the oil industry − and oil producers, in particular − function in a truly globalized, integrated world. In the foreseeable future, the challenges stem from uncertainties surrounding such important areas as world economic growth, the expansion of non-OPEC supply (especially in the medium term), technological advances, inter-fuel competition, the energy policies of consuming countries and the oil price path. Many analysts believe that the most pressing market challenge is the short-to-medium-term one of reducing bottlenecks across the supply-chain, most especially downstream.

In addition, at a structural level, there are new challenges for OPEC's national oil companies, as they seek to enhance competitiveness and take a more pro-active approach to developing their domestic petroleum sectors, as well as, in many cases, diversifying their activities abroad. This should occur at the same time as accommodating multilateral issues that range high on the international agenda, such as sustainable development, environmental harmony and world trade. It can involve strengthening cooperation with other NOCs within the Organization, as well as with the international oil companies, in accordance with Member Country governments' individual policies and objectives, as well as sovereignty over their indigenous natural resources.

This chapter begins by examining the prevailing market outlook near the end of 2005, notably the rising prices and volatility that have dominated it in 2004 and 2005. It then covers the medium-term prospects, up to 2010, before venturing onto the main topic, which is the long-term outlook through 2020. This latter aspect is based on a scenarios approach, and these have been designed to improve understanding of the nature and impact of the most important drivers affecting the future. However, it is important to remember that these scenarios should not be seen as forecasts. Rather, they constitute a useful tool that helps explore the factors that drive change, and provide an accessible framework for the analysis of long-term issues.

2.   Characteristics of the Oil Market near the End of 2005

When looking at the state of the market towards the end of 2005, it is clear that an unusual combination of factors has been at work over the preceding two years, leading to persistent price rises and volatility. Moreover, there has also been a shift in emphasis from upstream to downstream, with regard to the underlying causes of the present malaise.

The exceptionally high, synchronized world economic growth — reaching 5.1 percent in 2004 — particularly in China and other emerging developing economies, has led to much higher-than-expected oil demand. The year 2003 began with strong oil demand growth, which, in 2004, accelerated to its highest rate for the last 25 years. This has also led to widespread concern about future supply disruptions that may result from increased geopolitical tensions.

While growth has remained at healthy, although more modest and sustainable levels during 2005, the challenge of providing adequate crude supply has been met successfully. This has been largely due to OPEC's actions, in raising its production by more than 4.5 million barrels a day since 2003, as part of its market-stabilization measures. This has, in turn, led to a steady rise in OECD commercial oil stocks, which now exceed their five-year average.

OPEC has also been acting on a second front, with regard to these market-stabilization measures. Its Member Countries have sought to accelerate their plans to bring on-stream new production capacity, with the aim of re-establishing a comfortable level of spare capacity, to help calm markets at a time when demand is expected to continue growing strongly. The expanding capacity, together with the current spare capacity now at 2 mb/d, will be more than adequate to cover oil demand growth throughout this winter and in 2006. Furthermore, much of the new capacity from both OPEC and non-OPEC will, in 2006, consist of lighter crudes, which are much in demand in the market. More increases in capacity have been planned for the rest of the decade, and these are already being implemented.

With regard to supply side developments, the recent non-OPEC production trend has also impacted the market. For the third year in a row, growth in non-OPEC supply has been falling behind that of world oil demand, reversing earlier trends of matching or exceeding demand growth. Nevertheless, while some non-OPEC producers have been experiencing absolute declines in production in recent years, others have seen their output grow. On balance, it is forecast that average annual growth will be around 1 mb/d in non-OPEC supply up to 2010, slowing thereafter.

Combined with the planned growth in OPEC capacity from 32.5 mb/d to more than 38 mb/d by 2010, together with an additional 1.5 mb/d increase in natural gas liquids over the same period, this means that cumulative world oil production capacity will rise by around 12 mb/d by the end of the decade, and this will be well above the expected cumulative rise in demand of 7–8 mb/d. In other words, the expansion of global production capacity will more than cover the forecast growth in demand up to 2010 (Figure 1). Also, the progressive recovery of Iraq should contribute significantly to raising OPEC's overall production capacity.

The second distinct aspect of the recent volatility and high prices

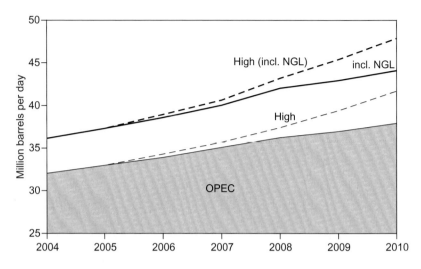

Figure 1: OPEC Production Capacity. Million Barrels per Day

occurs downstream. While the situation upstream has been stabilized and is expected to continue to improve in the coming years, there is a very different picture further along the supply-chain.

The continued serious downstream bottlenecks in some major consuming countries − due mainly to a lack of timely investment and increasingly stringent product specifications − had seen refineries operating at near capacity (Figure 2) to keep pace with rising demand, putting pressure not only on product prices, but also on crude prices. This has been particularly apparent in the post-Hurricane Katrina market, with OECD refinery utilization at over 90 percent, and, in some regions, close to 95 percent. It is clear, therefore, that the industry at large − especially in consuming countries − must pay more attention to the downstream part of the supply-chain, in the interests of overall market stability. In particular, concrete measures should be taken on the part of consuming countries, to create an enabling environment to encourage rapid, sizeable investments in the refining sector, especially in conversion capacity, which has persistently lagged behind market requirements.

However, without adequate and timely measures in the main consuming countries, today's high and volatile oil prices are likely to remain a feature of the market. Writing near the end of 2005, it does not appear that the growth in refinery capacity will match demand growth before 2007 (Figure 3).

Additionally, there has been widespread concern about possible

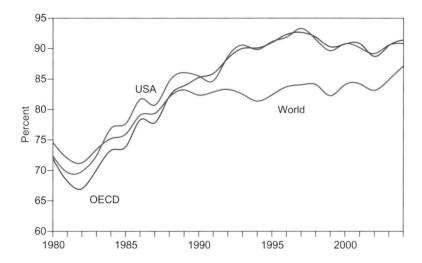

Figure 2: Refinery Utilization Trend 1980–2004. Percent

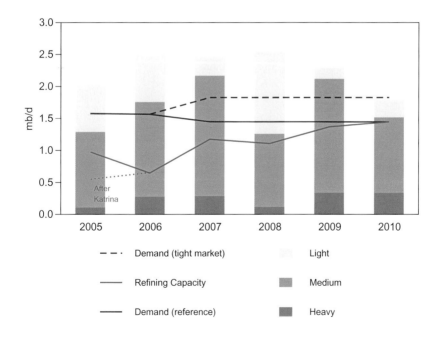

Figure 3: Incremental Product Demand Compared with Crude and Refining Capacity Expansion. Million Barrels per Day

future supply disruptions in areas of the world that are subject to increased geopolitical tensions. These factors have been reflected in increased speculation in futures markets, particularly through a rise in activity by non-commercials, notably pension and trust funds. A clear and strong correlation indeed exists between oil prices and both 'open commitment' and trade volumes in NYMEX oil futures in the past few years. Without doubt, perceived capacity constraints have made the market very nervous and over-responsive to external impulses.

3. Crude Oil Prices in an Historical Context

It is important to put the present prices in their proper historical context. Although prices reached record-breaking levels in nominal terms during 2005, in real terms they have been still well below those witnessed 20 years ago (Figure 4). This is in spite of the fact that oil demand in 2005 has been much higher and the world — that is, the richer part — much wealthier. The world is decreasingly dependant on oil for its economic growth. Globally, oil intensity has fallen by around 50 percent since 1970, due to such factors as advancing technology, improved efficiency, government policies and changing consumer behaviour.

On the other hand, at times of deteriorating exchange rates for the dollar, the producers' income from the sale of oil becomes much

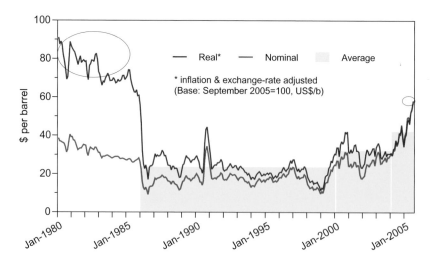

Figure 4: OPEC Reference Basket Price (nominal and real terms, $/b)

lower, especially for those trading with other major world currencies. Similarly, due to existing wide price differentials among the various traded crudes, the heavy crude streams of many OPEC Member Countries attract even lower prices than other widely traded crude benchmarks.

There are also other downsides to excessively high oil prices – in the context of the international oil market. Notably, these can distort the distribution of returns from oil sales. In many countries, particularly in Europe, consumer governments benefit the most, through the receipt of higher levels of extra revenue, from taxation, than either the producers or refiners. Averaged across the last five-year period (2000–04), the oil tax revenues of the G-7 countries were significantly above OPEC's oil export revenues.

OPEC recognizes the need for oil prices to be at reasonable levels that are acceptable to producers and consumers alike. In practice, this is a comfort zone for prices, where they are high enough to give fair returns on the production side and allow adequate reinvestment in future capacity, and low enough to contribute positively to world economic growth.

This was reflected in the price band mechanism being pursued by OPEC in the opening years of this century. Here, the aim was to ensure that OPEC's overall output remained at levels that would keep prices within the limits of the band – set at US $22–28 a barrel – thus coping with the delicate task of riding-out normal market fluctuations, at the same time as maintaining price levels considered acceptable to all parties. This approach was well received by the market and was very successful as a stabilizing force in 2000–03. However, the market conditions from 2004 on – with sufficient supply and rising prices – have rendered the band mechanism inoperable, and therefore it was suspended in early 2005.

Excessive market volatility can have an adverse effect on activity at all levels of the industry and, ultimately, this can have negative repercussions for both producers and consumers. It can, in particular, be highly detrimental to investment in future production capacity, as indeed it can for domestic development in oil-producing countries. If prices are so high that they cannot be sustained, they may well be followed by very low prices and reduced revenues not long afterwards – just as occurred in the 1980s, with dramatic consequences for OPEC Member Countries and their aspirations to sound, sustained economic and social development.

4.   Long-term Oil Outlook

Turning to the longer term, the uncertainties surrounding the fundamental interlinked drivers are even greater. Quantification efforts are important to identify key issues that need to be addressed. To support this, the OPEC Secretariat has adopted a scenario-based approach, depicting contrasting possible futures of the global energy scene. The drivers of change that are shaping each scenario include, among others, the world economy, energy policies and technology.

The Dynamics-as-usual scenario (DAU), envisages a future where the oil market drivers continue past patterns. Economic growth is clearly a major factor behind expected growth in oil demand. The DAU assumes real GDP growth rates similar to median growth observed over the past 15 years, averaging globally 3.6 percent over the period 2004–2020, on a purchasing power parity basis (Table 1). OECD economic growth over the medium-to-long term averages 2.3 percent p.a. and will be driven by relatively modest increases in the capital stock and total factor productivity. On the other hand, with a low capital stock base, considerable technological and labour quality catch-up potential, and continued population expansion, the GDP growth rates in developing countries are expected to be higher than for OECD countries, averaging 5.0 percent p.a. over the next two decades.

Table 1:   Real GDP Growth Rates in the Dynamics-as-usual Scenario. Percent per Annum

	2006–2010	*2011–2015*	*2016–2020*	*2004–2020*
OECD	2.4	2.3	2.2	2.4
DCs of which:	5.1	4.8	4.6	5.0
China	6.5	6.1	5.8	6.5
FSU	3.8	3.3	3.2	3.9
World	3.5	3.4	3.4	3.6

China is expected to increasingly act as an engine of world growth. Over the period 2003–2005 GDP grew by an average of around 9 percent, making it the fastest growing economy in the world. The Chinese government has the target of quadrupling the 2000 GDP by 2020, but this will only be possible if structural problems are solved. Moreover, the competitive advantage, provided by cheap abundant labour and an affordable currency, will be eroded over time. However, although these recent high rates are not expected to be sustained, China is still seen as remaining the fastest growing region. South Asia, with

India and Pakistan accounting for most of the GDP and population in this region, is also expected to remain one of the fastest growing regions of the world, as further reform of the Indian economy brings considerable increases in productivity and capital growth.

Together with these economic growth rates, past trends in efficiency improvements and energy taxation are assumed to continue, with no additional environmentally- or energy security-driven policy measures that impact oil demand growth in the Dynamics-as-usual scenario. Global climate change and local pollution will continue to be dominant concerns but without major impacts on oil demand. Oil prices are assumed to gradually decline from the record levels observed in 2005, settling at real values over the longer term that are higher than average levels seen over the past two decades.

With these assumptions, oil demand is expected to rise in the DAU by 22 mb/d from 2005 to reach 106 mb/d by 2020 (Table 2), representing average annual growth of 1.5 mb/d. OECD countries will continue to account for the biggest proportion of world oil demand as, although growth is slow, it builds on a high base. However, close to 80 percent of the increase in global oil demand will come from developing countries. In per capita terms, there remains a wide gulf in demand patterns, with, for example, North America consuming in 2020 almost seven times more per head than China.

Table 2: Regional Oil Demand in the Dynamics-as-usual Scenario. Million Barrels per Day

	2005	2010	2015	2020
North America	25.6	26.7	27.6	28.4
Western Europe	15.6	16.0	16.3	16.5
OECD Pacific	8.6	8.7	8.8	8.9
OECD	49.8	51.4	52.7	53.8
Latin America	4.6	5.2	5.8	6.4
Middle East and Africa	2.9	3.3	3.7	4.3
South Asia	3.2	4.2	5.3	6.7
South-East Asia	4.4	5.3	6.2	7.1
China	6.9	8.6	10.5	12.4
OPEC	7.0	7.8	8.7	9.5
DCs	29.0	34.4	40.3	46.4
FSU	3.9	4.2	4.4	4.6
Other Europe	0.9	0.9	1.0	1.0
Transition economies	4.8	5.1	5.4	5.7
World	83.6	90.9	98.4	105.9

The transportation sector is particularly important for this growth. The huge potential in developing countries is clear, particularly in China and India; in these two countries, there are just 10−20 vehicles per 1,000 inhabitants, compared with over 500 vehicles per 1,000 inhabitants in the OECD region. Asian countries, home to more than half the world's population, with strong expected economic growth and this large potential for expansion of oil demand, will remain the key source of demand increases in the developing world.

Up to 2020, the region, which has emerged as a key player in global markets for commodities, manufacturing and services, is expected to account for a rise of 12 million b/d in oil demand over the next two decades, which represents two-thirds of the total in all developing countries. Nevertheless, there is substantial uncertainty attached to these figures, for all regions, in particular related to economic growth, policy developments and the rate of development and implementation of new and existing technologies.

In the broader context of total energy demand growth, not only has oil been in the leading position in supplying the world's growing energy needs for the past four decades, but also there is a clear expectation that this will continue at least for the next two decades. The DAU projections foresee, for example, oil's current share of around 40 percent falling to 37 percent by 2020, when it will still be the single most important fuel satisfying the world's demand for energy, which grows at an average of just over 2 percent p.a. over the period 2000−2020 (Figure 5).

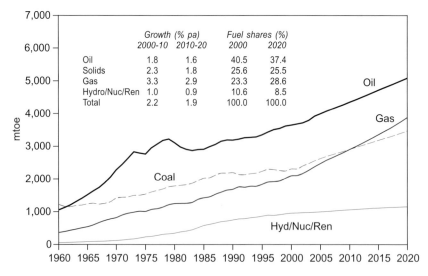

Figure 5: World Energy Demand by Fuel Type. Mtoe

The strongest growth by fuel-type is for natural gas, whose demand increases by an average of more than 3 percent p.a. over the 2000–2020 period, with its share in total energy demand rising from 23 percent in 2000 to 29 percent by 2020; indeed, around the end of the first decade of the twenty-first century, gas is due to become the second most important fuel after oil in satisfying world energy requirements. On renewables, these constitute an important option that needs to be developed at every opportunity, but, in reality, there are still many hurdles to overcome. Starting from a low base, it is likely to be a long time – decades, rather than years – before renewables acquire a significant share of the world energy mix, even with the most ambitious double-digit growth rates. In some consuming countries, there is a rethink about nuclear, but notable increases are not expected for a long time, given the need for enhanced safety and improved safeguards, as well as cost.

From the supply perspective, the first key observation to make is that the resource base is sufficient to satisfy such demand growth. Estimates of ultimately recoverable reserves of oil have been increasing over time with advancing technology, enhanced recovery from existing fields, and the development of new reservoirs. Indeed, at each point in time of estimates published by the US Geological Survey since the mid-1980s, the cumulative global production, as a percentage of the estimated resource base, has been relatively stable, at just under 30 percent (Figure 6). Technology also continues to blur the distinction

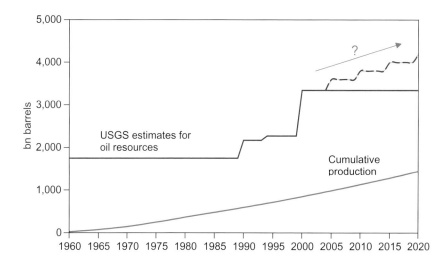

Figure 6: The Resource Base and Cumulative Production

between conventional oil, non-conventional oil and, in the future, other fossil fuels.

Turning to the prospects for non-OPEC supply, along with upward trends from the Former Soviet Union (FSU), three hot spots will stand out: deepwater offshore areas in the US Gulf of Mexico; West Africa, led by Angola; and Latin America, led by Brazil. On the other hand, production from areas such as the North Sea, that has provided the bulk of non-OPEC supply and most of its growth over the past two decades will now plateau and start long-term decline.

Total non-OPEC oil supply is expected initially to continue to increase, rising to a plateau of 56–57 mb/d (Table 3) in the post-2015 period, benefiting from continued advances in technology, increased exploration success and expanded access to new frontier areas. Although counterbalanced with strong decline rates, such as in the North Sea, offshore resource production, particularly in deep water, as well as non-conventional oil from Canadian oil sands, should offer opportunities for growth. The strongest regional decline is expected in Western Europe, with output halving over the next 20 years. Developing countries production reaches around 19 mb/d, while FSU output rises to 15 mb/d. Nevertheless, the evolution of non-OPEC supply will be subject to considerable uncertainty over the medium-to-long term, depending upon a number of factors shaping production profiles, such as technological advances, oil price movements, investment patterns, and exploration success rates.

Table 3:   Non-OPEC Supply in the Dynamics-as-usual Scenario. Million Barrels per Day

	2005	*2010*	*2015*	*2020*
OECD of which:	20.9	20.9	20.2	19.9
USA and Canada	10.6	10.9	10.8	11.0
W. Europe	5.9	5.3	4.4	3.7
DCs of which:	16.1	18.2	19.1	19.4
Latin America	4.2	4.9	5.3	5.6
Middle East and Africa	5.6	6.7	7.2	7.2
Russia & Caspian	11.7	13.1	14.3	14.9
Non-OPEC supply	50.5	54.3	56.0	56.8

Thus, the increases over the medium term from non-OPEC are expected to exceed OPEC output expansion. However, in the longer term, a distinct reversal in this relative contribution to increases in production is expected, with OPEC in the period 2010–2020 supplying over 80 percent of the incremental barrel, leading to a share in world supply more in line with its share of world reserves. In the DAU, the amount of oil required from OPEC, including natural gas liquids and non-conventionals, will gradually increase to around 37 mb/d in 2010 and 49 mb/d in 2020 (Table 4).

Table 4: Oil Supply and Demand in the Dynamics-as-usual Scenario. Million Barrels per Day

	2010	*2015*	*2020*
World demand	90.9	98.4	105.9
Non-OPEC production	54.3	56.0	56.8
OPEC supply (incl. NGL & non-conventional oil)	36.6	42.3	49.1
OPEC market share (%)	40.2	43.0	46.3

Beyond the upstream, it is nevertheless crucial, when looking at future oil market developments, to consider the required expansion along the entire oil supply chain. In this regard, as discussed in the previous section, the tightness of markets in 2005 could be largely linked to insufficient refining capacity. Nevertheless, the DAU assumes that such downstream tightness eases, although this is naturally contingent upon the appropriate investment being made, in particular in the United States, by expanding secondary processing capacity, as well as the development of new grassroots capacity in Asia. This, however, is a major issue, and the uncertain availability of suitable refinery capacity could prove to be a major source of future price volatility.

It is sometimes thought that a major challenge for refiners in the future could stem from the average crude slate becoming heavier and sourer. Indeed, crude oil quality will play an important role in determining the future of the refining sector. A heavier crude slate requires higher conversion capacity to produce the same volume of light products while simultaneously reducing the volumes of residual fuel oil generated. Higher sulphur content puts pressure on necessary investments in desulphurization processes. However, although the biggest increase in crude supply over the next ten years is expected to come from crudes from the Middle East, which are typically slightly heavier and sourer than the current average slate, this will be compensated for by increases in supply of sweet, light crudes from Africa, Brazil, Caspian, as well as synthetic crudes. Although, regionally, significant changes might be expected, on average, no dramatic change in the crude slate is therefore expected.

However, the real challenge for the downstream sector results from the fact that the product demand structure will change, with gasoil/

diesel expected to exhibit by far the largest demand growth. Essentially, all demand growth over the projection period is for light products, with the largest challenge being to produce additional distillate. Additionally, substantial fuel quality improvements are expected in all regions. Gasoline sulphur content should be reduced to levels of 10−50 ppm in OECD regions and 80−150 ppm in most developing countries. Even more dramatic changes are expected for automotive diesel oil.

The question is how the refining system will develop in response to this outlook. For primary processes, the global refinery distillation capacity additions over the next decade are estimated at almost 10 mb/d (Figure 7). Asia will be the main area where expansion will be required to cope with the expected demand growth of that region.

Expansion of secondary capacity will be dominated by further desulphurization, not only because of increasingly stringent product specifications, but also because of the rising volumes of cleaner products and, partially, the sour nature of what will increasingly be the incremental barrel of supply. In addition, more than 6 mb/d of upgrading capacity will be required in order to meet the demand for light products. This includes all major processes like vacuum distillation, hydro cracking, cat cracking and coking. There are, therefore, large investment challenges in the downstream sector in volume terms, and it is possible that this sector could remain a factor of volatility and tightness if these investments do not take place in a timely manner.

This dynamics-as-usual scenario should be regarded only as a reference case from which to assess the uncertainties concerning key drivers,

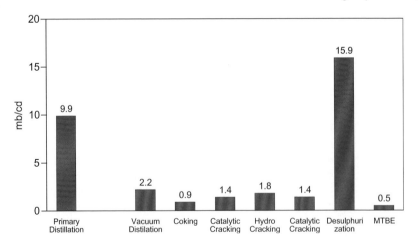

Figure 7: Estimation of Primary and Secondary Processing Capacity Additions. 2015 vs. 2004

as well as identifying potential pressure points and important areas requiring close attention over the longer term. To this end, a further plausible scenario, named Protracted market tightness (PMT), has been developed, which envisages a future where the oil market is characterized by higher demand for OPEC oil as well as higher oil prices.

This future is shaped, in particular, by geopolitical and economic conditions which are growth-inducing. Average economic growth in this scenario is 0.5 percent p.a. higher than in the DAU, at 4.1 percent p.a., leading to stronger oil demand growth, reaching over 111 mb/d by 2020 (Table 5).

Initially slower non-OPEC supply growth compared to the Dynamics-as-usual scenario, contributes to this tighter market, as a result of a more pessimistic view on the application of technologies, access to acreage, fiscal conditions, and so on, although longer-term upward reactions to the higher oil price are clearly possible. As a result of these developments, OPEC output would need to be of the order of 54 mb/d by 2020, almost 5 mb/d higher than in the DAU, demonstrating the potential for a higher increase in the amount of OPEC oil required over the medium-to-long term than the Dynamics-as-usual case portrays. This scenario also serves to highlight even further the potential for the downstream sector to be a major cause of price volatility, given the higher demand levels and consequent higher refinery capacity requirements.

Another scenario that has been developed is called the Prolonged soft market (PSM) scenario, depicting a future where the identified drivers of change are compatible with a relatively low growth in oil demand, resulting in low OPEC supply and prices. In this scenario, the world economy grows at a moderate rate of 3.1 percent. Oil demand is assumed to be strongly managed through enforceable domestic policies

Table 5: World Demand and OPEC Production in three Scenarios. Million Barrels per Day

	2010	2015	2020
World demand			
Dynamics-as-usual	90.9	98.4	105.9
Protracted market tightness	93.1	101.9	111.2
Prolonged soft market	89.2	94.4	99.0
OPEC supply (incl. NGL & non-conventional)			
Dynamics-as-usual	36.6	42.3	49.1
Protracted market tightness	40.4	46.6	53.9
Prolonged soft market	32.7	37.4	42.6

and plurilateral agreements in developed as well as in major developing countries.

The strong downward pressures on demand from lower economic growth and policy developments swamp any upward pressure stemming from the lower prices, to produce the net negative impact upon demand, even in the long term. On the other hand, lower prices could have some negative long-term impacts upon both OPEC and non-OPEC production, as the question of the sufficiency of investment funds becomes a key issue. By 2020, OPEC production is only 43 mb/d in this scenario, more than 6 mb/d lower than the DAU. Even lower levels for the demand for OPEC oil can be considered possible. This demonstrates the tremendous uncertainties confronting the Organization in making investment decisions in an industry characterized by long lead times and uncertainty.

What do these outlook figures imply for investment requirements over the coming two decades? There is clearly a need for expanded production capacity to satisfy the increases in demand, as well as maintain sufficient spare capacity to cope with sudden, unexpected shortages in supply. On top of these investments, however, the natural decline in existing capacities will need to be compensated for to the extent that is economically acceptable, with in-fill drilling, work-overs, reservoir and production management as well as routine maintenance.

Technological developments contributed to finding and development costs falling substantially through the 1980s and early 1990s. This technical progress drove down costs by opening up frontier areas that were previously inaccessible, such as deep offshore, increasing exploration efficiency, as well as making more efficient the development and production from existing areas. Although costs began rising from the end of the 1990s, there is still room in every aspect of the upstream industry to improve costs through continued technological progress. It is also important to distinguish between the costs observable in cyclical tight markets from longer-term expected trends. For example, the cost of expanding capacity in OPEC countries increased by approximately 50 percent over 2000–2005, as a result of increased cost of service industry prices, but also of raw materials, steel in particular.

The investment in OPEC Member Countries is expected to be much more productive in terms of yield in additional capacity. For this reason, the future global investment challenge is expected to remain similar to the past, with a rising share of funds being devoted to OPEC countries. The average global annual upstream investment over the long term, given the DAU developments, is expected to be close to $120 billion in today's prices (Figure 8).

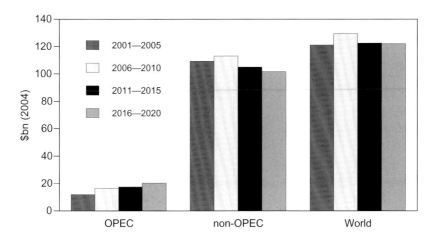

Figure 8: Annual Upstream Investment Requirements 2000−2020. $bn (2004)

However, as the scenarios have demonstrated, there is considerable uncertainty regarding the demand for oil from OPEC, and hence the possible investment needs. The level of OPEC production in 2020 in these three scenarios ranges from 43 mb/d in the PSM to 54 mb/d in the PMT, with a central DAU estimate of 49 mb/d. Moreover, the unit cost of expanding capacity will depend upon the type of market that evolves. The uncertainty concerning investment needs, even to 2010, is therefore very large, with the DAU estimate of $100 billion over the period 2005−2010 conceivably reaching $30 billion higher or lower (Figure 9).

By 2015, the range of possible cumulative investment requirements from 2005 is $130−240 billion. These are very significant differences, with important implications: under-investment will lead to a shortage of crude and over-investment will lead to excessive, costly idle capacity, as well as represent a huge waste of scarce resources in Member Countries. Neither situation is desirable: both can have a serious, detrimental impact on market stability and prices.

The long lead-times involved in investment in the petroleum sector, especially in this climate of uncertainty, represent a major challenge in the pursuit of long-term market stability. Clearly, one of the prerequisites for long-term stability is ensuring that timely and appropriate investments take place, and one of the crucial conditions for this is the prevalence of a reasonable price level that enables adequate sources of investment to be secured. Moreover, the issue of security of demand is central to the concerns of producer countries, and is the flip side to the

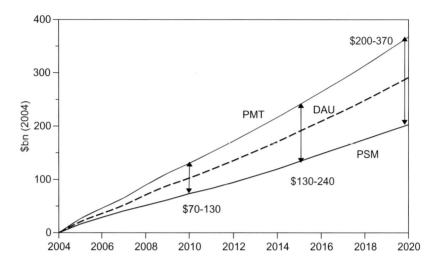

Figure 9: Cumulative OPEC Investment Requirements

idea of security of supply. Improved data, both in terms of quality as well as timeliness, is also a crucial element in improving the prospects for stability, and the Joint Oil Data Initiative (JODI), developed under the auspices of the International Energy Forum (IEF), is a good example of the way forward. More generally, ongoing dialogue, between producers as well as between producers and consumers is necessary to improve understanding of the challenges ahead, as well as to aim for a fairer distribution of burdens of risk.

One key area for supporting longer-term market stability is the downstream sector. On the basis of the developments outlined in the DAU, projected cumulative capital costs through 2015 are likely to reach almost $120 billion (Figure 10).

Of the refining investment, approximately 60 percent is primarily related to improvements in quality specifications of refined products, mainly desulphurization, while another roughly 40 percent is due to volume expansion. Geographically, Asia is still expected to be the dominant region in terms of downstream investment requirements, both for volume expansion and quality improvements, accounting for more then $30 billion of downstream investments. However, investment requirements elsewhere are also substantial, mainly because of the high expenditure implied by the need to move towards cleaner products. The investments in North America, Europe and Pacific industrialized regions are associated principally with product quality regulatory

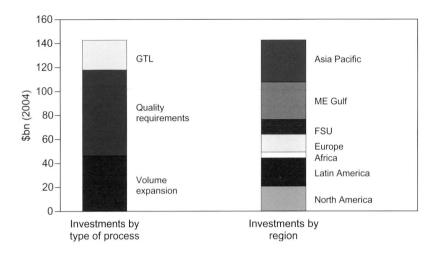

Figure 10: Estimation of Global Refining and GTL Investments. 2015 vs. 2004

compliance and with upgrading residuum. Those in other world regions are driven in addition by expected demand growth. Uncertainties over future investment needs are increased due to potential trends in policies regarding product specification in consuming countries.

5.   Technology and the Future of Oil

It is inevitable that petroleum will continue to play the crucial role in satisfying the world's energy requirements over at least the next two decades. For petroleum, technological advances will impact the ability to satisfy the world's growing needs for energy in a way that is affordable and cleaner, both in production and use.

In the upstream, technology development is expected to play a significant role in the future prospects for non-OPEC supply. Non-OPEC output might continue to increase at different rates, depending, inter alia, on the pace of technology development. In OPEC Member Countries, the application of advanced upstream technologies can potentially reduce costs further, increase recovery rates, and open up frontier areas. Focus is needed, in particular, on the continued development and use of advanced oil and gas production technologies suited to the special characteristics of present and future large oil fields, typically found in some OPEC countries, where most oil reserves and resources are located.

Future technological progress would allow the development of significant amounts of unconventional oil at lower cost and may enable the extension of the availability of oil supply in the long run. In addition, gas-to-liquids technology is making available increasing amounts of clean fuels. Costs have gradually declined as technology has improved; however, impacts over the rest of this decade are rather limited.

Technology will also play a crucial role from the demand perspective. This will be particularly the case for the transportation sector, which has been the main source of oil demand growth over the past 20 years, a trend that is expected to continue for the next two-to-three decades. New technologies will continue to improve efficiencies as well as address concerns over urban air pollution and the potential long-term impact of greenhouse gas emissions. For example, as well as improvements to conventional engines, hybrid vehicles may witness strong growth over the coming years, as performance, economics and safety continue to improve. New lightweight composite materials are being developed which in the longer term could allow the production of safe, sturdy vehicles of all types with even higher fuel efficiencies. More efficient vehicles could contribute to a longer-term more sustainable use of oil, and give the world sufficient time to make a smooth transition to alternative sources of energy in the very long term.

Longer term, the hydrogen fuel cell could represent a significant alternative to oil in the transportation sector. However, fuel cell vehicles are currently very expensive, and it is likely that, even if they are produced in large volumes, they would continue to cost more than conventional vehicles. Besides, hydrogen fuel, which is an intermediate energy carrier and not an energy source, is likely to be produced from hydrocarbon sources in the early stages.

In the face of potential climate change and the consequent calls to limit or reduce net emissions, the use of carbon dioxide capture and storage technology is a promising mitigation option that will allow the continued use of fossil energy resources to fuel development needs in a carbon-constrained world. Indeed, the potential for CO_2 reductions from carbon capture and storage is substantial. Injection of CO_2 in depleting oil reservoirs has been successfully used for enhanced oil recovery (EOR) in projects over the past three decades.

In sum, OPEC favours promoting, encouraging and participating in the development and rapid deployment of technologies that improve the environmental footprint of oil production and use, while supporting the widely accepted role that oil will be expected to play in fuelling the development needs of the world. Indeed, these technologies are the basis of several collaborative R&D programmes that have been recently initiated among OPEC Member Countries.

6.   Conclusion

Today, fuels, of which oil represents the major share, make up almost one-tenth of world trade by value. Also, around four-fifths of all produced oil is exported. With this international outlook, the oil industry has already adjusted to many changes and grown used to the risks and benefits that a global market provides. Recent developments have presented new challenges to the way the oil industry, and oil producers, in particular, function in a truly globalized world.

When viewing the market situation towards the end of 2005, it is clear that a combination of factors has been at work over the past couple of years, leading to the persistent price rises and volatility. Moreover, there has also been a shift in emphasis from upstream to downstream, with regard to the underlying causes of the present malaise. An examination of the short-term oil market reveals oil's centrality to global economic growth and mankind's associated energy needs, its dynamic, constantly changing nature and its inherent volatility. At the same time, it underlines OPEC's important role in seeking to achieve market order and stability, as has been successfully demonstrated and recognized in recent times. With capacity expansion plans under way in both OPEC and non-OPEC countries, ample crude supply should be available over the medium term. For the downstream, expansion of refining capacity, including upgrading, is not expected to catch up with product demand growth before 2007.

Over the longer term, the next few decades are expected to see fossil fuels continue to account predominantly for increases in energy demand, with oil set to maintain its major role as a source of energy. There is also a clear expectation that the oil resource base is sufficiently abundant to satisfy this demand growth. Yet, the expected ranges of demand growth reveal large uncertainties in the absolute magnitudes of demand. Thus timely deliverability, rather than just availability, is the key challenge facing producers and consumers in the early twenty-first century − in particular, capacity expansion and spare capacity management across the supply-chain. Moreover, although non-OPEC production is seen as continuing its recent expansion over the medium term, it is generally agreed that OPEC will gradually be relied upon to supply the incremental barrel, with its market share eventually set to rise. The absolute size of investment required by the oil industry in this outlook is huge, but not necessarily much different in magnitude to that observed in the past.

Maintaining stable prices will be complicated by the uncertainties about the future, with regard to how demand will evolve, as well as how non-OPEC supply behaves. Uncertainties over future economic growth, consuming governments' energy and environmental policies, and the rate of development and diffusion of newer technologies, raise questions over the future scale of required investment. This uncertainty, coupled with long lead times, inevitably complicates the task of maintaining market stability. The key to ensuring that sufficient investment takes place is oil market stability, whereby prices at reasonable levels need to be maintained to secure adequate sources of investment.

Technology development is of profound importance to the future of oil, for both supply and demand. It can contribute to demand growth, just as it can reduce demand through improved efficiency. Given that fossil fuels are set to meet most of the increases in energy demand over the next two decades, there is a need for cleaner oil technology. Sustainable patterns need to take advantage of promising technologies, such as carbon sequestration, reductions in gas-flaring and clean oil in electricity generation. The commercial potential of CO_2 sequestration, combined with enhanced oil recovery, presents an important opportunity for cooperation on technological issues, with the promise of great benefits for all.

All parties in the oil market share the responsibility of ensuring its steady, orderly evolution in the opening decades of the twenty-first century. This should be to the benefit of the world community at large, to people from all nations, rich and poor. Genuine dialogue and cooperation have an important role to play in this, together with a steadfast constructive approach to meeting the challenges that face us in the years ahead.

CHAPTER 3

THE ORIGINS AND EVOLUTION OF THE CURRENT INTERNATIONAL OIL PRICING SYSTEM: A CRITICAL ASSESSMENT

Bassam Fattouh

1. Introduction

The adoption of the market-related pricing system in 1986 opened a new chapter in the history of oil price determination. It represented a shift from a system in which prices were first administered by the large multinational oil companies and then by OPEC to one in which prices are set by 'markets'. But what is really meant by 'market prices'? The concept of the 'market price' of oil associated with this pricing regime has been surrounded with confusion. Crude oil is not a homogenous commodity. There are various types of internationally traded crude oil with different qualities and characteristics which have a bearing on refining yields.[1] Thus, different crudes fetch different prices. In the current system, the prices of these crudes are usually set at a discount or a premium to a marker or reference price according to their quality. This however raises an important question. How is the price of the marker or reference crude determined? A simple answer to this question would be 'the market' for these marker crudes. But this raises additional questions. What are the main features of the spot physical markets for the marker crudes? Do these markets have enough liquidity to enable them to discover adequately the marker price? And if they do not, can we rely on alternative markets such as the futures market and other derivatives based on these marker crudes? Does the distinction between the different layers of the market matter? And if the distinction matters, what do prices in different markets reflect? It is clear from all these questions that the concept of 'market price' may imply different things to different people and thus needs to be precisely

---

1 The two most important characteristics that determine oil quality are specific gravity and sulphur content.

defined. The argument that the market determines the price of oil has little explanatory power.

This chapter provides a critical assessment of these and other related questions in an attempt to understand the main features of the current oil pricing regime and its associated oil price concepts. In Section 2, we discuss the major transformations in the oil market during the last 50 years or so, and the different pricing regimes that have been associated with the various market structures. In this section, we will also discuss the complex structure of oil markets consisting of spot, physical forward, futures, options and other derivative transactions that emerged around Brent and West Texas Intermediate in the mid 1980s. In Section 3, we discuss the main features of the pricing formulae that constitute the basis of the market-related pricing system. In Section 4, we discuss the main characteristics of the markets of the marker crudes, some of the problems facing these markets, and their implications for price determination. In Section 5, we analyse the main reasons for the recent shift to the futures markets and provide a critical evaluation of this shift. In Section 6, we discuss the relationship between OPEC and the oil market. The last section concludes.

2.   Market Structure and Previous Oil Pricing Regimes

The emergence of the current oil price regime can not be understood in isolation from previous ones. The current system has emerged in response to major developments that transformed the oil market in the 1970s and 1980s and to crises that caused the abandonment of previous regimes. It is important to stress that every oil pricing system has been associated with a specific price concept.

2.1   History

Since the beginning of the oil era in the Middle East until the early 1970s, the oil-producing countries of the OPEC region did not participate in production or pricing of crude oil but simply received a stream of income through royalties and income taxes. In fact, the host governments acted only as competing sellers of licences or 'concessions' to produce crude oil. The oil pricing regime associated with the concession system that prevailed in the Third World until the mid 1970s was centred on the concept of 'posted' price.[2] The posted price

2   The first posted price in the Middle East was made by Mobil for the Iraqi Kirkuk crude in October 1950.

served only as a fiscal parameter which was used to calculate the royalty and income tax per barrel of crude oil going to the host governments (Mabro, 1984).

Spot prices, transfer prices and long-term contract prices could not play such a fiscal role. Until the late 1950s, the international oil industry outside the USA, Canada, the USSR and China has been characterized by the dominant position of the large multinational oil companies (known as the Seven Sisters or the majors) which had majority control of reserves, extraction and production, transportation, and marketing.[3] Each of these Seven Sisters was vertically integrated and had control both of upstream operations (exploration, development and production of oil) and to a significant but lesser extent of downstream operations (transportation, refining and marketing). At the same time, they controlled the rate of supply of crude oil through joint ownership of companies that operated in different countries. This type of inter-linking enabled them to control the bulk of oil exports from the major oil-producing countries and to prevent large amounts of crude oil accumulating in the hands of sellers and thus minimizing the risk of sellers competing to dispose of unwanted crude oil to independent buyers (Penrose, 1968). This vertically and horizontally integrated industrial structure meant that oil trading became to a large extent a question of inter-company exchange with no free market operating outside these companies' control, which resulted in an underdeveloped spot market. Transfer prices used in transactions within the subsidiaries of an oil company were not acceptable for calculating the income tax due to the host countries. These prices did not reflect a market price but were merely used by multinational oil companies to minimize their worldwide tax liabilities by transferring profits from high-tax to low-tax jurisdictions. Because some companies were crude long and others crude short, transactions used to occur between the multinational oil companies on the basis of long-term contracts. However, the prices used in these contracts were never disclosed by oil companies who considered this piece of information to be a commercial secret. Oil-exporting countries were not particularly keen on using contract prices because these prices were usually lower than posted prices.

Income tax and royalty calculations thus had to be based on posted prices. Being a fiscal parameter, the posted price did not respond to the usual market forces of supply and demand and thus did not play any allocation function. The vertically integrated oil companies were

---

3   In 1950 the majors controlled 85% of the crude oil production in the world
    outside Canada, USA, Soviet Russia and China (Danielsen, 1982).

comfortable with the system of posted prices because it maintained their oligopolistic position, and until the late 1960s OPEC countries were too weak to change the existing pricing regime.

By the late 1950s, however, the dominance of the vertically integrated companies was challenged by the arrival of independent oil companies who were able to invest in upstream operations and obtain access to crude oil outside the Seven Sisters' control.[4] In the mid 1950s, Venezuela granted independents (mainly from the USA) some oil concessions, and by 1965 non-majors were responsible for 15% of total Venezuelan production (Parra, 2004).[5] Oil discovery in Libya increased the importance of independents in oil production as Libya chose as a matter of policy to attract a diversity of oil companies and not only the majors.[6] Competition to majors also appeared elsewhere. In the late 1950s, Iran signed two exploration and development agreements in the Persian Gulf offshore with non-majors and in 1951 Saudi Arabia entered in an agreement with the Japan Petroleum Trading Company to explore and develop Saudi Arabia's fields in the Neutral Zone offshore area.[7] Crude oil from the Former Soviet Union also began to make its way into the market (Parra, 2004).[8] These and other developments led to the creation of a market for buying and selling crude oil outside the control of the Seven Sisters. However, the total volume of crude oil from US independents and other companies operating in Venezuela, Libya and the Gulf offshore remained small and the growth of Russian exports came to a stop after 1967 and started to decline in 1969 and 1970 (Parra, 2004). These factors limited the scope and size of the market and by the late 1960s the majors were still the dominant force both in the upstream and downstream parts of the industry (Penrose, 1968). [9]

---

4   The state-owned Italian company ENI is a good example.

5   This share though declined from 1966 onwards.

6   In 1965, production by independents in Libya totalled around 580 thousand barrels per day increasing to 1.1 million barrels per day in 1968 (Parra, 2004).

7   The volume of oil produced from these concessions did not constitute a major threat to majors, but the conclusion of the agreements placed pressure on the majors from host governments to improve the terms on existing concessions.

8   The discovery and development of large fields in the Soviet bloc led to a rapid growth in Russian oil exports from less than 100 thousand barrels per day in 1956 to nearly 700 thousand barrels per day in 1961 (Parra, 2004).

9   This does not undermine the impact of these developments on the global oil scene. In fact, the competitive pressures from other oil producers were partly responsible for the multinational oil companies' decision to cut the posted price in 1959 and 1960. Another important factor was the US decision to impose mandatory import quotas which meant an increase in competition for outlets outside the USA and put pressure on prices. The formation of OPEC

Between 1970 and 1973, global demand for oil increased at a fast rate with most of this increase being met by OPEC countries; this resulted in an increase in OPEC's share in global crude oil production. Other developments in the early 1970s, such as Libya's production cutbacks and the sabotage of the Saudi Tapline in Syria, tightened the demand-supply balance further. This created a strong sellers' market and significantly increased OPEC governments' power relative to the multinational oil companies. On 2 September 1970 the Libyan government reached an agreement with Occidental wherein this independent oil company agreed to pay income taxes on the basis of posted price which was increased by 30 cents and to make retroactive payment to compensate for the lost revenue since 1965.[10] Soon after, all other companies operating in Libya submitted to these new terms. As a result of this agreement, other oil-producing countries invoked the most favoured nation clause and made it clear that they would not accept anything less than the terms granted to Libya. The negotiations conducted in Tehran resulted in a collective decision to raise the posted price and increase the tax rate from 50 to 55 percent. The agreement was for five years until 31 December 1975. The twenty-first OPEC Conference in Caracas in December 1970 confirmed and consolidated what had been achieved in the Tehran agreement.

In September 1973, OPEC decided in its thirty-fifth Conference in Vienna to reopen negotiations with the companies to revise the Tehran Agreement and seek large increases in the posted price. The bargaining position of OPEC had been strengthened by tight market conditions that prevailed in the period between 1970 and 1973 and because of a lack of a short-term alternative to OPEC as a source of incremental supply to satisfy the growing global oil demand. Oil companies refused OPEC's demand for an increase in the posted price and negotiations collapsed.[11] As a result, on 16 October 1973, the six Gulf Members

---

in 1960 was an attempt to end the erosion in the posted price (Skeet, 1988).

10 Occidental was the ideal company to exert pressure on. Unlike the majors, Occidental relied heavily on Libyan production and did not have much access to oil in other parts of the world.

11 Oil companies did not have the incentive for negotiations to succeed for they did not want to be seen as responsible for a major rise in the world oil prices. This can be implied from the negotiation tactics adopted by the oil companies' representatives. On 8 October 1973 the oil companies' representatives asked for the negotiations to be adjourned for two days to allow consultation among companies and with their governments. When negotiations resumed on 12 October, the oil companies requested a further adjournment of two more weeks for further consultations. This was not accepted by OPEC and negotiations were called off.

of OPEC unilaterally announced an immediate increase in the posted price of the Arabian Light crude from $3.65 to $5.119. On 19 October 1973, members of the Organization of Arab Petroleum Exporting Countries (OAPEC) less Iraq announced production cuts of 5% of the September volume and a further 5% per month until 'the total evacuation of Israeli forces from all Arab territory occupied during the June 1967 war is completed and the legitimate rights of the Palestinian people are restored' (Skeet, 1988, quotations in original). In December 1973, OPEC raised the posted price of the Arabian Light further to $11.651. This jump in price was unprecedented. The year 1973 represented a dramatic shift in this balance of power towards OPEC. For the first time in its history, OPEC assumed a unilateral role in setting posted prices (Terzian, 1985). Before that date, OPEC had been able only to prevent oil companies from reducing them (Skeet, 1988).

The oil industry witnessed a major transformation in the early 1970s when some OPEC governments stopped granting new concessions[12] and started to claim equity participation in the existing concessions, with a few of them opting for full nationalization (Algeria, Iraq and Libya).[13] Demands for equity participation emerged in the early 1960s, but the multinational oil companies downplayed these calls from the start. Oil companies became wary in the late 1960s when they realized that even moderate countries such as Saudi Arabia had begun to make similar calls for equity participation.[14] In the twenty-fourth OPEC Conference in July 1971, a Ministerial Committee was established to construct a plan for the effective implementation of the participation agreement. OPEC's six Gulf Members (Abu Dhabi, Iran, Iraq, Saudi Arabia, Qatar, and Kuwait)[15] agreed to negotiate collectively the participation agreement with oil companies and empowered the Saudi oil Minister

---

12 As early as 1957, Egypt and Iran started turning away from concessions to new contractual forms such as joint venture schemes and service contracts. In 1964, Iraq decided not to grant any more oil concessions (Terzian, 1985).

13 Nationalization of oil concessions in the Middle East extends well before that date. Other than Mossadegh's attempt at nationalization in 1951, in 1956 Egypt nationalized Shell's interest in the country. In 1958, Syria nationalized the Karatchock oilfields and in 1963 the entire oil sector came under the government control. In 1967, 'Algerization' of oil companies had already begun and by 1970 all non-French oil interests were nationalized. In 1971, French interests were subject to Algerization with the government taking 51% of French companies' stake (Terzian, 1985).

14 In 1968, Yamani advocated the concept of participation and in a speech at the American University of Beirut in that year he argued that participation is the only way for the majors to avoid nationalization.

15 Iran announced its withdrawal early in 1972.

Zaki Yamani to speak in their name. On October 1972 and after many rounds of negotiations, oil companies agreed to an initial 25% participation which would reach 51% in 1983.[16]

Equity participation gave OPEC governments a share of the oil produced which they had to sell to third-party buyers. This led to the introduction of new pricing concepts to deal with this new reality (Mabro, 1984). As owners of crude oil, governments had to set a price for third-party buyers. Here entered the concept of official selling price (OSP) or government selling price (GSP). However, for reasons of convenience, lack of marketing experience and inability to integrate downwards into refining and marketing in oil-importing countries, most of the governments' oil was sold back to the companies that held the concession and produced the crude oil in the first place. These sales were made compulsory as part of the equity participation agreements and used to be transacted at buyback prices. The complex oil price regime of the early 1970s centred on three different concepts of oil prices (posted, OSP, and buyback prices) was highly inefficient. It meant that a buyer could obtain a barrel of oil at different prices (Mabro, 2005). Lack of information and transparency also meant that there was no adjustment mechanism to ensure that these prices converge. Thus, this regime was short-lived and by 1975 it ceased to exist.

The oil pricing regime that emerged in 1974−75 after the short lived episode of the buyback system was radical in many aspects, not least because it represented a complete shift in the price determination of crude oil to OPEC. OPEC countries became the dominant producers and assumed responsibility for the new pricing system. The system was centred on the concept of reference or marker price with the Arabian Light 34° API being the chosen marker crude. In this administered price system, individual members retained the Government Selling Prices (GSP) for their crudes, but these were now set in relation to the chosen reference price. More specifically, the GSP for any particular crude was equal to reference price plus or minus a differential which takes into account the differences in quality among crudes (sulphur

---

16 Out of the six Gulf States, Saudi Arabia, Abu Dhabi and later Qatar signed the general participation agreement. Iran announced its withdrawal early in 1972. Iraq opted for nationalization in 1972. In Kuwait, the parliament fiercely opposed the agreement. In 1974, the Kuwaiti government took a 60% stake in the capital of the Kuwait Oil Company and called for 100% stake before 1980. 100% equity participation in Kuwait was achieved in 1976 and Qatar followed suit in 1976−77. For other countries, Libya proposed 51% participation to oil companies at the beginning but after the October war of 1973, equity participation accelerated sharply resulting in full nationalization.

content and gravity) and their location. Crudes have different properties which have a bearing on refining yields and thus they fetch different prices. The differential relative to the marker price used to be adjusted periodically depending on a variety of factors such as the relative supply and demand for each crude variety and the relative price of petroleum products among other things. The flexibility of adjusting differentials by oil-exporting countries complicated the administering of the marker price and in the slack market of 1983 in face of falling global demand, OPEC opted for a more rigid system of setting price differentials, but without much success.

Equity participation and nationalization of oil resources profoundly affected the structure of the oil industry and led to the emergence of a market for crude oil. As a result of these developments, multinational oil companies lost large reserves of crude oil and they found themselves increasingly net short and dependent on OPEC supplies. The degree of vertical integration between upstream and downstream considerably weakened. During the 1950s and 1960s the crude oil obtained from upstream was used for downstream operations. However, vertically integrated companies were not always completely 'in balance' with respect to their downstream requirements (Penrose, 1968, p.153). If a vertically integrated oil company ended up with more crude oil than required downstream, it would sell the excess oil to other companies through a very long-term contract to another major who happened to be crude short. Nationalization of oil concessions from 1975 onwards changed the position. Oil companies retained both their upstream and downstream assets, but their position became more imbalanced and in one direction: the companies no longer had enough access to crude oil to meet their downstream requirements. This encouraged the development of an oil market outside the inter-multinational oil companies' trade and pushed companies to diversify their sources of oil supply by gaining access and developing reserves outside OPEC.[17]

During the years 1975–78, OPEC countries consolidated their

---

17 This process began well before the 1970s. The North Sea attracted oil companies from the early 1960s and the first rounds of leasing were awarded in 1964 and 1965. In 1969, oil was found in the Norwegian sector and in 1970 a major find (the Ekofisk field) was confirmed. In the UK sector, in 1969 Amoco found some oil but it was deemed to be non-commercial. In 1970, BP drilled the exploratory well that found the Forties field. One year later, Shell-Esso discovered the Brent field (Parra, 2004). It is important to note that all these major discoveries preceded the large rise in oil prices. Seymour (1990) shows that half the increase in non-OPEC supply over the 1975–85 period would have materialized regardless of the level of oil prices.

control over production, prices and investment. However, they remained dependent on multinational oil companies to lift and dispose of the crude oil and initially sold only low volumes through their national oil companies to firms other than the old concessionaires. Thus, at the early stages of the OPEC-administered pricing system, the major companies continued to have preferential access to crude. This narrowed the scope of a competitive oil market.

The situation changed in the late 1970s with the emergence of new players on the global oil scene. National oil companies in OPEC started to increase the number of their non-concessionaire customers. The appearance of new players (such as independents, Japanese and Wall Street refiners, state oil companies, trading houses and oil traders) permitted such a development which accelerated during and in the aftermath of the 1979 Iranian crisis. The new regime in Iran cancelled any previous agreements with the oil majors in marketing Iranian oil and they became mere purchasers as any other oil companies. In Libya, there was an attempt to switch away from the main term contract customers, including majors, to new customers – primarily governments and state oil corporations. Soon all OPEC countries followed suit.

During the 1979 crisis, spot crude prices rose faster than the official selling prices. The long-term contract represented an agreement between the buyer and the seller which specified the quantity of oil to be delivered while the price was related to the OPEC marker price. Since contracts obliged producers to sell certain quantities of oil to the majors at the marker price, this meant that oil companies would capture the entire differential between official selling prices and the spot prices by buying from governments and selling in the spot market or through term contracts with other companies who did not have direct access to producers. This was unacceptable to governments and hence they began to sell their crude oil directly to third-party buyers (Stevens, 1985). Faced with a large number of bidders, small OPEC producers such as Kuwait began to place an official mark-up over the marker price. By abandoning their long-term contracts, the producers had the freedom to sell to buyers who offered the highest mark-up over the marker price.[18] The result was that the majors lost access to large amounts of crude oil that were available to them under these

---

18 Saudi Arabia was a major exception to this behaviour. They maintained their long-term contracts with the four Aramco concessionaires (Exxon, Chevron, Texaco and Mobil). These continued to obtain oil at the OPEC official price and enjoyed what their competitors referred to bitterly as the 'Aramco advantage'.

long-term contracts. This had the effect of worsening dramatically the imbalance within oil companies and reduced the degree of integration between downstream and upstream with the latter becoming only a small fraction of the former. Faced with virtual disruption of traditional supply channels, multinational oil companies were forced into the market. This had a dramatic effect on oil markets as de-integration and emergence of new players expanded the external market where buyers and sellers engaged in arm's-length transactions. The crude market became more competitive and the majority of oil used to move through short-term contracts or the spot market. Prior to these developments, the market had consisted of a small number of spot transactions usually done under distressed conditions, for disposing of small amounts of crude oil not covered by long-term contracts or for price discovery purposes.

The other development that further consolidated the market for crude oil was the increase in non-OPEC supply. New discoveries in non-OPEC countries responding to higher oil prices, and taking advantage of new technologies, meant that significant amounts of oil began to reach the international market from outside OPEC. Between 1975 and 1985 non-OPEC countries managed to increase their share of world total oil production from 48% in 1973 to 71% in 1985 (EIA, 2005). Most of the increase in oil production came from Mexico, the North Sea, and the Soviet Union (Parra, 2004). The increase in non-OPEC supply had three main effects. First, non-OPEC countries were setting their own prices which were more responsive to market conditions and hence more competitive. Second, the number of crude oil producers increased dramatically. The new suppliers of oil who ended up having more crude oil than required by contract buyers secured the sale of all their production by undercutting OPEC prices in the spot market. Buyers who became more diverse were attracted by the competitive prices on offer which were below the long-term contract prices. Third, the emergence of two interlinked centres of price determination in OPEC and non-OPEC countries in the early and mid 1980s increased the importance of the time dimension. By the time the oil cargoes reached the buyer, OPEC f.o.b. prices were much less competitive than non-OPEC crudes. Also, under the competitive pressures from non-OPEC suppliers, OPEC had to resort to price cuts on many occasions during the early 1980s. This meant that OPEC f.o.b. prices at the time of the conclusion of the contract would usually be higher than the oil price when the cargo reached its destination in Europe or the USA. In such an environment, there was a need by market participants to hedge against price fluctuations.

The OPEC-administered oil pricing regime lasted until 1985 but not without serious challenges and tensions. Disagreements within OPEC created tensions with some implications on the pricing regime. Saudi Arabia used to lose market share with every increase in the marker price and hence opposed them while others pushed for large increases. At times, disagreements within OPEC led to the adoption of two-tiered price reference structure. Such a structure emerged first in late 1976 in the forty-eighth conference in Doha when Saudi Arabia and UAE set a lower price for the marker crude than the rest of OPEC.[19] This was repeated in 1980 when Saudi Arabia used $32 per barrel for the marker while the other OPEC members used the per barrel marker of $36. Thus, two new concepts were introduced: the actual marker price which was fixed by Saudi Arabia and the deemed marker price which was fixed by the rest of OPEC (Amuzegar, 1999).

The growth of non-OPEC crude oil production represented a major challenge to OPEC's administered pricing regime and was mainly responsible for its demise. The decline in global oil demand and the increase in non-OPEC supply[20] created competitive pressures on OPEC. It became clear by the mid-1980s that the OPEC-administered oil pricing regime was unlikely to hold for long and OPEC's, or more precisely Saudi Arabia's, attempts to defend the marker price would only result in a dramatic reduction in its oil exports and loss of market share as other producers could offer to sell their oil at a discount to the Arabian Light. As a result of these pressures, the demand for OPEC oil declined from a yearly average of more than 30 million barrels per day (mb/d) in 1979 to a yearly average of 16 mb/d in 1985 with Saudi Arabia's production falling from a yearly average of 9.9 mb/d in 1980 to around a yearly average of 3.4 mb/d in 1985. [21] With this continued decline in demand for its oil, OPEC saw its own market share in the world's oil production fall from 52% in 1973 to less than 30% in 1985.

Saudi Arabia adopted the netback pricing system in 1986 in order to restore both the $26 price level and the country's market share.[22] The netback pricing system provided oil companies with a guaranteed refining margin even if oil prices were to collapse. It involved a general formula in which the price of crude oil was set equal to the *ex post*

---

19 This two-tier pricing system lasted until July 1977 when Saudi Arabia and UAE announced the acceptance of the price $12.70 for the marker crude.
20 For example, North Sea production increased to 3.5 mb/d in 1985 from less than 2mb/d in 1975.
21 Source of data is EIA.
22 For a detailed analysis of the netback pricing system and the 1986 price collapse, see Mabro (1986).

product realization minus refining and transport costs. A number of variables had to be defined in a complex contract including the set of petroleum products that the refiner can produce from a barrel of oil, the refining costs, transportation costs, the time lag between loading and delivery. The netback pricing system resulted in the 1986 price collapse, from $26 a barrel in 1985 to less than $10 a barrel in mid-1986. Out of the 1986 crisis, and not out of conviction, the current 'market related' oil pricing regime was born.

2.2   *The Development of a Complex Oil Market Structure*

After the collapse of the OPEC-administered system and the short experiment with netback pricing, oil-exporting countries adopted a kind of market-related pricing. First adopted by the Mexican national oil company PEMEX in 1986, market-related pricing received wide acceptance among many oil-exporting countries and by 1988 it became and still is the main method for pricing crude oil in international trade. Its structure is based on formula pricing where the marker or reference price is derived from the market rather than being the administered OPEC marker price.[23] The oil market scene was ready for such a transition. As indicated above, the end of the concession system and nationalization which disrupted the oil supplies of the multinational oil companies especially in the late 1970s established the basis of arm's-length deals and exchange outside these companies. The emergence of many suppliers and many buyers further enhanced the importance of such arm's-length deals. This led to the development of a complex structure of oil markets which consist of spot, physical forward, futures, options and other derivative markets. Such a complex structure emerged in the North Sea around Brent.[24] In the early 1980s, the Brent market only consisted of the spot market and the informal forward physical market (the 15-day market). By the late 1980s, the Brent market had become quite complex including also a futures contract traded on the International Petroleum Exchange (IPE), options, swaps and other trading instruments. In North America, other complex layers of trading

---

23 This raises a large number of issues which will be discussed in detail in Section 3.

24 The Brent Blend is a mixture of oil produced from separate oil fields and collected through two main pipeline systems, the Brent and Ninian to the terminal at Sullom Voe in Shetland, UK and then transferred through tankers to European refiners or when arbitrage allows across the Atlantic. The Brent and Ninian systems were separate but in 1990, the two systems were commingled and Ninian ceased to trade as a separate crude.

instruments emerged around the West Texas Intermediate (WTI).[25] Although these various markets are linked, they are not the same. The question then becomes: Which of these markets should determine the marker price? This question is dealt with later in detail, but first we examine briefly the current complex oil market structure, focusing mainly on Brent and WTI.

The incentive for oil companies to engage in tax spinning through the forward market was the main factor responsible for the emergence of the forward 15-Day Brent market (Mabro et al. 1986; Horsnell and Mabro, 1993; Bacon, 1986). The valuation of oil for UK fiscal purposes was based on market prices. In an arm's-length transaction, market prices were obtained from the realized prices on the deal.[26] If oil was merely transferred within a vertically integrated system, then the fiscal authorities would assign an assessed price to the transaction based on the prices of 'contemporary and comparable' arm's-length deals. Until 1984, these followed the official British National Oil Corporation (BNOC) price. Because of the differential rates of taxation between upstream and downstream (the tax rate being lower in the downstream), the impact of the fiscal regime was not neutral and affected a vertically integrated oil company's decision to sell or retain crude oil.[27] When the spot price was lower than the official BNOC price, integrated oil companies had the incentive to sell their own crude arm's-length and buy the crude needed for their own refineries from another seller. When the spot price was higher than the assessed price, oil companies had the incentive to keep the oil for use in their own refineries. In doing so, the oil companies would achieve higher after tax profits. After the abolition of BNOC, the assessment process of transactions within the firm became more complex. The market value of non arm's-length transactions was to be based on the average price of contracts (spot

---

25 WTI is a blend of crude oil produced in the oil fields of Texas, New Mexico, Oklahoma and Kansas. WTI deliveries are made at the end of the pipeline system in Cushing, Oklahoma. Cushing serves as a major crude oil marketing hub in the USA where, in addition to WTI, it receives imported crude oil from the Gulf Coast. Crude oil in Cushing is then transported through a network of pipelines to refineries in central parts of the USA.

26 The fiscal authorities specified a number of conditions before a contract can qualify as arm's length including the condition that the deal is not made back to back.

27 Tax spinning refers to this situation in which for fiscal reasons oil companies would resort to buying and selling crude oil in the market though it would have been more convenient and cheaper to internalize the transaction (Horsnell and Mabro, 2003, p.63).

and forward) preceding the deal.[28] This encouraged oil companies whether vertically integrated or not to engage in tax spinning through the forward market.[29]

Although tax spinning continued to provide a motive for trading in these markets its importance has declined as tighter regulations introduced later in 1987 made it more difficult and much less predictable. But by then, the 15-day forward market was well established and expanding fast as various market participants including oil companies, traders, and refiners began to trade actively in this market for risk management and speculative purposes.

The previous 15-day Brent market (which has now become the 21-day Brent market)[30] largely evolved in response to the peculiar nature of the delivery schedule of Brent. The companies producing crude oil in the 15-day Brent system nominate their preferred date for loading at the relevant month by the 5th of the preceding month. The loading programme is then organized and finalized by the 15th of the preceding month. Until this happens producers do not know when their crude oil will be available for delivery. Once the schedule is finalized, a producing company or any other entity that enters the forward Brent market must give the buyer a notice of the first date of a three-day loading window. Thus, contracts for 15-day Brent are forward contracts which specify the delivery month but not the particular date at which the cargo will be loaded. Under the 15-day contract, sellers were required to give at least 15 days notice. Under the 21-day Brent contract, the seller is required to provide the purchaser at least 21 days notice as to when the cargo will be loaded. For instance, if the loading date is in the first three days of the delivery month, then the seller has to nominate the buyer on the 11th day of the previous month. Depending on the market conditions at the time of nomination, the purchaser may or may not want the actual possession of the cargo. In fact, it is likely that the original cargo purchaser has already sold another 21-day contract, in which case he must give notice to the new buyer to take the cargo at least 21 days in advance.

Two types of futures contracts, the **NYMEX** light sweet crude oil

28 For instance, for a deal made in December, one should take into account the average of spot and forward prices between 1 November and 15 December.
29 For details on how tax spinning can be transacted through trading in the forward market, see (Horsnell and Mabro, 1993, Chapter 6 and Bacon, 1986).
30 In 2002, Forties and Oseberg grades were added to Brent and as a result the 15-day Brent contract was replaced by the 21-day BFO (Brent-Forties-Oserberg) contract. However, due to historical reasons, the forward market is still referred to as the 21-day Brent market.

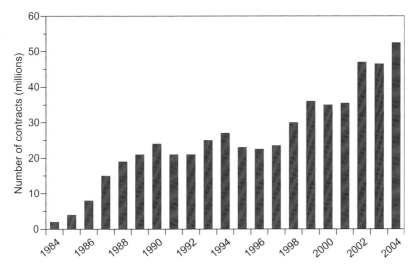

Source: NYMEX Website (www.nymex.com)

Figure 1: Annual Number of Light, Sweet Crude Oil Futures Contracts
(1984−2004)

futures contract (referred to as WTI futures contract) and the IPE
Brent futures contract, dominate oil futures trading. The WTI crude
oil futures contract specifies 1,000 barrels of WTI to be delivered at
Cushing, Oklahoma at a specified time in the future.[31] This futures
contract has been trading on the New York Mercantile Exchange
(NYMEX) since 1983, and has grown since then to become the largest
traded commodity futures contract. The annual volume of contracts
traded increased dramatically over the last twenty years, reaching almost
53 million contracts in 2004 compared to approximately less than 2
million in 1984. The IPE Brent futures contract was launched on the
International Petroleum Exchange (IPE) in London in June 1988 after
a number of failed attempts. The IPE Brent crude oil futures contract
specifies 1,000 barrels of Brent crude oil at a specified time in the
future. The IPE Brent crude oil futures market has grown dramatically
in the last two decades with the year 2004 representing a record year
for the IPE Brent crude futures contracts where around 25.5 million
contracts were traded in that year alone.

---

31 NYMEX also allows the delivery of domestic types of crude (Low Sweet Mix,
New Mexican Sweet, North Texas Sweet, Oklahoma Sweet, South Texas Sweet)
and foreign types of crude (UK Brent and Forties, Norwegian Oseberg Blend;
Nigerian Bonny Light, Qua Iboe, and Colombian Cusiana) against the light
sweet crude oil futures contract. (Source: NYMEX Website).

In November 1986, NYMEX launched the light sweet crude oil option contract. In May 1989, IPE launched the IPE Brent option contract. For both option contracts, the underlying financial asset is a futures contract. Thus, the option market represents an additional layer of the Brent and WTI markets which are based on futures which in turn are based on forward (only in the case of Brent) and these in turn are based on the spot market. The NYMEX and IPE Brent option contracts are of the American type i.e. they give buyers the right to buy (a call option) the underlying asset (i.e. transform the contract into a Brent or WTI futures contract) at a specified price (the strike price) over a specified period of time and can be exercised at any time before the expiry of the contract.[32] The number of light sweet crude oil option contracts traded in NYMEX has witnessed a dramatic increase during the past few years, rising from around 3 million contracts in 1987 to more than 11.5 million in 2004. This is in contrast to the volume of IPE Brent crude option contracts which has been shrinking in the last ten years. Reaching a peak of over 1 million contracts in 1993, there has been a decline over the years and in 2004 the number was only 28,688 contracts. It is important to note that these figures include only option contracts traded on the exchange and do not take into account those traded over the counter (OTC). Option contracts traded over the counter are customer-tailored and hence give more flexibility to market participants, for instance by allowing for a longer maturity. The volume of option contracts traded in the OTC market is not known but is believed to be much larger than those traded on the exchange.

In addition to forward, futures and option markets, the 1990s saw the emergence of an active crude swaps market. Crude swaps provide additional tools for hedgers and speculators to trade a specific type of risk: the risk associated with the change in the price differential between the spot price at the time of settlement of the paper contract and the price of the first forward month at the time of settlement of physical trade. This type of risk assumes a special importance in the Brent market due to its peculiar logistics. There has been growing reliance on Contracts for Differences (CFD) to manage the risk arising from differential price moves between the dated and forward market. CFD

---

32 All open contracts are marked-to-market daily. Due to futures style margining option premiums are not paid/received at the time of the transaction. Rather margins are paid/received every day according to the changing value of the option and the total value to be paid/received is only known when the position is closed (by an opposing sale/purchase, exercise or expiry). It is a fundamental principle of option trading that the buyer never pays more than the premium.

is essentially a swap contract that locks in the price of dated Brent in relation to first month forward Brent deliveries in return for a fixed insurance premium. The CFD market has grown rapidly since the mid 1990s in response to the very volatile relationship between dated Brent and first forward 15 and 21 day contracts.[33] An active swaps market known as Dated to Front Line (DFL) also emerged for trading the difference between dated Brent and the IPE front line futures contract. This has been supplemented by the dated to Brent Weighted Average (BWAVE) market i.e. a swap market that trades the difference between dated Brent and the daily trade-weighted Brent average reported by the IPE. These crude swaps along with Brent futures and forwards are tools for risk management for trade in crude oil.[34]

3.   Formula Pricing in the Current Oil Pricing Regime[35]

The emergence and expansion of the market for crude oil allowed the development of market-referencing pricing off spot crude markers such as spot WTI (initially Alaska North Slope), dated Brent (which is considered as the spot market for Brent) and Dubai. Formula pricing which constitutes the basis of the current 'market-related' oil pricing regime is a relatively recent development in the oil market. First adopted by the Mexican national oil company PEMEX in 1986, formula pricing received wide acceptance among many oil-exporting countries and by 1988 it became and still is the main method for pricing crude oil in international trade.

   Long-term contracts constitute the main method for arranging physical delivery of oil in this system. These contracts are negotiated directly between the parties and they specify among other things, the volumes of crude oil to be delivered, the actions to be taken in case of default, the tolerance level[36] and the method used in calculating crude oil prices.

---

33 For detailed analysis of the Contracts for Differences in the Brent market, see Barrera-Rey and Seymour (1996).

34 Source: Platts,   *http://www.platts.com/Oil/Resources/Risk%20Manage ment/crudeswaps.html*

35 This section is based on Chapter 15 of Horsnell and Mabro (1993).

36 Operational tolerance, which is spread throughout the industry, refers to a situation in which a buyer of a cargo is given the choice of loading more or less crude oil than the agreed volume depending on the agreed tolerance rate. For instance, in the Brent market, the contract defines a cargo of Brent as 500,000 barrels and gives the buyer a plus or minus 5% operational tolerance, i.e. on the loading date the buyer has the choice to load either 475,000 barrels

The formula used in pricing oil is straightforward. The price of a certain variety of crude oil is set as a differential to a certain marker or reference price. Specifically, for crude oil variety X, the formula pricing is: $P_X = P_R \pm D$ where P_X is the price of crude X, P_R is the reference or marker price and D is the value of the price differential.

3.1   The Differential and Time Lags

The differential, also called the coefficient of adjustment, determined independently by each of the oil-producing countries, is usually set in the month preceding the loading month, and adjusted monthly or quarterly. Theoretically, the differential is supposed to reflect differences in the quality of crude. Specifically, the differential should reflect the differences in the Gross Products Worth (GPW) obtained from refining the reference crude R and the crude X. The differential between the different varieties of crude oil is not constant over time and changes continuously according to relative demand and the relative supplies of the various crudes which in turn depend on the relative prices of petroleum products. Furthermore, an oil-exporting country will not only consider the differential between its crude and the reference crude, but has also to consider how its closest competitors are pricing their crude in relation to the reference crude. This implies that the timing of setting the differential matters, especially in a slack market. Oil-exporting countries that announce their differentials first are at a competitive disadvantage of being undercut by their closest competitors. This can induce them to delay the announcement of the differential or in the case of multiple transactions compensate the buyers with adjusting downwards the differential in the next rounds.

The 'equivalence to the buyer' principle, which means in practice that the price of crudes has to be of a c.i.f. nature so that equivalent crudes have equivalent prices at destination, adds another dimension to the differential. The location in which a comparison of prices should be made is not the point of origin but must be closer to the destination where the buyer receives the cargo. Since the freight costs vary depending on export destination, some formulae also take into account the relative freight costs. Specifically, they allow for the difference between the freight costs involved in moving the reference crude from its location to a certain destination (e.g. Brent from Sollum Voe to Rotterdam) and the costs involved in moving crude X from the oil

---

or 525,000 barrels. Depending on whether the market price is favourable or not, the buyer will maximize or minimize the loading volume.

country's terminal to that certain destination (e.g. Arabian Light from Ras Tanura to Rotterdam).

There is also a time lag problem. When there is a lag between the date at which a cargo is bought and the date of arrival at its destination, there is a price risk. Oil-exporting countries usually share this risk with their buyers through the pricing formula. Agreements are sometimes made for the date of pricing to occur around the delivery date. For instance, in the case of Saudi Arabia's exports to the United States, the date of pricing can vary between 40 to 50 days after the loading date.[37] The price used is the market price quotes averaged over 10 days around the delivery date. Since the point of sale is closer to the destination than the origin, this is closer to c.i.f. rather than f.o.b. pricing. For oil sold to the Far East, the pricing is the average of the reference price over the calendar month in which loading took place.[38] For instance, for loading on the 3rd of January, the market price used is the average of quoted prices over the entire month of January. One important implication of formula pricing is that the same crude is priced differently (on an f.o.b. basis) depending on its destination.[39]

It is important to note that although the pricing date can be as long as 50 days after the loading date and hence the purchase price around the delivery date will reflect the market price of oil, the pre-setting of the adjustment factor before loading month means that the differential is bound to be out of date. For instance, if the differential is set at the beginning of the month preceding the loading date based on information from the previous two to three weeks, the adjustment coefficient could be seven weeks out of date by the time of loading. These problems are usually addressed through bargaining between buyers and sellers. As such, the ex-post changes in the adjustment factor can not account only for the differences in quality of crudes or freight costs, but also the bargaining positions of the different parties.[40]

---

37 This does not completely eliminate price risk because the actual delivery may not correspond to the conventional time lag used in the formula. For instance, the cargo can reach its destination 60 days after the loading date while the pricing formula sets the date of pricing as 50 days after loading.

38 At times, Asian oil sales involve a premium referred to as the Asian premium which involves an element representing the difference in freight between Asia on the one hand and Europe and USA on the other hand (Ogawa, 2002; Soligo and Jaffe, 2000).

39 Fesharaki and Vahidy (2001) argue that to exclude the possibility to make profits out of this differential, the contract should forbid the resale of the crude. If resale was allowed, arbitrage would result in a situation where crude is sold at the same f.o.b. price regardless of its destination.

3.2   The Benchmark Crudes

The next important element of formula pricing is identification of the reference or benchmark crude. Brent, WTI and Dubai/Oman are the main crude oil benchmarks of the current oil pricing regime. Nearly all oil traded outside America and the Far East is priced using Brent as a benchmark. Brent blend is a combination of crude oil from different oil fields located in the North Sea. Its API gravity is 38.3 degrees while it contains about 0.37% of sulphur making it 'light sweet' crude oil, but less so than WTI. WTI is the main benchmark used for pricing oil imports into the USA.[41] Although more crude oil is priced off Brent, the standard WTI futures contract is the most widely traded commodity futures contract in the world. West Texas Intermediate (WTI) crude oil is of very high quality. Its API gravity is 39.6 degrees and it contains only about 0.24% of sulphur making it 'light sweet' crude oil. Dubai-Oman is used as a benchmark for Gulf crudes (Saudi Arabia, Iran, Iraq, the UAE, Qatar and Kuwait) sold in the Asia-Pacific market. Dubai's API gravity is 31 degrees and has a high sulphur content (2 percent). Initially, Dubai became the main price marker for the region by default as it was one of the few Gulf crudes available for sale on the spot market.

With a large fraction of traded crude oil being priced off Brent, Brent price takes a privileged position within the current oil pricing regime. A number of special features favoured its choice as a reference. In the early 1980s the volume of production in the Brent market was quite large and ensured enough physical liquidity. But similar bases of physical liquidity could be found in other regions of the world, especially in OPEC which constituted (and still does) the largest physical market for crude oil to question the choice of Brent price as a marker. Thus, the volume of production though important is not the determining factor for the choice of a marker. Horsnell and Mabro (1993) identify additional determinants, the most important of which is ownership diversification. The commodity underlying the forward/futures

---

40 If the coefficient of adjustment is one month and this turns out to be too large or too small, subsequent adjustment can remedy this situation. If the volume purchased by a company is more or less constant from month to month, over-pricing in one month could be compensated for by under-pricing in a subsequent month. In such cases, on average, there is no loss for either party.

41 Mexico's formula for sales to the USA is much more complex. It may include the price of more than one reference crude (WTI, ANS, West Texas Sour (WTS), Light Louisiana Sweet (LLS), dated Brent) and may be linked to fuel prices.

contracts should be available from a wide range of sellers and not from a single seller. Monopoly of production increases the likelihood of squeezes and manipulation, increasing the risk exposure of buyers and traders who will thus be reluctant to enter the market in the first place. In fact, Newbery (1984) argued that if a dominant producer has enough market power, then even if market participants have rational expectations, the dominant producer has the incentive to destabilize the spot market and engage in destabilizing speculation in the futures market. Every country in OPEC is a single seller and hence OPEC crudes did not (and still do not) satisfy this criterion of ownership diversification. Monopoly of production also prevented the development of a complex market structure in other markets with a larger physical base such as Mexico. This is in contrast to the Brent market which has always been characterized by a large number of companies with entitlement to the production of Brent. The co-mingling of Brent with Ninian in 1990 and the development of new fields reinforced this aspect and resulted in an even higher degree of ownership diversification.[42]

Because the international benchmarks are differentiated by the type of crude oil and its location, they fetch different prices. For instance, WTI is lighter and sweeter than Brent and as a result of these gravity and sulphur differences WTI typically trades at a dollar or two dollars premium to Brent. The price differential between the international benchmarks, also known as spreads, however are not fixed and can widen or narrow in response to factors other than differences in the intrinsic properties of the crude. Movements in oil price differentials can result from non-parallel movements of either of the underlying benchmark prices or both (for instance in the WTI-Brent differential, movements in the differential can occur either because of a movement in Brent, or WTI or a non-parallel movement of both). These non-parallel movements are possible because each type of crude oil can be influenced by local conditions, the set of traders in the market, their perceptions of the oil crude market. This may lead sometimes to wide spreads between the benchmarks (Milonas and Henker, 2001). For instance, demand for oil has a seasonal component which is not likely to affect two types of crude oil in the same way. In addition, squeezes in the Brent market can cause the price of Brent to rise sharply

---

42 The WTI market is also characterized by a large number of independent producers who sell their crude oil to gatherers based on posted price. Brent however has additional advantages over WTI. First, Brent is waterborne and thus is not subject to problems of pipeline scheduling. Second, Brent is exportable which makes it more flexible to respond to the trading conditions in the western hemisphere (Horsnell and Mabro, 1993).

relative to the price of WTI. For instance, during September 2000 due to a successful squeeze of the Brent market, the price of Brent rose almost three dollars above that of WTI. This increase was not due to a fundamental change in the underlying supply and demand for global oil, but rather to a local event that affected Brent but did not have an effect on WTI.[43]

Although these markets for crude oil are subject to local conditions, they are linked together by arbitrage where arbitrage is broadly defined as the simultaneous purchase of an instrument in one market and the sale of the same instrument in another market in order to exploit price differential. If too large differentials occur between the international benchmarks, arbitrage will ensure that markets return to a state where they move within narrow bands. In this respect, arbitrage between WTI and Brent markets determines the volume of trans-Atlantic trade in crude oil. Accounting for quality differences, if the price of Brent plus transportation costs across the Atlantic is less than WTI, this opens the trans-Atlantic window − i.e. refiners in the USA will import Brent and crudes priced off Brent. When the WTI-Brent price differential is small enough such that it does not cover the transportation costs across the Atlantic, the trans-Atlantic window will close and refiners will rely on crude oil produced in North and South America.[44] The Brent-Dubai differential also determines whether to import European oil or African crudes to Asia.

4.   Features of the Benchmark Crudes and Implications for the Current Oil Pricing Regime

In the early stages of the current oil pricing regime, the marker prices were spot WTI, dated Brent and Dubai. The spot markets of the

---

43 Unlike the WTI-Brent differential which reflects the relative market conditions in Europe and the USA, Horsnell and Mabro (1993) argue that the Brent-Dubai differential does not usually reflect the trading conditions of Asian markets except in some rare occasions such as the Iraqi invasion of Kuwait. In normal times, Dubai crude is more responsive to trading conditions in Europe and the USA than the Far East. Specifically, the authors argue that the Brent-Dubai differential reflects better the relationship between prices of sweet and sour crudes. In support of this hypothesis, they argue that when OPEC decides to cut production, these cuts affect the production of heavy sour crudes. As a result, the price of these crudes will strengthen relative to sweet crudes leading to the strengthening of Dubai prices relative to Brent.

44 Note that because WTI can not be exported arbitrage only involves physical movement of Brent.

marker crudes however suffer from serious problems that have raised doubts about their ability to generate a marker price that reflects accurately the price at the margin of the physical barrel for oil. In this section, we highlight three problems with the spot market for marker crudes. The first problem is that they have become very thin i.e. very little actual trading occurs. The second problem is that while WTI and Brent are used as benchmark crudes, neither of them comes close to representing the marginal barrel any longer as the marginal barrel has become heavier. The third problem arises in the context of the Brent market where some observers have argued that the spot market for Brent does not exist. Each of these problems is discussed below.

4.1 Benchmark Markets Are Thin

The physical markets for the international reference crudes, Brent, WTI and Dubai, have been in fast decline. The Brent system witnessed a dramatic decline in production in the early 1990s. The co-mingling of the Brent system with the Ninian system at the end of 1990 alleviated this problem and the combined production from the Ninian and Brent systems (still known as Brent) increased to around 900,000 barrels per day in 1992. Thereafter, the production of Brent has been falling to around 350,000 barrels per day in 2002. In terms of cargoes, this represents around 20 cargoes per month or less than one cargo per day. To avoid potential distortions and squeezes in the thin Brent market, Platts broadened its definition of Brent to include Forties[45] (UK North Sea) and Oseberg (Norway) as of 10 July 2002. This new benchmark has been given the name BFO. The inclusion of these two grades increased the number of cargoes to around 60–70 cargoes per month. This remedy however created some problems of its own: the seller alone decides which type of crude to deliver i.e. whether to deliver Brent, Forties or Oseberg to buyers. This implies that buyers have to bid for a BFO contract at a price that reflects the value of the lowest quality of crude in the BFO set.[46]

For the last ten years or so, the production of WTI has also been in rapid decline with production reaching less than 400,000 barrels per

---

[45] Forties is a mixture of oil produced from separate oil fields and collected to the terminal Forties Voe in Hound Point.

[46] A large number of companies are responsible for the production of physical Brent, Forties and Oseberg although some consider that Statoil's share in Oseberg might be considered higher than ideal for a forward contract (Bossley, 2004). Since the 21-day BFO contract allows for the delivery of Brent, Forties or Oseberg, this should not pose an immediate problem.

day in 2002. This decline reflects a sharp fall in US crude oil production in the past few years. During 2003, the United States produced around 7.8 million barrels per day of hydrocarbon liquid of which 5.7 mb/d was crude oil. US total oil production in 2003 declined sharply from the 10.6 mb/d averaged in 1985.[47]

Initially, the Dubai benchmark only included crude oil produced in Dubai. The volume of Dubai crude production has dropped from a peak of 400,000 b/d in the period 1990−95 to under 120,000 b/d in 2004, with production hovering around 100,000 b/d in 2005, i.e. there are no more than seven cargoes of Dubai available every month (Montepeque, 2005).[48] Due to this rapid fall in Dubai's oil production, in 2001 Platts introduced a new assessment mechanism that permitted the delivery of Omani crude oil into the Dubai contract.[49] This counteracted the problem of low liquidity and effectively doubled the volume of crude oil in the Dubai-Oman benchmark for price formation. The addition of Oman in the assessment of Dubai however has created problems of its own. In the Dubai-Oman benchmark, Oman crude has lower sulphur content and higher gravity than the Dubai crude. The increase in demand for lower sulphur and higher gravity crude oil relative to heavier and sour crude caused mainly by underinvestment in hydro-cracking capacity in refineries has widened the price gap between the two types of crude. Since September 2004, the Dubai-Oman markets have diverged with the spread between the two grades reaching $2.80 per barrel in October 2004 (Montepeque, 2005).[50] As a result of this divergence, many observers are calling for another crude to be included in the Dubai assessment process which is closer to Dubai than Oman.[51]

In March 2004, Platts introduced another innovation, the partials

---

47 EIA (2005), United States Country Analysis Brief.
48 As a result of low liquidity, the Dubai benchmark witnessed a squeeze which sent the price of Dubai above the price of Oman.
49 Platts has also published a proposal to incorporate Abu Dhabi's Upper Zakum into the Dubai assessment.
50 It is important to note that such quality differences do not only arise from the commingling of two different crudes. The introduction of existing fields into one of the crudes can have similar effects. This is not particularly relevant in the Dubai context, but can play a role in other markets. For instance, the introduction of the Buzzard field into Forties blend can affect the quality of this crude and hence change the size of differential relative to Brent and Oseberg.
51 The pricing of a crude off Dubai-Oman requires setting two coefficients of adjustment (one off Dubai and one off Oman) and then taking some average between the two coefficients.

mechanism, to counteract the problem of Dubai's low production. The partials mechanism has the effect of cutting the cargo into small parcels that can be traded. In the Dubai context, the smallest trading unit was set at 25,000 barrels. Since operators do not allow sale of cargoes of that volume, it meant that a seller of a partial contract would not have been able to meet his contractual obligation. Thus, delivery will only occur if both buyer and sellers trade 19 partials totalling 475,000 barrels. Any traded amount less than 475,000 barrels is not deliverable and should be cash settled (Platts, 2004).[52] Montepeque (2005) argues that this innovation has produced encouraging results, increasing the trading liquidity and hence improving the efficiency of spot price determination, reducing the bid/offer spreads, and attracting new players to the market.

In thin and illiquid markets where actual deals are far apart and irregular, the number of price quotations for actual transactions is quite small. But for a crude to act as a reference or benchmark, price quotations should be generated on a regular basis. Here lies another important feature of the current pricing system. In order to obtain a regular flow of price quotes and daily price assessments of reference crudes, markets rely on oil price reporting agencies such as Platts and Petroleum Argus for price discovery. The prices generated are assessed prices and not market prices. In their assessment, oil price reporting agencies obtain information from market participants about the deals concluded and the bids and offers made. In order to provide reliable price assessments, reporters should observe plenty of arm's-length deals between market participants. Since market participants have different interests and different positions (short or long), it is very unlikely that they would reveal the actual price used in the deals. In a liquid market, this can be less of a problem since by pooling information from different participants biases could cancel out (but not necessarily). Furthermore, the reporters would then be observing actual trades taking place and thus are able to assess more accurately the gathered information.

Some observers have argued that in principle, there is not a certain level of production below which the integrity of the market is threatened. Before its substitution by WTI, the Alaskan North Slope (ANS) continued to generate market prices although the physical base was very narrow. The prices were derived completely from oil price reporting agencies' assessments of traders' perceptions about what the price

---

52 Settlement of cash differences that result from undeliverable partials uses the last price assessment of the trading month.

would be if there were actual trade in cargoes. Thus, they argue an oil pricing regime can survive without a physical base as long as there is confidence in it and as long as players want it to survive (Rehaag, 1999). This argument is unconvincing because confidence is unlikely to survive for very long in these conditions. In effect, markets become thinner and thinner, squeezes and distortions become more widespread and as a result prices become less informative and more volatile thereby distorting consumption and production decisions (Pirrong, 1996).

4.2   *The Marginal Barrel is Becoming Heavier*

While WTI and Brent are used as benchmark crudes, neither of these two come close to representing the marginal barrel any longer. In face of increasing global demand, the marginal barrel is heavier and sourer than these benchmark crudes. Furthermore, the marginal refining capacity in the world cannot process heavy, sour crudes. The conversion of existing refining capacity to deal with this new reality is expensive and time-consuming and thus is unlikely to take place in many parts of the world very soon. Faced with this refining bottleneck, the relative demand for sweet oil continues to rise creating a large wedge between sweet crude and sour crude prices. This represents a fundamental change in the oil industry and pressures to change the world benchmarks to heavier more sour crudes will become stronger.

4.3   *A Missing Spot Market for Brent*

The foremost important issue in formula pricing is the identification of the marker price. Ideally, the marker price should be generated in a physical market where supply and demand interact. In such a market, the price of a barrel of oil will be determined at the margin. The concept of referencing against international spot market crude however does not refer to such a construct. This will be explained in the context of the Brent market.

Generally speaking, a spot market transaction is a transaction in which oil is bought or sold at a price negotiated at the time of agreement and for immediate delivery. Although dated Brent is considered as the spot market for Brent, dated Brent contracts contain an important element of forwardness in them. Contracts for 21-day Brent are forward contracts that specify the delivery month but not the particular date at which the cargo will be loaded. Under the 21-day Brent contract, the seller is required to provide the purchaser at least 21 days notice as to when the cargo will be loaded. Once the notice period is expired, the Brent oil to be loaded on a specific date is traded on the dated Brent market. Thus, although the dated Brent is referred to as a spot transaction, a dated cargo is not for immediate delivery as there is a 21-day period between the time when the cargo becomes dated and the actual loading date. In other words, a dated Brent contract is a contract to buy or sell a cargo of Brent for loading up to 21 days from the date of agreement of the contract.

This has some implications. First, during this period, dated Brent can continue to be traded even after the expiration of the notice period. Second, the time gap implies considerable price risk between the price of dated Brent and the price of the cargo at the loading date. Various over-the-counter (OTC) markets developed to manage this form of price risk. These OTC markets became an integral part of the Brent market complex. Finally, it implies that a spot market for Brent is missing. As Caumon and Bower (2004) argue, a 'parcel of that particular commodity crude oil grade (Brent) has no value as a spot commodity at that location. However, it will have value on the short term forward market, called the dated Brent market....A true spot market for Brent and other North Oil sea grades is therefore missing.'

The significance of the forward element is also reflected in the important shift in the pricing of dated Brent. While in the early stages of the current oil pricing regime most deals were done on an outright basis, by 1991 all dated Brent was priced as differential to another crude oil price. The parties no longer agree on the absolute price of a cargo (for instance $40 per barrel), rather they agree on a price differential to the first month forward Brent (for instance 30 cents above the first month Brent forward on the loading date). In other words, the forward Brent market sets the level of prices while the spot market sets the differential.

5.   The Shift to the Futures Market

The declining liquidity of the physical base of the reference crudes and the narrowness of the spot market have caused many oil-exporting and oil-consuming countries to look for an alternative market to derive the price of the reference crude. The alternative was to be found in the futures market. In what follows, we discuss the main reasons for the shift to the futures market and assess the benefits and costs of such a shift.

Although the basic blocks of the current system have remained stable, the market-related oil pricing regime has undergone some major transformations in the last few years, the most important of which is the shift of price determination from the spot to the futures market. When formula pricing was first used in the mid 1980s, the WTI and the IPE futures contracts were in their infancy. Since then the futures market has grown to become not only a market that allows producers and refiners to hedge their risk and speculators to take positions, but is also at the heart of the current oil pricing regime where the price of oil is determined.

Instead of using dated Brent, several major oil-producing countries such as Saudi Arabia, Kuwait and Iran rely on the IPE Brent Weighted Average (BWAVE) as the basis of pricing crude exports to Europe. [53] The BWAVE is the weighted average of all futures price quotations that arise for a given contract of the futures exchange (IPE) during a trading day. The weights are the shares of the relevant volume of transactions on that day. Specifically, this change places the futures market, which is a market for financial contracts, at the heart of the current pricing system.

A futures contract is an agreement between a buyer and seller in which the seller (the short) agrees to deliver the buyer (the long) a specified quantity and grade of an identified commodity at a fixed time in the future and at a specified price agreed when they enter the contract. [54] The only feature of a futures contract that is in continuous adjustment is its price which changes according to the demand and supply of these contracts.

Unlike long-term contract and spot market transactions, the main purpose of the futures market is not to provide a mechanism for actual delivery, but rather a mechanism that allows market participants to spread the risk of price volatility. [55] The WTI and IPE Brent futures

---

53 Thus, the settlement price index (referred to as the IPE Brent Index) against which contracts should settle does not converge to spot prices but follows closely the prices of 21-day BFO. Specifically, the IPE Brent Index is calculated as the average of first month trades in the 21-day FBO market, the second month trades in the 21-day FBO market plus or minus a straight average of the spread trades between first and second months and an average of assessed prices published by various reporting oil price agencies.

54 A key feature of these futures contracts is standardization which facilitates their trading on the exchange. Thus, a single futures contract can be traded so many times up to the expiration of the contract.

55 One of the major advantages of a futures contract traded on recognized exchanges rather than a forward contract is that every futures exchange has a clearing house that guarantees performance on all the contracts traded on the exchange. To guarantee contract performance brokers trading on behalf of their customers must place a deposit or initial margin related to the value of

markets are no exception. Because it is not a mechanism for actual delivery, the volume of crude oil traded under these contracts can exceed by far the actual volume of oil production. For instance, in the case of the WTI futures market, a US Senate subcommittee notes that 'over the 7 years that the December 2001 NYMEX light sweet crude oil contract was traded, 5 billion barrels were traded, but only 31,000 barrels were actually delivered on those contracts' (p.37).[56]

The IPE Brent futures market is also rarely used for delivery although the IPE states that 'the IPE Brent crude futures contract is a deliverable contract based on EFP (exchange of futures for physical) delivery with an option to cash settle'. This statement is somehow overstating the importance of physical delivery in the futures contract. In practice, most open future positions are settled on the basis of cash against the IPE Brent Index. Even if a trader opts for the EFP option, the closing of the position using the EFP method involves the transformation of a futures contract into a physical forward contract.[57] This is not the same as delivering physical oil against a futures contract. Specifically, if the holder of a Brent futures contract wishes to ensure physical settlement, then he has to enter an OTC agreement with an equity producer of Brent. Although this conversion has to be registered with the IPE and London Clearing House, the LCH does not guarantee the settlement. Also the contract can not be cash settled two days after the trading ceases for such contracts. The implication of this arrangement is that there is a separation between the physical commitment and the determination of price since the parties will have to agree on the new price. There is an additional aspect of EFP that should be highlighted. When the cargo allocation period closes on the ninth day of the delivery month and if it happens that there are still unallocated EFPs, then these have to be cash settled outside the system.

---

the positions taken on their trades and sufficient to cover customer losses due to changes in futures prices. Thereafter, the position is daily 'marked to the market'. The clearing house computes daily for each member the net gains or net losses on the futures position that resulted from price movements during that day. The clearing house then collects the money from accounts that have suffered losses and pays them over to accounts that have gained value. These daily payments are known as variation margins and should be made before the market is open for trading the day after. In this manner, the risk that the other party will default is dramatically reduced.

56 Untitled Document from the Senate Committee on Homeland Security & Government Affairs downloaded from *http://www.senate.gov/~gov_affairs/ psisec3pricingofcude.pdf*

57 If the price agreed does not suit the contract holder then he will opt for cash settlement.

The futures market has attracted a wide range of participants. Table 1 shows the ownership structure of outstanding NYMEX crude oil futures contracts by sector. As can be seen from this table, the players in the futures market are various and include oil traders, integrated oil companies, refiners, marketers, financial institutions, investment funds, investors and floor traders. Oil traders are the major participants in the NYMEX crude oil futures accounting for 40% of outstanding NYMEX crude oil futures contracts. It is interesting to note that financial institutions, investment funds and investors account for 25% of the outstanding NYMEX crude oil futures. On the other hand, oil producers seem to be the least active in the futures market accounting for only 1% of outstanding NYMEX oil futures contracts.[58]

Table 1:   Ownership of Outstanding NYMEX Crude Oil Futures Contracts by Market Parties (January–October 2000)

Oil Traders	40%
Financial Institutions	13%
Integrated Oil Companies	13%
Refiners	9%
Investors	7%
Investment Funds	5%
Floor Traders	5%
Marketers	5%
Producers	1%
End Users	1%

Source: The Senate Committee on Homeland Security & Government Affairs downloaded from *http://www.senate.gov/~gov_affairs/psisec3pricingofcude.pdf*

---

58 The Commodity Futures Trading Commission (CFTC) reports the open interest position of 'commercials', which are usually treated as hedgers, and 'non-commercials' usually treated as speculators. This distinction is rather artificial. Hedging and speculation define agents' operations and not the agents themselves as agents usually undertake both hedging and speculation. Also in a recent survey conducted on the heating oil futures markets Ederington and Lee (2002) find that although non-commercials can be treated fairly accurately as speculators, many players classified within the 'commercials' category do not have any energy assets and hence it is inappropriate to treat all 'commercials' as hedgers. The survey also reports some interesting findings regarding the trading strategy of market participants. Instead of being either long or short, large traders hold spread positions such as inter-crude or calendar spreads. This particularly holds for floor traders and marketers.

5.1   Reasons for the Shift

The shift to this new stage was justified by the perceived inadequacies of the spot market. Many have argued that the natural decline in oil production in Brent reduced the overall market liquidity and hence increased the crude market's vulnerability to manipulation, distortions and squeezes. A squeeze refers to a situation in which a trader goes long in a forward market by an amount that exceeds the actual physical cargoes that can be loaded during that month. A corner is an extreme case of a squeeze when one trader acquires the entire cargoes of a particular month. If successful, the squeezer will claim delivery from sellers who are short and will obtain cash settlement involving a premium. One consequence of a successful squeeze is the increase in the price of the first forward month relative to dated Brent and relative to the price of the second forward month. Naturally, the price of the particular crude that has been squeezed will rise relative to that of other marker crudes. Sometimes a reversal of the sign of the usual differential may occur (for example, the Brent price in a squeezed market may exceed the WTI price although WTI is of better quality).

Squeezes are made possible by two features: the anonymity of trade and the huge volume of trading compared to the underlying physical base (Mollgaard, 1997). After all, squeezes are much easier to perform in a thin market such as Brent (Telser, 1992). This is in contrast with the futures markets where the volume of transactions is quite large and thus there is less room for squeezes and manipulation. Although it is easier to squeeze a thin spot market, it is important to note that the futures market is not totally immune from being squeezed or manipulated.[59]

The decline in liquidity and the peculiarities of the Brent market have also increased the difference between the Brent spot price and that of the nearest futures i.e. the basis risk. In recent years, the basis

---

59 The challenge of the U.S. Federal trade commission to the BP Amoco-Arco merger was partly based on the fear that by controlling the physical infrastructure, the WTI futures market can be squeezed. The Federal trade commission notes that 'the restriction of pipeline or storage capacity can affect the deliverable supply of crude oil in Cushing and consequently affect both WTI crude cash prices and NYMEX futures prices' (p.7). Then it states that 'a firm that controlled substantial storage in Cushing and pipeline capacity into Cushing would be able to manipulate NYMEX futures trading markets and they enhance its own positions at the expense of producers, refiners and traders' (p. 7) (United States of America Before Federal Trade Commission in the Matter of BP AMOCO P.L.C. and Atlantic Richfield Company downloadable from *http://www.ftc.gov/os/2000/08/bparco.pdf*

has fluctuated so widely that both oil-producing and oil-consuming countries grew increasingly dissatisfied. This has led to the development of new financial instruments such as CFDs to hedge this basis risk. Ironically, while CFDs provide price insurance against a volatile Brent market, it has also encouraged speculators and increased the attractiveness of squeezes. This has led one observer to argue that,

> the array of instruments available to traders enables a small number of powerful and sophisticated players to operate squeezes or launch other operations which cause prices to move in directions that do not always reflect the actual state of the supply/demand balance. Whether these games whose frequency has been increasing in recent years affect price trends over the medium term is debatable. It is certain however that they cause higher price volatility and that they rob prices from their most important function which is to signal at every movement the state of the supply/demand balance (Mabro, 2000).

Another reason for shifting to the futures market for price determination is that a futures price is after all determined by actual transactions in the futures Exchange and not on the basis of some assessed prices by oil reporting agencies. Furthermore, the timely availability of futures prices enhances price transparency. Futures contracts are traded for delivery of commodity at various points in the future which results in the establishment of many different prices for that commodity at different points in time. It is often argued that these prices reflect market participants' perceptions about the current supply and demand conditions in the oil market. In the case of price for future delivery, market participants also bring their expectations about prospective demand and supply at various points in the future. Futures prices that are sufficiently higher than current spot prices reflect the market expectation of a relative shortage of the supply of a commodity in the future. On the other hand, futures prices that are sufficiently lower than current spot prices reflect the market expectation of a relative surplus of the supply of a commodity in the future. Through futures trading, information about market participants' expectations will be summarized in a single futures price. These futures prices are continuously updated and are disseminated to the public. At any time, a seller or a buyer can look at the prevailing price and use it in the spot and term contract. The volume of daily transactions and open positions is additional useful information to gauge the liquidity of the market.[60]

---

60 One observer doubts that the reduction in liquidity is the major factor responsible for the shifting of price determination towards the futures market. They argue that talks of frustrations of some OPEC Members about the ongoing

Although some OPEC members have expressed dissatisfaction with frequent manipulation of the spot markets and thus shifted to the futures market for price to be used in formula pricing, this shift is fraught with dangers. As we shall see in Section 6, it means that OPEC's influence on prices is now dependent on the expectations of participants in the futures markets. The large number of players and the large variety of participants (floor traders, fund managers, refiners, producers, financial institutions, and speculators) have certainly complicated the process of decision making within OPEC. The list of factors that OPEC has used to explain its decisions since 1999 has expanded to include not only the level of oil and product stocks and supply/demand situation but also the open interest position of 'non-commercial' traders on the futures market (Garcia, 2005; Mabro 2005).[61]

However, substituting an illiquid market for physical oil with a highly liquid market for financial contracts should not be considered as a triumph of the market. In fact, 'it is a major defeat in that it implies that self-regulation has failed in the area of pricing spot crude oil' (Horsnell, 2000). The question is why has self-regulation not worked? Part of this answer might lie in producers' behaviour. Mollgard (1997) argues that 'at any rate, if the producers (i.e. the North Sea equity producers) think that the market (i.e. Brent market) will not lose customers, their incentive to assuage the effects of squeeze is of course non-existent' (p.111).

---

manipulation of dated Brent should not be taken at face value as national oil companies such as Aramco benefited whenever squeezes raised dated Brent prices. Thus, Rehaag (1999) states that in fact 'on average Aramco has probably gained as much from past manipulative plays on physical Brent as it has lost' (p.xiv). He argues that the issue is not that of low liquidity and manipulation. Instead, he claims that Aramco's move from a spot market to a futures benchmark could be the result of the addition of new participants in the Brent market which made it more competitive than any time in its history. Thus he raises the following questions: 'Is this the cause of Aramco's concern that North sea has become too aggressive, too independent, too competitive. Is it too hard to influence and exert any measure of control?'(p.xviii) This argument however is misleading and neglects the fact that moving to the futures pricing introduces new players over whom OPEC's national oil companies have even less control and whose actions are very difficult to influence. If control is an issue, then it is infeasible to control behaviour in the futures markets.

61 In addition, other factors have been mentioned such as speculation, basket price range, geopolitical factors and US$ exchange rate (Garcia, 2005).

5.2   *Assessment of the Shift*

Thus, the current oil pricing regime is market related, but it is the market of futures contracts that determines the most important international oil price. This raises the following questions: What do prices in the futures market reflect? Do these prices always reflect the fundamentals of the underlying physical market for crude oil? What about speculation? Does speculative behaviour destabilize oil markets and exacerbate oil price volatility? Providing answers to these questions is important to evaluate the performance of the current oil pricing system.

Relationship between the Futures and Spot Markets

Garbade and Silber (1983) developed a model that analyses the relationship between the futures market and the underlying spot market. Specifically, they are interested in testing whether price formation in the futures market influences and improves the pricing in the spot market. This aspect is referred to as the price discovery function of the futures market. That spot prices influence the futures price is quite expected as futures markets are derivative markets which are based on the spot market. The price discovery aspect emphasizes the opposite channel, i.e. a situation in which the formation of prices in the futures market influences the prices in the spot market. Futures markets can influence spot markets by allowing market participants to hedge, providing liquidity, affecting producers' production decisions and providing information about prices. It is important to note that the effectiveness of the price discovery function depends on the close relationship between the prices of futures contracts and spot prices.

Under the assumptions of no taxes or transaction costs, no limitations on borrowing, no costs to storing a long cash position, no limitations on the sale of the commodity in the cash market and that the cash market follows a Gaussian diffusion process, Garbade and Silber (1983) show that the futures price will be equal to the cash price plus a premium reflecting the deferred payment on a futures contract. These assumptions also imply that arbitrage will be infinitely elastic such that if the futures price is different from the spot price plus the premium, then the hedger can make an immediate risk-less profit. In the extreme situation when all these assumptions hold and there is perfect arbitrage, Garbade and Silber (1983) show that the futures contracts will be a perfect substitute for the spot market and in fact there will be no meaningful distinction between the two markets. In the other extreme case when there is no arbitrage, then the futures contract will be a poor substitute for the cash market and there will be no tendency for the futures prices and spot prices to converge. In this extreme case, the futures market ceases to perform its dual roles of price discovery and risk shifting. In the intermediate case when there is less than perfect arbitrage, prices in the two markets will follow an intertwined random walk. This intermediate case is the most realistic one as arbitrage transactions between the futures and spot markets are not risk-less and hence arbitrage is not perfect as the basis can change due to a number of factors such as the grade and location of delivery of the commodity, constraints on storage and availability of capital.[62] In this intermediate case, the higher the degree of arbitrage, the more correlated are the unexpected changes in futures and spot prices and hence the price divergence that might occur between the two markets will be more rapidly eliminated.

Garbade and Silber (1983) perform empirical analysis for seven commodities (crude oil is not included) and find that for all the seven commodities, markets are well integrated over one or two months, but there is considerable divergence between cash and futures markets over shorter time intervals, especially for grains. As regards the price discovery function, they find that in general futures markets dominate. Spot markets act just as satellites of the futures markets. Most of the new information is incorporated first in the futures markets and then flows to the spot market.

Garbade and Silber's method has not been applied to the crude oil markets. However, it is possible to derive some interesting conclusions from the above study. Their results suggest that prices determined in the futures market will not always reflect the underlying conditions or fundamentals of the spot market. Divergence between the spot and futures markets can occur from time to time. Whether the divergence occurs regularly and persists for a long time depends on the specific features of the storable commodity and is an empirical issue. When such divergence occurs, the question becomes: which market should be used for oil price determination? In the current oil pricing system, the futures market seems to have the upper hand. As to the price discovery function, one would expect (as in other commodities) the futures market to dominate the spot market for crude oil. As argued above, the futures market determines the most important price in the oil market and it would be hard to imagine any transaction in the spot market taking

---

62 Shleifer and Summers (1990) identify other risks that can result in limited arbitrage. The authors argue that 'the assumption of limited arbitrage is more general and plausible as a description of markets for risky assets than the assumption of perfect arbitrage which market efficiency relies on' (p.30).

place without parties taking into account information about prices flowing from the futures market.

Speculation
There has been a long debate on whether futures trading improves the efficiency or destabilizes prices in the underlying spot markets (see Mayhew (2000) for a recent review). This debate is closely linked to the role of speculators in the futures market. Futures prices summarize the collective expectations of market agents about prospective demand and supply and hence provide valuable information that can contribute to more efficient inter-temporal prices. When a futures market is introduced, a speculator can buy a commodity when prices are low and simultaneously lock in the profit by selling a futures contract. If carrying costs are predictable, then price smoothing through storage becomes an arbitrage activity.[63] In fact, based on this analysis, many models predict that the futures market has an unambiguously stabilizing effect on the spot market (see for instance Peck, 1976). A survey of the empirical literature by Mayhew (2000) provides support for this view. His survey reveals that the futures market has either no effect or in fact can contribute to the stability of the underlying spot market.

Contrary to the preceding studies, some have argued that futures markets can destabilize the spot market in such a way that they move further from the underlying fundamentals. Newbery (1984) for instance argues that the futures market allows producers to hedge for risk and as a result may encourage them to adopt riskier production technologies with the effect of destabilizing prices. Figlewski (1981) argues that less experienced participants in the futures market can transmit noise to the more informed participants in the spot market. Stein (1987) provides support for this view where he shows that the entry of new speculators can change the information set for spot traders and destabilize prices even when speculators are rational profit maximizers. However, this destabilizing effect can not persist for a long time. If the futures markets

---

63 Notice that speculation in this context is very specific and is associated with arbitrage activity. Speculation can have different though related meanings in various contexts. For instance, suppose that all investors receive the same information set but traders interpret this information differently. Then one might consider speculators as those traders who take positions on the opposite side of the market. Speculation can also be related to the concept of offering insurance to hedgers. In return for providing insurance, speculators will obtain a financial gain. There is also the popular view of speculators. In this view, speculators are perceived to be irrational and uninformed gamblers trading on noise and sentiments rather than any fundamentals. Noise traders are discussed in the next subsection.

continue to provide wrong signals, then spot market participants would stop basing their decision on the signals from the futures market, unless information in the underlying physical market is also unreliable.

A related issue is whether heavy trading in the futures market is good or bad for price discovery. Again the evidence here is mixed. Some take the viewpoint that a high volume of trade indicates a higher rate of information flow and thus there is a monotonic positive relationship between volume of trading and efficiency of price (Easley, Keifer and O'Hara, 1997). From a slightly different angle, some have argued that heavy trading indicates the extent of different opinions among market participants, in which case heavy trading helps in price discovery through the aggregation of heterogeneous beliefs, interpretation of the same information set or valuation models (Brandt and Kavajecz, 2004). Another strand in the literature argues the opposite. Investors may be overconfident about the precision of their information and as a result they trade too much (Odean, 1999; Benos, 1998) and overreact to private and under-react to public signals (Daniel, Hirshleifer and Subrahmanyam, 1998). Various studies have shown that this overconfidence can persist (See for instance, Wang, 2001).

Noise Trading
In a number of articles in energy and non-energy publications and international policy reports, many observers have argued that a large component of transactions in the crude oil futures markets is being driven by speculation. It is often argued that speculators, attracted to oil markets by higher returns, trade on noise and sentiment rather than on fundamentals with adverse effects on the functioning of oil markets. The *BIS Quarterly Review* (2004) notes that, 'the rapid increase in oil prices in recent months has focused attention on the role of speculators in the oil market. With prices in most major equity, bond and credit markets moving sideways or even declining, investors in search of higher returns have reportedly turned to commodity markets, oil in particular' (p.6). Then the report notes that the changes in non-commercial traders' net long position have tended to coincide with changes in the oil price. Thus, 'it is possible that the larger presence of non-commercial traders in the oil market contributed to herd-like behaviour. Their presence, coupled with the upward trend in oil prices, might have made traders wary of positioning against further increases in oil prices, thereby effectively reinforcing the upward trend' (p.6). The Federal Reserve Chairman Alan Greenspan (2004) also tends to believe that speculation might have played a role in affecting oil prices. In his September 2004 testimony before the House Budget Committee, Greenspan identified

that high oil prices might be caused by 'speculators and investors who took larger net positions in crude oil futures'.

Most observers of oil markets tend to adhere to the view that speculators are merely noise traders who trade on bad information or even no information at all. Black (1986) defines noise traders as agents who sell and buy assets on the basis of irrelevant information. The noise traders transact on the basis of extrapolating past trends (technical analysis) or irrational investors' sentiments (herding) rather than on market fundamentals or the arrival of new information (see for instance Shiller (1981) and Shleifer and Summers (1990) among many others). These are usually contrasted with arbitrageurs, rational speculators or 'smart money' that trade on the basis of information and thus tend to push prices towards fundamentals.

Although there is a general agreement that noise traders exist in financial markets, the traditional view has been that these noise traders can be ignored in models of price formation. The traditional view's argument is that irrational investors or speculators trading on noise will lose their money and thus will not survive in the market. If noise traders hold overvalued assets, then arbitrageurs should sell these assets to noise traders and push down their prices. As prices fall towards the fundamental value, then noise traders will lose money and over time will go bankrupt. If on the other hand, the trader holds over-valued assets, then arbitrageurs should purchase the asset from the noise trader and raise the price towards its fundamental value making a profit in the process. The net effect is that noise traders will lose money and would sooner or later exit the market. This argument was forcefully made by Friedman (1953) who states that 'people who argue that speculation is generally destabilizing seldom realize that this is equivalent to saying that speculators lose money since speculation can be destabilizing in general only if speculators on average sell low and buy high' (p.175).

This traditional view has been challenged recently. Many papers have shown that the survival of noise traders is possible. Shleifer and Summers (1990) argue that on average noise traders may be more aggressive than arbitrageurs, either because they are more optimistic or overconfident, and thus are likely to bear more risk. If higher risk is rewarded in the market, then noise traders can earn higher expected returns on average and hence as a group they need not disappear from the market. Specifically, no individual rational investors will know with certainty that all other rational investors will push the price of the financial asset towards its fundamental value in the period she is holding the asset. If this is the case, then arbitrage is not risk-less because there is always the chance that noise traders will continue to push prices even further away from its fundamentals causing in this case a loss for the rational speculators. Thus, it is not necessarily true that fully rational speculators or arbitrageurs can always arbitrage away the impact of noise traders. If noise traders hold large shares of assets subject to noise-trader risk, they can earn an above average return and survive in the market. In a more recent paper, Kogan, Ross, Wang and Westerfield (2003) find that irrational traders can affect prices significantly even if trading decreases their wealth over time. In other words, the price impact of irrational traders does not rely on their long-run survival. They also show that the ability of irrational traders to impact prices even when their wealth is diminishing can significantly affect their chances for long-run survival. That is, in the long run, a price impact can occur regardless of whether the irrational traders survive or not.

Even if noise trading in financial markets can persist, the main issue is whether changes in demand due to noise trading are big enough to affect prices and destabilize the market. The answer has to do with the potential of herding. If the shifts in demand are correlated across noise traders and do not cancel each other out then noise trading is capable of influencing market prices. Furthermore, the potential for herding behaviour implies that arbitrage is not riskless and hence it is not necessarily the case that arbitrageurs will always be able to arbitrage away the noise trade. In fact, the arbitrageurs may not have the incentive to counter shifts in demand by noise traders and may instead decide to ride the wave in the hope that they can dispose of the assets near the top before the noise traders. In other words, arbitrageurs can amplify the price bubble. Hart (1977) showed how a sophisticated speculator can take advantage of the naïve forecasting techniques of less sophisticated speculators and then destabilize the futures market. In an interesting study, Brunnermeier and Nagel (2004) show that rational traders may have the incentive to trade in the same direction as irrational traders in the short run (i.e. herd themselves) if convergence is expected to be slow.

Herding results from investors' decisions to follow the trading strate-gies of others. The literature has identified a number of reasons why investors may decide to engage in herding.[64] One approach explains herding in terms of investors' irrational behaviour (De Long et al,

---

64 This should be distinguished from 'spurious herding' where investors facing the same information set decide to undertake the same investment decisions. Spurious herding is efficient whereas intentional herding may be inefficient. In practice it is very difficult to distinguish between the two forms of herding (Bikhchandani and Sharma, 2000).

1990; Froot et al, 1992). Another approach tries to explain why fully rational profit maximizing investors may be influenced by other investors' decisions and decide to reverse their investment strategy and follow the herd. Several reasons are given, the most important of which are informational asymmetries (Bikchandani et al, 1998), compensation structures of money managers (Roll, 1992) and reputation concerns (Scharfstein and Stein 1990).[65] These are discussed in the context of institutional investors.

The Role of Institutional Investors
In a number of articles in energy and non-energy publications and international policy reports, observers have argued that a large component of transactions in the crude oil futures markets is being driven by institutional investors' noise trading. They lament the entry of new large players such as the short-term technical funds, commodity pools, hedge and investment funds, and market markers into the oil futures markets with limited knowledge of the petroleum industry. Given the large size of institutional investors, their trading strategies are likely to influence equilibrium price setting and tend to push prices away from fundamentals. There is a common belief that these institutional investors engage in psychology trading,[66] herding and irrational feedback trading behaviour and as a result contribute to the destabilization of financial markets and induce oil price volatility beyond what would have been generated in a market composed only of individual investors.

A number of theoretical studies have suggested that institutional investors have a tendency to engage in rational herding or momentum trading. For instance, Lakonishok, Shleifer and Vishny (1992) attribute herding behaviour to the agency problems in the money management industry. Due to agency problems, the manager may act on behalf of her own interests (for instance maximizes her commission income) rather than acting on behalf of the fund she is responsible for. These principal agent problems may result in short-term horizon investments and as a consequence institutions may adopt similar trading behaviour even if their own information suggests otherwise. For instance, Roll (1992) argues that if a manager's compensation

---

65 For a review of this approach see Bichchandani and Sharma (2000).
66 Some of these psychological factors were exaggerated in recent years. For instance fears of shortage or disruption of future oil supplies due to the fragility of the ruling family in Saudi Arabia and the possibility of religious extremists taking over or affecting the operations of the richest oil fields received considerable media attention. It remains unclear whether such and similar extreme and implausible views carried any weight in market agents' expectations.

depends on how her performance compares to other managers, this can distort the managers' investment decision resulting in holding of inefficient portfolios and causing herd behaviour. In Scharfstein and Stein's (1990) model, herding occurs because by conforming to other investment professionals, an investment manager is likely to preserve her reputation when there is uncertainty about a manager's ability to pick the right stocks. The reputation of an unsuccessful investment manager will be less damaged if all other managers have also made unprofitable investments. Other models have emphasized irrational herding or momentum trading. According to these models, institutional investors can engage in feedback trading in which institutions tend to sell when prices are low and buy when prices are high and move prices away from fundamentals.

Herding is likely to affect markets when no investors take offsetting positions. Some have argued that institutional players are diverse both in their time horizon and trading strategies and while some investors such as technical funds may use trading strategies based on technical analysis and herding and have a very short-term horizon, others such as hedge funds and commodity pool operators invest in obtaining market information, specialized skills and may even have a long-term horizon. Badrinath and Wahal (2002) argue that it is possible for the momentum trading by some institutional investors to be offset by other institutional investors' contrarian trading and therefore institutional investors as a whole may have no impact on price formation. This argument however does not take into account that not all institutions have the choice to act in a contrarian manner. For instance, mutual funds must adhere to asset allocation strategy and hedge funds which are expected to take a contrarian position may be forced to herd, for example if credit is freely available in the upturn of asset price and sharply withdrawn in the downturn (Davis, 2003). Furthermore, because of the risk reward structure of incentives, it is also possible that fund managers will not have the incentive to deviate from the benchmark and thus will not be willing to take contrarian positions. In other words, there might be preference for a fund manager to follow other managers just because the consequences of taking a contrarian position that turns out to be wrong are much more severe than the consequences of a manager performing badly along with all her other peers.

Despite the richness of these theoretical discussions, the empirical evidence on herding lags behind. Empirical studies from other financial markets reveal that generally institutional investors feedback trade more than individuals (Nofsinger and Sias, 1999). However, they also show that the extent of such feedback trading is rather modest. In reviewing the literature, Sias (2005) notes that out of the eleven studies reviewed, four did not find any momentum trading, five did find weak evidence of institutional momentum trading, while two found strong evidence of momentum trading. He concludes that 'as a whole, extant evidence of institutions momentum trading is, at best very weak' (p.2).[67]

Despite the weak empirical evidence, the general view from the oil industry is that the entry of institutional investors to the oil futures market has influenced both oil price levels and oil price volatility. In the absence of empirical evidence from the crude oil market, it is very difficult to evaluate the extent of the impact of institutional investors on oil prices. However, the popular debate on the role that institutional investors play in the oil market has so far neglected a few important aspects. First, the argument that institutional investors only trade on noise and sentiment is oversimplified. There is some evidence from other markets that institutional investors may in fact have superior information.[68] In other words, it is not necessarily true that institutional investors will only bring 'hot' or 'dumb money' and that their entry is always associated with higher noise trading. Second, the observation that changes in non-commercial traders' net long position have tended to coincide with changes in the oil price does not establish that speculators influence market prices. This observation could be the result of change in fundamentals that affect both oil prices and the futures position of speculators. That is why the BIS report when commenting on the role of speculators in the oil market has been careful in noting that 'it is also possible that shifts in activity in the futures market were driven by changing perceptions of fundamental imbalances in the supply of and demand for oil, including the changing perceptions of commercial traders' (p.6). In the case when speculators react and change their position on the basis of the arrival of new information, speculation is not necessarily destabilizing. In fact, if the speculators have superior information that enables them to respond quickly to the arrival of new information, then they may even improve the functioning of the market by speeding the price adjustment process. Finally, institutional investors enhance the liquidity of the futures market and reduce the chances of successful squeezes. In fact, as argued above, this has been the prime

---

67 This weak empirical evidence may be due to the fact that herding takes place periodically and is not a continuous phenomenon. Herding also occurs in stress rather than normal times. These factors make it difficult for standard econometric techniques to detect herding based on average volatility (Davies, 2003).

68 For instance, Trzcinka (1998) shows that Initial public offerings chosen by institutional investors tend to outperform those chosen by the general public.

reason for oil-producing countries to shift to the futures market for price determination.

Empirical Evidence from the Crude Oil Markets
It is clear from the above brief review that the theoretical literature provides different predictions regarding the price discovery function of the futures market and whether price formation in the futures market enhances efficiency or destabilizes prices. Empirical studies on the crude oil market provide mixed evidence. Using a Generalized Autoregressive Conditional Heteroskedastic (GARCH) model before and after introducing the futures market, Antoniou and Foster (1992) find that the introduction of the futures markets did not have any effect on the underlying Brent crude oil spot market. For the petroleum derivatives, Nainar (1993) examines the relationship between futures trading, market information and spot prices and finds that past prices are significant in explaining spot prices in periods of futures trading. This is interpreted as implying that futures trading contributed to increased market information in the spot market.

In a more recent paper, Fleming and Ostdiek (1998) examine the effect of derivatives trading on the crude oil market for the period 1989–1997. They first examine whether the introduction of crude oil futures and subsequent introduction of energy derivatives increased crude oil price volatility. Their results indicate that large unexpected abnormal volatility occurred after the introduction of the futures contracts. The authors however argue that one can not infer from this evidence that the introduction of the futures market was responsible for the increased volatility. The increase in volatility might have been induced by deregulation in the US energy markets which in turn might have contributed to the increase in derivative trading. To support this argument, Fleming and Ostdiek (1998) find that the subsequent introduction of crude oil derivatives did not seem to have an effect on crude oil price volatility. The authors then turn to examine the influence of futures trading activity and the depth and liquidity of the market. They found an inverse relationship between open interest in crude oil futures and spot market volatility. They interpret these findings as evidence that trading in futures improves the depth and liquidity of the underlying market and thus contradict the argument that derivatives markets destabilize the underlying spot market.

Examining the volume-volatility relationships for crude oil futures, Foster (1995) shows that both trading volume and volatility are contemporaneously correlated in the crude oil market. He suggests that movements in both of these variables are influenced by the arrival of new information. He also finds that lagged volumes explain price variability and takes this as evidence that the market is characterized by some degree of inefficiency.

Lautier and Riva (2004) examine whether the presence of futures markets induces volatility. They find that trading variances are higher than overnight variances which in turn are higher than weekend variances. They propose two explanations for this observation: either information arrives only during the exchange trading hours or there is excess volatility and mis-pricing caused by trading. To distinguish between these two effects, the authors identify the presence of noise trading and then test whether this noise trading is associated with higher volatility during trading hours. To identify noise trading, they test auto-correlation of daily returns and conclude that there is noise trading in the crude oil market with part of this noise actually being generated by the electronic trading system. Then the authors make a comparison between daily variances and long-run variances and find that a significant fraction of daily variances disappears in the long run. Thus, a significant proportion of daily variance is caused by pricing errors, but pricing errors have relatively little effect on six-month holding period returns. Finally, the authors examine whether noise trading is responsible for the high variance recorded during trading hours. Lautier and Riva (2004) find that higher volatility recorded during trading hours is mainly due to the arrival of new information. The decline in the weekend or overnight variance is due to the reduction of arrival of new information. They note that trading plays an important role in incorporation of information: the greater the number of trades, the more rapid is the incorporation of information in prices.

Using controlled experiments, Ballinger, Dwyer, and Gillette (2004) provide evidence that the futures market for West Texas Intermediate crude increased the variability of the posted cash price of crude.[69] They explain the increased variability in terms of the ability of the futures markets to aggregate information into futures prices that is not reflected in cash prices.

The impact of futures markets was also examined by testing the excess-comovement hypothesis i.e. whether the prices of commodities move together beyond what can be explained by fundamentals. Pindyck and Rotenberg (1990) examine the prices of seven commodities (wheat, cotton, copper, gold, crude oil, lumber and cocoa). Although the commodities are unrelated, the authors find that the prices of

---

69 These are prices posted by refiners for lifting of crude oil from producers.

these commodities have a persistent tendency to move together. This comovement is in excess of what can be explained by common macroeconomic effects such as interest rates, inflation, exchange rates and changes in aggregate demand. The authors suggest that commodity price movememt can be explained by herd behaviour in these markets. Recent evidence however tends to contradict Pindyck and Rotenberg's findings (see for instance Cashin, McDermott and Scott, 1999; Booth and Ciner, 2001).

Using a different empirical approach, some studies have examined the degree of integration in prices across markets using the methodology of co-integration. The idea is that the futures markets should help investors to arbitrage price differentials across markets and thus bring price convergence across markets. Using monthly data from the period 1974−1996, Ewing and Harter (2000) show that ANS and Brent crude oil prices follow a random walk and that both these oil markets share a long-run common trend. The authors interpret these results as evidence that the two markets are 'unified' and that there is price convergence. Kleit (2001) suggests a modelling technique based on the theory of arbitrage which is more appropriate to answering the question of whether markets have become more unified. Studying seven types of crude oils, he finds substantial evidence for the integration hypothesis. Although this approach provides some useful insights, it does not directly test the impact of the futures markets. It remains unclear whether the evidence of integration across geographically dispersed markets is due to arbitrage in the futures market.

5.3 *Backwardation in the Futures Oil Market*

One of the main features of the crude oil futures market is that it is in backwardation most of the time. As noted by Litzenberger and Rabinowitz (1995), 80−90% of the time the oil forward curve is in backwardation, i.e. futures prices are often observed to be below spot prices.

The theory of storage implies that the futures price at time t of a contract that expires at time T can be explained by the cost of carrying the commodity from the present to the delivery period. Thus, the futures price for delivery at time T can be written as

$$F_{t,T} = S_t e^{(r+g)[(T-t)/365]} \tag{1}$$

Where S_t is the spot price of commodity at time t, r is the nominal risk-free interest rate and g is the cost of storage and $T\text{-}t$ is the fraction of the year until expiration of the futures contract. Notice in the above equation that since r and g are positive (the sum of the two is referred to as the cost-of-carry), then this implies that futures prices should be above the spot prices. In this case, the market is said to be in 'contango'. The underlying logic behind this equation is that arbitrage will ensure that the futures price of oil will be equal to the cost of borrowing funds, buying oil in the market and storing it over the same period. For instance, assume you sell a futures contract in which you agree to deliver oil in six months at an agreed price. If the futures price is above the spot price, then one strategy is for you to borrow money today, buy the oil in the spot market, store it for six months and then deliver it at the agreed upon price. Thus, there are three factors that tie up the futures price: the spot price, the interest rate and storage costs.

However, the observation that most of the time the oil forward curve is in backwardation implies that the cost-of-carry oil is not the only determinant of futures prices and that for some commodities one has also to consider in addition to storage costs, the benefits derived from storing the commodity. In other words, there is an implied yield or return from holding the commodity. Equation (1) can be modified to reflect the existence of a convenience yield

$$F_{t,T} = S_t e^{(r+g-y)[(T-t)/365]} \qquad (2)$$

where y is the convenience yield. For sufficiently high value of y, it is clear from equation (2) that the futures price can be below the spot price.

This implied yield is referred to as the 'convenience yield', a concept developed by Kaldor (1939) and Working (1948). Brennan and Schwartz (1985) define the convenience yield as 'the flow of services that accrues to an owner of physical asset but not to an owner of a contract for future delivery of the commodity'. The yield may not necessarily be a pecuniary return, but an implicit and not directly measurable return that a refinery or an end user of oil places on its ability to use inventories to meet contractual obligations, not to interrupt supplies to long-standing customers, and to respond to demand shocks.[70] Lovell (1961) suggested that inventories serve to smooth production, but empirical support is mixed (see for instance Blinder (1986), Blinder and Maccini (1991)).

---

70 Lovell (1961) was among the first to suggest that inventories serve to smooth production.

Litzenberger and Rabinowitz (1995) provide an explanation of backwardation using the analogy that ownership of oil reserves can be seen as holding a call option. If discounted futures prices are higher than spot prices and if extraction costs grow by no more than the interest rate, then all producers have the incentive to defer production and leave the oil in the ground. This gives the producers the option of leaving oil there rather than extracting it and incurring high storage costs, risks of sabotage, and so on. But if every oil producer waits, then there will be a shortage of oil today causing the price to rise. The net result is backwardation in which the oil price rises today to offset the advantage of waiting to see the price before oil extraction. Thus, weak backwardation is a necessary condition for current production.[71]

The level of oil inventories plays an important role in the futures market, especially if falling inventories are perceived by market participants as a decline in the supply of oil relative to demand while an increase in inventories is interpreted by market participants as an increase in the supply relative to demand.[72] In a contango, participants have the incentive to build up their inventories. If this is interpreted by the market as a sign of excess supply, then spot and futures prices will go down. But this in turn will induce participants to augment their stock further, which will cause further decline in prices. There are limits on the extent of the contango. The 'contango' is bounded by three elements: the spot price, the interest rate and storage costs. If the futures price is too high relative to the spot price, then participants would undertake arbitrage, driving the spot price up and the futures price down. A recent example of when the oil market was trapped in reinforcing contango occurred in 1998. A warm winter that year and the Asian crisis led to a dramatic decline in oil demand. Around the same time, OPEC decided to raise its quotas and increase supply. These two key factors resulted in a dramatic fall in prices. This triggered the accumulation of crude oil which led to the decline in spot prices and encouraged contango in the market. Weak oil demand encouraged participants to stay net short in the futures market. At the same time, growing contango encouraged participants to buy wet barrels and sell futures. The arbitrage of buying oil and storing it for

---

71 This result may imply that production will stop when the market is in contango, which does not conform to the reality of oil production.

72 This view however contains a fallacy. For instance, a reduction in the level of inventories can be driven by factors unrelated to supply conditions. There may be plenty of crude oil supplies in the market but because of 'just in time inventories' policy, inventories may remain at very low levels.

future delivery did not stop the accumulation of inventories and the market went into a deeper contango. It took large cuts in production led by Venezuela, Mexico and Saudi Arabia to help reduce the extent of the contango.

On the other hand, there are no limits to the extent of backwardation. When a market is in backwardation, participants have no incentive to accumulate inventories as buyers of oil for the future will buy paper rather than buy oil at a higher price today and store it. As a result, physical stocks or inventories will go down. Lower inventories send signals to traders that there is a shortage of supply relative to demand. This has the effect of pushing spot prices higher. But this in turn will induce companies to hold lower inventories, which will in turn send signals of shortage driving prices up. Thus, backwardation could lead to prices rising and rising. According to Mabro (2000), 'only big shocks can stop these movements. But big shocks do not only arrest the price movement. They can reverse it, recreating the problems of relentless rise or fall until the next shock.'

This circular view in which backwardation tends to reduce inventory level which in turn would result in further backwardation implies that price relationships and inventories are the causes of movements in oil prices. However, there is a more likely possibility that both backwardation and inventory levels are the outcome of shocks to market fundamentals. For instance, an unexpected supply shock can both reduce the level of inventories and increase current prices above futures prices.

The case when spot prices are higher than futures prices may also be explained in terms of the predictive power of futures prices. If spot prices are expected to fall, then futures prices will be below the spot price. In other words, the futures price is the expected spot price. However, this approach is quite problematic not only because it ignores other factors such as interest rates and storage costs but also it does not fit the data. The oil market shows that backwardation with spot prices exceeding futures prices dominates most of the time. If futures prices predict spot prices, then spot prices should fall quite often. This however is not the case. For instance, in 2002 when the futures market was predicting a fall in prices, oil prices reached a record level. The futures price is not a predictor of the price in the future. In fact, evidence suggests that the average error (i.e. the average difference between the predicted and actual value) using oil futures prices is larger than the average error using the spot price. Moosa and Al-Loughani (1994) carried out a formal analysis on the crude oil futures market to test its predictability power. If futures prices are rational or efficient forecasts of future outcomes, then the forecast errors must have mean zero and must be uncorrelated with any variable in the information set at the time when the forecast was made. The evidence of forecast errors with mean zero is consistent both with market efficiency and the unbiasedness property of the forecaster. There are a number of ways in which forecast errors might be found to be predictable. Under standard equilibrium asset pricing theory futures prices are the sum of an expectations component and a risk premium component. Thus, rejection of forecast efficiency could simply mean that the risk premium component is important. In the presence of risk premia futures prices diverge from market expectations. Moosa and Al-Loughani (1994) find that futures prices are neither an unbiased nor an efficient forecaster of spot prices. They also find evidence of time varying risk premium which they model as a GARCH process.

Despite the strong evidence that futures prices are not a good forecaster of spot prices, it is still used by some institutions. For instance, the Bank of England *Inflation Report* states that 'the MPC (Monetary Policy Committee) judges that the futures curve continues to offer the best indication of the prospective path for oil prices' (cited in Farrell, Kahn and Visser, 2004).

6.   The Relationship between OPEC and the Market

On the face of it, OPEC seems to play a very limited role in the formation of oil prices in the current oil pricing regime. OPEC countries, like other exporters, just take the marker price from the spot, forward and more recently from the futures market and plug it in the pricing formula to arrive at the price at which they sell their oil. But this simple characterization does not provide an accurate description of the whole picture. After all, OPEC countries account for the largest share of oil in international trade, and as such are bound to have an influence.

In the current oil pricing system, the links running from OPEC's large liquid oil market to the illiquid markets for reference crudes do exist but their impact is very difficult to define. First, in certain instances when oil markets are subject to abrupt disruptions and severe shocks, OPEC and in particular Saudi Arabia can play the role of a swing producer filling the oil gap and moderating oil prices. Saudi Arabia has done this at various times such as during the Iran-Iraq war when output from both countries collapsed; during and after the first Gulf war, when output from Iraq and Kuwait disappeared; in 2003, when civil strife in

Venezuela and Nigeria curbed output from both countries; and when the US invasion of Iraq had disrupted Iraqi oil output. But the effectiveness of this role is sometimes undermined, especially if participants in the futures market have doubts about the volume of these supply increases. On other occasions different doubts arise as regards the size of OPEC's spare capacity and its ability to act as a swing producer. Events in 2004 highlight this fact. Saudi Arabia had gone to unusual lengths to provide assurances to markets and analysts that it can increase production to meet growing global demand. These assurances however were not able to calm oil prices. This led *The Economist* to comment that, 'Ali Naimi, the Saudi oil minister, usually moves markets when he speaks. Yet when he promised a few days ago that more oil is on the way, traders ignored him and the rally continued apace.'[73]

Second, OPEC can influence prices through adjusting its production quotas. These quota decisions are essentially signals to the market about OPEC's preferred prices. This mechanism may or may not succeed, depending on how the market interprets these signals. Market participants may ignore or misinterpret the signals and in that case their response will not be the expected one. The effectiveness of the signal will depend on whether the market believes that OPEC is united in its decision and will therefore make the necessary output adjustment. Although OPEC has on many occasions succeeded in defending the oil price, adjusting output has sometimes been problematic both in the face of falling and growing global oil demand though for different reasons. In other words OPEC's response is asymmetric to global demand conditions. If global demand for oil falls, non-OPEC suppliers will continue to produce. Those who are willing to co-operate with OPEC, as happened on occasion with Norway, Mexico, or Oman, will usually wait for OPEC to negotiate with them and plead the common interest. When the required cuts are significant the small OPEC Members usually find it difficult to reduce their production on a pro-rata basis (the usual system adopted by OPEC over the years) and commentators begin to doubt the effectiveness of the policy and thus influence the market's perceptions in a negative way. In the case of growing global demand, although agreements to increase quotas are easier to reach and implement, OPEC may not respond quickly to this upward trend, especially in an environment of imperfect information. After all, the decision to wait and not increase output is more profitable than to increase output if the trend turns out to be false. Thus, although

---

73 *The Economist*, 'Unstoppable? How OPEC's Fear of $5 Oil Led to $50 Oil', August 21, 2004.

OPEC sends signals to the futures markets their impact depends on participants' understanding of their implications and their perceptions of the credibility of the policy.

In 2000, OPEC adopted the price band as a signalling mechanism. This mechanism sets a target range for the OPEC basket price between US$22 and US$28 per barrel of oil. If prices are below the lower band for ten consecutive days, OPEC will automatically cut production. If prices are above the upper band for twenty days, OPEC will automatically increase production. But as argued above, OPEC may not respond to prices above the upper band unless there are fears that price rises will weaken demand. Furthermore, in a very tight market when spare capacity is very low, the upper band becomes irrelevant as OPEC will not be able to defend it. Thus the most important feature of the price band is the lower band, as in principle OPEC will be able to defend it by cutting production. Although on certain occasions, as was the case in 1998, the positive impact of the production policy on prices took a long time to emerge, the importance of this signalling mechanism is that it indicates that when prices fall below the lower band OPEC will respond by cutting production. The difficulty of managing such a price band in a tight market led OPEC to suspend its five-year old oil price band mechanism in January 2005.

Some observers have argued that the extreme movements in oil prices can be explained in terms of OPEC market power and behaviour. For instance, Adelman (1999) notes that crude oil prices have been more volatile than other commodity prices although in principle they ought to be less volatile. On the demand side, oil consumption has been stable relative to GDP while on the supply side it is easier and faster to adjust output to demand conditions than in other industries. From these observations, he hypothesizes that oil prices must be more volatile for a special reason and that reason is OPEC output control. Thus, Adelman (1999) argues that high oil price volatility is due to the fact that 'the thermostat of a competitive market has been turned off' and then predicts that 'price volatility will continue higher than in other markets' (p.1).

Explaining volatility in terms of OPEC's 'price fixing' is not warranted. As we have previously seen, OPEC abandoned fixing the reference price in 1986 in favour of a system in which OPEC sets production quotas based on its assessment of the market's call on OPEC supply. Oil prices fluctuate in part according to how well OPEC does this calculus. Through the process of adjusting its production quotas OPEC can only hope to influence price movements. However as explained earlier, this adjustment process can prove quite problematic, at times inducing undesired price volatility. For instance, in face of falling demand, expectations of cuts induce speculation about OPEC's ability to adhere to these output cuts and the size of the cut. These expectations can cause swings in net speculative positions, and reversal of such positions if the cut is less than expected or if the cut does not materialize. In the case where global demand for oil rises, OPEC may not respond quickly to this upward trend, especially in a situation of imperfect information. The slowness of the response to an upward trend can contribute further to volatility by undersupplying the market.

OPEC's response has been complicated by the lack of reliable data about production, consumption and inventory levels. Although oil prices have become more transparent over the years, information about crude oil consumption and production has not improved either in quality or in the speed of information dissemination.[74] On the demand side, data on the consumption of oil, even for OECD, are uncertain, subject to major revisions, and published with a considerable lag. This problem is becoming worse with the increasing importance of developing countries such as India and China as major oil consumers but with even less reliable data. On the supply side, the dominance of less transparent national oil companies and uncertainty about OPEC production levels (OPEC partly relies on 'secondary sources' to find out what its own members are producing) increases the uncertainty about oil production. Furthermore, the advent of many small oil producers on the scene makes it even more difficult to collect reliable and timely information. The quality of inventory data is also subject to uncertainties and revisions.[75] Since in a tight market every barrel matters, doubts about supply and demand contribute to the volatility of oil markets both through magnifying the oil gap and increasing speculation. In addition, unreliable data may induce OPEC to misinterpret the market's direction and pursue policies that add to the uncertainty and compound volatility.

OPEC's ability to deal with shocks that cause increases in demand has been more difficult from the late 1990s onwards. In the mid-1980s, OPEC countries were left with a huge surplus capacity estimated at around 11 million barrels a day (on an annual basis) when new discoveries in non-OPEC countries, responding to higher oil prices and taking advantage of new technologies, increased their production capacity by a comparable amount. What is not well recognized by some observers

---

74 This discussion is based on Lynch (2002).
75 Lynch argues that such factors have magnified the 'missing barrels' problem where the term 'missing barrels' refers to the statistical discrepancy between oil consumption and production that is not reflected in the inventory levels.

is that during the mid-1980s the oil market reached a turning point in which demand for OPEC oil slowly started to increase. At the same time, the 1986 price collapse saw a production collapse in some old non-OPEC basins. As a consequence, OPEC's spare capacity began to decline gradually over the years until it was reduced to very low levels. Without this cushion of surplus capacity, OPEC's ability to stabilize oil prices in a tight market has considerably declined, contributing to price volatility. This problem has been compounded by market scepticism about OPEC's spare capacity and its ability to act as a swing producer.

There is some evidence to suggest there is an increase in volatility as OPEC meetings approach. It is possible to distinguish between the two ordinary conferences required by OPEC's charter and extraordinary meetings which are similar in structure and are organized in response to macroeconomic developments. In addition, there are various committees and sub-committees which hold meetings and make recommendations to be approved by the overall conference. The most important of these sub-committees is the Ministerial Monitoring Committee (MMC) which was established to monitor market conditions. Using implied volatility measures from option on crude oil futures, Horan, Peterson and Mahar (2004) find that volatility drifts upward as these meetings approach but drops by around 3% after the first day of the meeting. The most pronounced decline in volatility after the first day is associated with the meeting of the MMCs while the ordinary bi-annual conferences result in little drop in volatility. This does not necessarily mean that OPEC is causing an increase or reduction in volatility; it merely reflects market agents trying to be on the upside of a bet about what OPEC might or might not do.

Some have argued that OPEC may have contributed to volatility through its deliberate policy to keep inventories at very low levels (Weston and Christiansen, 2003). This view seems to be accepted by many oil market analysts. For instance, *The Economist* notes that 'Saudi Arabia targeted inventory levels: whenever oil stocks in the rich countries of the OECD started rising, OPEC would reduce oil quotas to stop prices softening.'[76] This view however ignores other factors that may have contributed to the current low levels of inventories. Under pressure to maximize shareholder value, international oil companies have undergone major cost-cutting exercises including cutting inventories to their lowest possible level and shifting to a 'just in time' inventories policy. In this shift of policy, oil companies have relied on OPEC's large holdings

---

76 *The Economist*, 'Oil in Troubled Water', April 28, 2005

and consuming countries' strategic petroleum reserves (SPR) and on a developed spot market for immediate deliveries. The main consequence was a downward trend in the level of inventories over the last decade which meant that any swing in demand could not be met immediately by changes in inventory levels but rather by fluctuations in crude oil prices.[77] Furthermore, the relationship between levels of inventory and volatility is not straightforward and can run in the opposite direction, i.e. volatility of oil prices can affect inventory levels. Pindyck (2001) argues that oil price volatility causes increased volatility in consumption and production and as a result market participants will want to hold greater inventories to buffer these fluctuations. Oil price volatility also increases the opportunity cost of producing now, such that producers will not be willing to extract oil unless the spot price is higher than the futures price, i.e. except when the market is in strong backwardation. Pindyck provides evidence that market variables such as inventory levels can not explain crude oil price volatility whereas volatility can influence some market variables such as production, although the effect is empirically very small.

7.   Conclusion

The adoption of the current market-related pricing system in 1986–8 represented a major chapter in the history of oil price determination. It brought in a departure from the administered oil pricing system that had dominated the oil market from the 1950s until the mid 1980s. As in all previous oil pricing regimes that have emerged in the eventful history of oil, the current system has introduced specific concepts of oil price. These concepts have not been static but have evolved over time in response to strains within the system. Initially, the marker prices used in formula pricing were spot prices. The rationale for using the spot market is that prices must be generated in a market where there is interaction of demand and supply for physical oil. The spot markets for marker crudes however suffer from a number of problems which prevent them from playing this price discovery role effectively. The most acute problem is the decline in the physical base of the marker crudes which increased the vulnerability of these markets to squeezes.

---

77 Weston and Christiansen (2003) however argue that 'OPEC still appears to have had an impact over and above this trend decline: measuring stock vs. the 3-year average effectively accounts for a gradually declining norm – and for almost 80% of the time since January 2000, total US inventories have been below this level (p.8).'

Remedies in the form of commingling various crudes and broadening the definition for price assessment presented problems. For this and other reasons, a group of oil-exporting countries have replaced the spot with the futures market for the determination of the reference price. The volume of transactions in the futures market is quite large and thus there is less room for squeezes and manipulation. Furthermore, futures prices are determined by actual transactions and not on the basis of some assessed prices by oil reporting agencies. Thus, it is true that the current oil pricing regime is still market related, but there has been a shift towards a new stage in which the most important international oil price is determined in a market for financial contracts and not in a market for transactions in physical cargoes.

Some observers consider that the shift to the futures market does not pose a problem as it represents a mere shift from a less efficient market to a more efficient one. Others are more sceptical about the role of the futures market in the determination of the reference prices. Within this group, some highlight the role of speculation arguing that although some shifts in investor demand for futures contracts are rational, not all demand shifts appear to be so. In some cases, demand changes may respond to changes in expectations or sentiments that are not based on reliable information. Others argue that factors not related to the oil market can determine transactions in the futures markets. For instance, price movements can occur when investors change the composition of their portfolios due to developments outside the oil markets. These and other arguments suggest that although the different layers of markets are linked, decoupling of price movements between the futures and the physical markets may occur.

Every pricing regime in the past reflected the balance of power at the time and this present regime is no exception. Despite its drawbacks, the implicit position of most market analysts, oil producers, oil companies, oil reporting agencies and especially traders is to maintain the current system in place as long as it works. Those who highlight the deficiencies of the current oil pricing regime do not propose a viable alternative. Thus, for the time being, the current system though imperfect seems to be safe from being toppled by another. However, as the eventful history of oil has taught us, an imperfect pricing system can continue to survive unchallenged for a long time until a powerful shock or a series of small shocks exposes its weaknesses and limitations and most importantly alters the balance of power (or perceived power) among the main players. It is only then that the consensus about the validity of the regime will collapse and a transition to a new (and probably a fundamentally different) pricing regime will occur.

References

Adelman, M.A. (1999), 'Oil Prices: Volatility and Long Term Trends', MIT Centre for Energy and Environmental Policy Research, August.

Amuzegar, J. (1999), *Managing the Oil Wealth: OPEC's Windfalls and Pitfalls*, London: I.B. Tauris.

Antoniou, J.H. and A.J. Foster (1992), 'The Effect of Futures Trading on Spot Price Volatility: Evidence from Brent Crude Oil Using GARCH', *Journal of Business Finance and Accounting*, 19(4):473−84.

Bacon, R. (1986), 'The Brent Market: An Analysis of Recent Developments.', Oxford: Oxford Institute for Energy Studies.

Badrinath, S. G. and S. Wahal (2002), 'Momentum Trading by Institutions', *Journal of Finance*, 57: 2449−78.

Ballinger, A., G. P. Dwyer Jr., and A. B. Gillette (2004), 'Trading Institutions and Price Discovery: The Cash and Futures Markets for Crude Oil', Federal Reserve Bank of Atlanta Working Paper 2004−28, November.

Bank for International Settlement (2004), *BIS Quarterly Review* September 2004: International Banking and Financial Market Developments, Basle: Bank for International Settlement.

Barrera-Rey, F. and A. Seymour (1996), 'The Brent Contract for Differences (CFD): A Study of an Oil Trading Instrument, its Market and its Influence on the Behaviour of Oil Prices', SP5, Oxford: Oxford Institute for Energy Studies.

Benos, A.V. (1998), 'Aggressiveness and Survival of Overconfident Traders', *Journal of Financial Markets*, 1:353−83.

Bikhchandani, S. and S. Sharma (2000), 'Herd Behavior in Financial Markets: A Review', IMF Working Paper WP/00/48.

Bikhchandani, S., D. Hirshleifer & I. Welch (1998), Learning from the Behavior of Others: Conformity, Fads and Informational Cascades', *Journal of Economic Perspectives*, 12:151−70.

Black, F. (1986), 'Noise', *Journal of Finance*, 41:529−43.

Blinder, A.S. (1986), 'Can the Production Smoothing Model of Inventory Behavior Be Saved?', *Quarterly Journal of Economics*, 101:431−53.

Blinder, A.S., and S.J. Maccini (1991), 'Taking Stock: A Critical Assessment of Recent Research on Inventories', *Journal of Economic Perspectives*, 5:73−96.

Booth, G. G., & C. Ciner, (2001), 'Linkages among Agricultural Commodity Futures Prices: Evidence from Tokyo, *Applied Economics Letters*, May, 8(5).

Bossley, L. (2004) 'BFO Futures to Be Affected By New Crude Streams, Warns CEAG', *MEES*, XLVII(17), 26 April.

Brandt, M. W. and K. A. Kavajecz (2004), 'Price Discovery in the U.S. Treasury Market: The Impact of Order flow and Liquidity on the Yield Curve', *Journal of Finance*, 59:2623−54.

Brenan, M.J. and E.S. Schwartz (1985), 'Evaluating Natural Resource Investments', *Journal of Business*, 58:153−7.

Brunnermeier, M. K., and S. Nagel (2004), 'Hedge Funds and the Technology

Bubble', *Journal of Finance*, 59:2013−40.

Cashin, P. C., J. McDermott and A. Scott (1999), 'The Myth of Comoving Commodity Prices' IMF Working Paper 99/169 (December).

Caumon, F. and J. Bower (2004), 'Redefining the Convenience Yield in the North Sea Crude Oil Market', WPM28, Oxford Institute for Energy Studies.

Daniel, K., D. Hirshliefer & A. Subrahmanyam (1998), 'Investor Psychology and Security Market Under- and Over-reactions,' *Journal of Finance*, 53:1839−86.

Danielsen, A. (1982), *The Evolution of OPEC*, New York: Harcourt Brace Jovanovich.

Davis, E.P. (2003), 'Institutional Investors, Financial Market Efficiency and Financial Stability', Discussion Paper PI-0303, London: The Pensions Institute, Birkbeck College.

De Long, J.B., A. Shleifer, L.H. Summers and R.J. Waldmann (1990), 'Noise Trader Risk in Financial Markets', *Journal of Political Economy*, 98:703−38.

Easley, D., N. Kiefer, and M. O'Hara (1997), 'The Information Content of the Trading Process', *Journal of Empirical Finance*, 4:159−86.

Ederington, L. and Jae Hae Lee (2002), 'Who Trades Futures and How: Evidence from the Heating Oil Futures Market', *Journal of Business*, 75:353−74.

EIA (2005), Non-OPEC Fact Sheet, Country Analysis Briefs, downloadable from *http://www.eia.doe.gov/emeu/cabs/nonopec.html*

Ewing, B. T. & C.L. Harter (2000), 'Co-movements of Alaska North Slope and UK Brent crude oil prices', *Applied Economics Letters*, August, 7(8):553−8.

Farrell, G.N., B. Kahn and F.J. Visser (2004), 'Price Determination in International Oil Markets: Developments and Prospects', *South African Reserve Bank Quarterly Bulletin*, March.

Fesharaki, F. and H. Vahidy (2001), 'Middle East Crude Oil Trade and Formula Pricing,' *Middle East Economic Survey* (MEES) 44(43), October.

Figlewski, S. (1981), 'Futures Trading and Volatility in the GNMA Market', *Journal of Finance*, 36:445−56.

Fleming, J. and B. Ostdiek (1998), 'The Impact of Energy Derivatives on the Crude Oil Market', *Energy Economics*, 21:135−67.

Foster, A. (1995), 'Volume-volatility Relationships for Crude Oil Futures Markets', *Journal of Futures Markets*, December, 15(8):929−52.

Friedman, M. (1953), *The Case For Flexible Exchange Rates: Essays in Positive Economics*, Chicago: Chicago University Press.

Froot, K. A., D. S. Scharfstein, and J. C. Stein (1992), 'Herd on the Street: Informational Inefficiencies in a Model with Short-Term Speculation', *Journal of Finance*, 67:1461–84.

Garbade, K.D. and W.L. Silber (1983), 'Price Movements and Cash Discovery in Futures and Cash Markets', *Review of Economics and Statistics*, 65:289−97.

Garcia, P.A.M. (2005), 'OPEC in the 21st Century: What Has Changed and What Have We Learned?' *Oxford Energy Forum*, Issue 60, Oxford: Oxford Institute for Energy Studies.

Greenspan, A. (2004), Testimony before the U.S House of Representatives'

Budget Committee. September 8, 2004. See *www.federalreserve.gov/boarddocs/testimony/2004/*.

Hart, O.D. (1977), 'On the Profitability of Speculation', *Quarterly Journal of Economics*, 91:579−97.

Hart, O.D. and D.M. Kreps (1986), 'Price Destabilizing Speculation', *Journal of Political Economy*, 94:927−52.

Horan, S. M., J. H. Peterson and J. Mahar (2004), 'Implied Volatility of Oil Futures Options Surrounding OPEC Meetings', *The Energy Journal*, 3:103−25.

Horsnell, P. (2000), 'Oil Pricing Systems', Oxford Energy Comment, Oxford Institute for Energy Studies, May.

Horsnell, P. and R. Mabro (1993), *Oil Markets and Prices: The Brent Market and the Formation of World Oil Prices*, Oxford: Oxford University Press.

Kaldor, N. (1939), 'Speculation and Economic Stability', *Review of Economic Studies*, 7(1):1−27.

Kleit, A. N. (2001), 'Are Regional Oil Markets Growing Closer Together? An Arbitrage Cost Approach', *Energy Journal*, 22(2):1−15.

Kogan, L., S. A. Ross, Jiang Wang and M. M. Westerfield (2003), 'The Price Impact and Survival of Irrational Traders', MIT Sloan Working Paper No. 4293-03; AFA 2004 San Diego Meetings; 14th Conference on Financial Economics & Acctg. (FEA); 5th Ann. Texas Finance Fest.

Lakonishok, J., A. Shleifer, and R. Vishny (1992), 'The Impact of Institutional Trading on Stock Prices', *Journal of Financial Economics*, 32:23−43.

Lautier, D. and F. Riva (2004), 'Volatility in the American Crude Oil Futures Market', *Cahier de recherche du CEREG* n° 2004-08, Université Paris-Dauphine.

Litzenberger, R. and N. Rabinowitz (1995), 'Backwardation in Oil Futures Markets: Theory and Empirical Evidence', *Journal of Finance*, 50:1517−45.

Lovell, M.C. (1961), 'Manufacturers' Inventories, Sales Expectations, and the Acceleration Principle' *Econometrica* 29:293−413.

Lynch, M. (2002), 'Causes of Oil Price Volatility', Background Paper for the 8th International Energy Forum, Osaka, Japan.

Mabro, R. (1984) 'On Oil Price Concepts', Oxford: Oxford Institute for Energy Studies.

Mabro, R. (1986), 'The Netback Pricing System and the Price Collapse of 1986', Oxford: Oxford Institute for Energy Studies.

Mabro, R. (2000), 'Oil Markets and Prices', Oxford Institute for Energy Studies, Oxford Energy Comment, August.

Mabro, R. (2001), 'Does Oil Price Volatility Matter?', Oxford Energy Comment, Oxford Institute for Energy Studies.

Mabro, R. (2005), 'The International Oil Price Regime: Origins, Rationale, and Assessment', *The Journal Of Energy Literature*, XI(1).

Mabro, R. et al (ed.) (1986), *The Market for North Sea Oil*, Oxford: Oxford University Press.

Mayhew, S. (2000), 'The Impact of Derivatives on Cash Markets: What Have We Learned?', Working Paper, Department of Banking and Finance, Terry

College of Business, University of Georgia (Athens, GA, February).

Milonas, N. and T. Henker (2001), 'Price Spread and Convenience Yield Behaviour in the International Oil Market,' *Applied Financial Economics*, 11:23−36.

Mollgaard, H.P. (1997), 'A Squeezer Round the Corner? Self-Regulation and Forward Markets', *The Economic Journal*, 107(440):104−12.

Montepeque, J. (2005), 'Sour Crude Pricing: A Pressing Global Issue', *Middle East Economic Survey*, XLVIII(14), April.

Moosa, I. and N.E. Al-Loughani (1994), 'Unbiasedness and Time Varying Risk Premia in the Crude Oil Futures Market', *Energy Economics*, 16(2):99−105.

Nainar, S.M.K. (1993), 'Market Information and Price Volatility in Petroleum Derivatives and Futures Markets', *Energy Economics*, 15:17−24.

Newbery, D.M.G. (1984), 'The Manipulation of Futures Markets by a Dominant Producer', Chapter 2 in Anderson, R.W. (ed.), *The Industrial Organization of Futures Markets*, Lexington, MA: Lexington Books.

Nofsinger, J., and R. Sias (1999), 'Herding and Feedback Trading by Institutional and Individual Investors', *Journal of Finance*, 54:2263−95.

Odean, T. (1999), 'Do Investors Trade Too Much?' *American Economic Review*, 89:1279−98.

Ogawa, Y. (2002), 'Proposals on Measures for Reducing Asian Premium of Crude Oil', Research Report, Homepage of the Institute of Energy Economics, Japan (IEEJ).

Parra, F. (2004), *Oil Politics: A Modern History of Petroleum*, London: IB Tauris.

Peck, A. E. (1976), 'Futures Markets, Supply Response and Price Stability', *Quarterly Journal of Economics*, 90:407−23.

Penrose, E. (1968), *The Large International Firm in Developing Countries: The International Petroleum Industry*, USA: Greenwood Press.

Pindyck R., & J. Rotenberg (1990), 'The Excess Co-movement of Commodity Prices', *Economic Journal*, 100:1173−89.

Pindyck, R. (2001), 'Volatility and Commodity Price Dynamics', M.I.T. Centre for Energy and Environmental Policy Research Working Paper.

Pirrong, S. (1996), *The Economics, Law and Public Policy of Market Power Manipulation*, Boston, MA: Kluwer.

Platts (2004), 'Review of the Partial Dubai and Oman Price Discovery Mechanism', Platts Discussion Paper, November. Downloadable from: *http://www.platts.com/Oil/Resources/Market%20Issues/*

Rehaag, K. (1999), 'Disappearing Benchmarks: The Demise of the Market Indexed Pricing?', IEA Energy Prices and Taxes, 4th Quarter, downloadable from: *http://data.iea.org/ieastore/assets/products/eptnotes/feature/4Q1999.pdf*

Roll, R. (1992), 'A Mean/Variance Analysis of the Tracking Error', *Journal of Portfolio Management*, 18:13–22.

Scharfstein, D. S. and J. C. Stein (1990), 'Herd Behavior and Investment', *American Economic Review*, 80:465–79.

Seymour, A. (1990), 'The Oil Price and Non-OPEC Supply', Oxford Institute for Energy Studies.

Shiller, R.J. (1981), 'Do Stock Prices Move Too Much to be Justified by

Subsequent Changes in Dividends?', *American Economic Review*, 71:421−36.

Shleifer, A. and L.H. Summers (1990), 'The Noise Trader Approach to Finance', *Journal of Economic Perspectives*, 4:19−33.

Sias, R. W. (2005), 'Reconcilable Differences: Momentum Trading by Institutions' (March 15). Working paper, USA: Washington State University.

Skeet, I. (1988), *OPEC: Twenty Five Years of Prices and Politics*, UK: Cambridge University Press.

Soligo, R. and A. Jaffe (2000), 'A Note on Saudi Arabia Price Discrimination', *Energy Journal*, 21(1):31−60.

Stein, J. (1987), 'Informational Externalities and Welfare-reducing Speculation', *Journal of Political Economy*, 95:1123−45.

Stevens, P. (1985), ' A Survey of Structural Change in the International Oil Industry 1945-1984', in David Hawdon (ed.), *The Changing Structure of the World Oil Industry*, USA: Croom Helm Ltd.

Telser, L.G. (1992), 'Corners in Organized Futures Market', in *Aggregation, Consumption and Trade: Essays in Honour of H.S. Houththaker* (eds) L.Phlips and L.D. Taylor, Dordrecht: Kluwer Academic Publishers.

Terzian, P. (1985), *OPEC: The Inside Story*, London: Zed Books.

Wang, F.A. (2001), 'Overconfidence, Investor Sentiment, and Evolution', *Journal of Financial Intermediation*, 10:38−70.

Weston, P.J. and M.R. Christiansen (2003), 'OPEC: Market Stabiliser or Disruptive Influence?' Downloadable from: *http://www.biee.org/downloads/conferences/WestonChristiansen%20Paper.pdf*

Working, H. (1948), 'Theory of the Inverse Carrying Charge in Futures Markets', *Journal of Farm Economics*, 30:1−28).

CHAPTER 4

THE INVESTMENT CHALLENGE

Bassam Fattouh and Robert Mabro

1.   Introduction

In a recent speech at the meeting of the World Petroleum Congress in Johannesburg, the Saudi Petroleum and Mineral Resources Minister Ali Naimi argued that the problem that the oil industry faces,

> is not one of availability; it is a problem of deliverability… Deliverability is a measure of our capacity as an industry to develop, produce, transport, refine and deliver to end consumers the energy they require in their daily lives. Currently, from the upstream to the downstream – all along the supply chain – the petroleum industry faces infrastructure constraints and bottlenecks that are causing market volatility and restricting its ability to bring oil from the ground to the consumer. We should not confuse this very real 'deliverability' challenge with resource 'availability', which is not a problem (Naimi, 2005).

Naimi's view on resource availability is shared among many – but not all – oil market analysts. Recently, Jeroen van der Veer (2005), the Chief Executive of Royal Dutch Shell group argued that 'there is still plenty of oil and gas still available to be found and produced' but he points out that 'oil and gas will be found in more and more hostile environments that present significant technical challenges'. Some of these potential sources of hydrocarbons are in isolated areas, in environmentally fragile regions and in countries perceived, rightly or wrongly, to be politically unstable. Their development will necessarily take time. Institutional arrangements as well as corporate rigidities can have a serious delaying effect. In similar vein, the International Energy Agency (IEA) (2003) and the International Monetary Fund (IMF) (2005) argue that global production growth is not yet limited by resource scarcity. There are ample opportunities, but their development requires large capital expenditures.

But if the main issue is deliverability and not availability, the central question becomes: how did the infrastructure constraints and bottlenecks in the oil industry mentioned by Minister Naimi arise? For many, the answer lies in long periods of insufficient investments in the oil sector. There is a general view that worldwide investment in the oil sector has reportedly been below levels needed to effectively meet the increase in actual and expected demand.

But if underinvestment is the problem, the question then becomes: what are the main factors responsible for this phenomenon? The dominant underlying model for analysing investment for international and national oil companies is a frictionless one where investment decisions are made in a world with complete information and where lags between planning, investment, development and production are not taken into account. This frictionless model underlies many politicians' calls for OPEC and oil companies to increase investment in oil production in order to relieve the current tightness of the market.[1] It also forms the basis as to what many observers consider a puzzle: despite robust oil prices in 2003–2005, why have so many national and international oil companies been hesitant to increase their investments in oil production?

The reality of the oil industry however is far from this 'perfect' world. Decision making and planning in manufacturing are much easier than in the extractive activities of the oil industry. The first is driven by margins while the second by economic rent generation and capture, where geology rules supreme. Compensating for the decline of mature basins poses an enormous challenge. Expanding capacity so as to cope with growing demand, even at relatively modest rates is also very challenging. Also, subsoil property rights over non-renewable resources pose a complex set of issues. Rent is affected by technological change and the capture of rent is a function of the upstream legal and fiscal regime. Geological features, completion and price risk management

---

1   Earlier this year, the UK Chancellor of the Exchequer, Gordon Brown, called on OPEC to boost investment in the energy sector: 'From the additional $300 billion a year in revenue OPEC countries are now enjoying and the additional $800 billion available to oil producers, there must be additional new investment in production and global investment in refining capacity' (HM Treasury Website: *http://www.hm-treasury.gov.uk/newsroom_and_speeches/press/2005/press_81_05.cfm*). Similar calls were also made to international oil companies. The EU energy commissioner Andris Piebalgs said in a statement after meeting with senior representatives from the energy industry that 'it is important that oil companies behave in the most responsible manner, as they play an important economic and social role by investing profits resulting from high prices in new production and – where needed – refining capacity.' *http://www.eubusiness.com/Energy/051013144748.1ho6ffou/view*

further add to the complexity of decisions. The oil industry is also driven by its own logistics, by lags between planning investment and production, and by the capital intensity of its assets. These factors and the interaction between them imply a very complex decision-making process which may often produce less than optimal investment decisions from a global perspective. The failure to take these factors into account will not only misguide our understanding of past investment decisions which shaped the current oil market structure but will also affect our assessment of future patterns of investments in the oil sector.

This paper addresses the underinvestment problem in both upstream and downstream oil production. In the corporate finance literature, underinvestment refers to a situation in which a company decides not to undertake a positive net present value project available to it (see Myers (1977) and Myers and Majlouf (1984) for instance). This underinvestment problem may be due to asymmetric information problems and/or agency costs that arise between shareholders and managers. Though this might be relevant in the context of current investment behaviour of international oil companies, the underinvestment problem in this paper refers to a situation in which the current or future capacity in the oil industry is unable to meet the current or expected increase in global demand for oil.[2]

This paper is organized as follows. In Section 2, we discuss the investment requirements in the energy and oil sectors, while in Section 3, the main factors that have caused underinvestment in the upstream are analysed. In Section 4, we discuss the underinvestment problem in the downstream. Section 5 is the conclusion.

2. Investment Requirements

The IEA has alerted us to the massive size of the requirements that need to be mobilized. Based on the Reference Scenario of the *World Energy Outlook 2002* where a 1.7 percent annual demand growth per year is projected, the IEA estimates that the total investment requirements for energy supply infrastructure are $16 trillion over the period 2001−2030. This investment is needed not only to expand existing capacity but also to replace existing and future supply facilities that will become obsolete. Although the figure is quite large in absolute amount, it is relatively small compared to the size of the world economy where the estimated

---

2   Although this is not the case in oil at present overinvestment, not a rare occurrence in many industries, causes problems as it can seriously depress prices.

investment requirements constitute only around 1 percent of world GDP. This however masks variation across regions where the volume of future investment is likely to be low in OECD countries while relatively high in developing countries. In fact, the IEA estimates that developing countries will require more than half of the global investment in the energy sector despite the fact that unit costs of capacity addition are lower there when compared to OECD. According to the IEA, the bulk of this investment is needed in the power sector. Power generation, transmission and distribution should account for around 60 percent of the total estimated energy investment requirements or around \$10 trillion. To meet the projected increase in oil demand of 45 million barrels per day (mb/d) from 75 to 120 mb/d between 2001 and 2030, \$3 trillion of total investments in each of the oil and gas sectors are needed. In the oil sector, most of this investment should be dedicated to exploration and development of conventional oil (See Figure 1).

The IMF (2005) also predicts oil demand growth, most of which to be met by OPEC whose members may have to double their output within 25 years. This requires unprecedented increases in investments which are '*conservatively* estimated by the IEA at around \$90 billion per year' (p.14, Italic added).

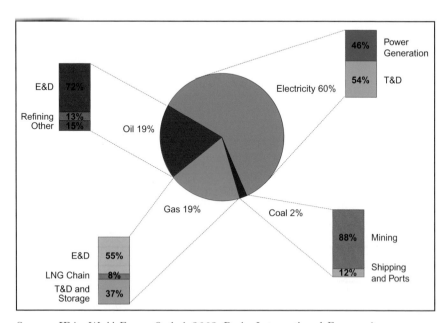

Source: IEA, *World Energy Outlook 2003*, Paris: International Energy Agency.

Figure 1: Investment Requirements in the Energy Sector

An OPEC scenario termed the Dynamics-as-usual scenario (a middle way scenario between a low and a high case) has an increase in world oil demand from 83.6 mb/d in 2005 to 105.9 mb/d in 2020. OPEC oil supplies (including natural gas liquids and non-conventional oil) could reach 49.1 mb/d in 2020. The cumulative investment requirement from 2005 to 2020 in this scenario is put at US$285 billion, half way between a low case of US$200 billion, and a high case of US$370 billion. (See Chapter 2)

3.   Factors Responsible for Underinvestment in Upstream Oil

The factors responsible for underinvestment in the upstream sector in the past two decades are various. In this section, we discuss six main factors. Although some of these are common both to national and international oil companies, the investment behaviour of national oil companies is quite different from that of international oil companies. Furthermore, although this section focuses on upstream investments, some of the issues discussed here are relevant for investment in the downstream.

3.1   Demand Pessimism and Low Oil Prices (1985–1999)

The underinvestment in upstream oil is determined to a large extent by developments in the oil markets during the 1980s and mid-1990s. In the mid-1980s, OPEC countries were left with a huge surplus capacity estimated at around 12 mb/d (on an annual basis) when new discoveries in non-OPEC countries, responding to higher oil prices and taking advantage of new technologies, increased their production by a large amount. In the presence of such huge surplus capacity, there was no reason for OPEC to invest as this will only increase the idle capacity extant at the time. As one observer has argued, 'for the resource holders, the presence of spare capacity has always acted as a barrier to new exploration. Why on earth would they want to invest in more idle capacity when their stake holders had a greater need to cash?' (Arnott, 2004) Oil-producing countries outside OPEC lacked such spare capacity and thus continued to increase investment in oil production.

What is not well recognized is that during the mid-1980s, the oil market reached a turning point in which demand for OPEC oil slowly started to increase. As a consequence, OPEC's spare capacity began to diminish. The persistent decline in the available surplus capacity over the 1990s should have given rise to a realization that the continuing downward trend will eventually eliminate the capacity cushion. Investment to increase the volume of surplus capacity was required early on to avoid this situation. Recognizing this critical juncture however has not been timely, nor have its implications been adequately understood, as badly informed optimism about non-OPEC supply has tended to prevail while there was widespread pessimism about global demand for oil. The growth of global oil demand was consistently underestimated by the various energy agencies, and since demand pessimism was associated with exaggerated expectations about non-OPEC supply, there was no incentive for OPEC to increase spare capacity. Thus the decision not to invest more in upstream oil can be understood in the context of the following paradox: consuming countries want surplus capacity to be provided by OPEC but expect demand increases to be met by non-OPEC.

This trend of demand pessimism was associated with real oil prices starting to fall in the mid-1980s and ending with the collapse of 1998 and early 1999. This threw the industry into a deep recession and reduced the attractiveness of both existing and new investment plans. As a result, both national and international oil companies considerably reduced their investments in upstream oil exploration and production throughout the 1990s at a time when spare capacity was in decline.

Thus, the current lower spare capacity in oil production, as well as the bottlenecks in refining and pipeline systems which reduced the flexibility of the oil sector to respond to shocks are partly due to past history of low oil prices and long periods of oil demand pessimism which created an environment that is not conducive to investment in upstream and downstream oil. An indicator of the decline in upstream activity can be seen from the fall in the worldwide rig count although part of it may be due to the increasing use of horizontal drilling which reduces the number of wells needed for producing a field (see Figure 2).

3.2   Constraints on National Oil Companies (NOCs)

Some NOCs face serious investment constraints though their nature differs from those faced by IOCs. For many but not all oil-producing countries, efficiency and competitiveness of their national oil companies are not the main drivers of their oil policy but rather maximization of oil revenue, which can then be used to achieve broad socio-economic objectives. In many countries, the capital expenditure budget of the

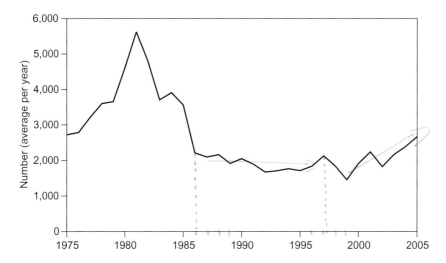

Source: Baker Hughes Website

Figure 2: Worldwide Rig Count

national oil company is determined by the finance ministry, the oil ministry or by some other authority. These capital budgets are not determined according to the availability of investment opportunities in the oil sector, but are subject to general government budgetary requirements. Given the competing and increasing demands for economic, social and infrastructure projects, this implies that the capital budget for national companies has been quite tight most of the time preventing them from undertaking investment, acquiring technological capabilities and enhancing their managerial expertise. There are some notable exceptions of course, with Saudi Aramco a case in point. As argued by Mr Claude Mandil, Executive Director of the IEA,

> unlike the major privately owned international oil companies that can mobilise large cash resources, the amount of earnings national oil companies can retain for investment purposes will necessarily be a derivative of the broader needs of national budgets. Financing new projects could become a problem where the national debt is already high and national considerations discourage or preclude private or foreign investment."[3]

The relationship between the owner of the natural resource (i.e. the government) and the operator and extractor of these reserves (i.e. the

---

3   IEA Press Releases (2004), *http://www.iea.org/textbase/press/pressdetail. asp?PRESS_REL_ID=126*

NOC) is highly complex. It causes inefficiencies in certain contexts, and yields sub-optimal rates of investment. The exception, at certain times, lies in those OPEC countries that carry spare capacity.

Many NOCs from resource-rich regions might be underestimating the magnitude of the cost and effort that capacity expansion requires. In many instances the main task since nationalization was to manage excess supply, not to grow capacity. Many of today's managers have little experience in dealing with large-scale capacity expansion projects. Even maintaining output capacity is now more expensive and more difficult than in the past, as reservoirs age. Cost inflation will also result from stronger pricing positions of key suppliers, given the widespread surge in demand for their products and services. This means that investment requirements could be greater than the initial estimates, projects will take longer than expected to complete and that output targets will be more difficult to achieve. These conditions will vary from company to company, and from region to region. But overall, the challenges will turn out to be greater than is commonly believed. This only calls attention to the need to redouble efforts now and formulate a new objective for the national oil companies and defining the role of the NOC vis-à-vis other state agencies.

The maturity of producing basins will induce many national oil companies to develop a two-track production and exploration strategy. The first one is to aim at improving the recovery factors of producing fields in order to reduce decline rates and increase reserves. This implies the application of new technology, more and better engineering, the adoption of best practice techniques, greater managerial discipline, rigorous economic analysis of alternatives and sustained efforts to reduce the costs of secondary and enhanced recovery projects. The challenges are not minor and several national oil companies are not well prepared for the task of designing and implementing these types of projects, although some have managed to successfully rejuvenate a number of basins and fields. The second track of the proposed strategy refers to the discovery and development of new oil and gas fields and basins. These projects are more risky and perhaps more capital intensive. The principal obstacle to the development of frontier projects that are being contemplated is not technology, which is readily available through service companies, where it is today embedded. The fundamental restriction however is a managerial one, further exacerbated when public sector type of bureaucratic rules and procedures discourage initiatives and hinder efficient execution.

3.3 Relationship between IOCs and Owners of Oil Reserves

International oil companies argue that difficulty of access to reserves in resource endowed regions is the main obstacle to these companies' rapid production growth. Of total global oil and gas reserves, only 14 percent are fully open to IOC without NOC participation. A further 17 percent are held by Russian companies while the rest (i.e. 69 percent) are held by NOCs. Out of this 69 percent, IOCs have some equity access of 11 percent (Zanoyan, 2004).

There are many strategic reasons why some producers have opted for restricting access to their oil resources. For example, Saudi Arabia had to maintain full control over its own level of production in order to effectively influence the overall management of supply. It does so by modulating the use of its excess capacity. This was essential in a world where excess supply tended to depress prices and reduce revenues, and will remain important in an oil world subject to sharp demand fluctuations and supply disruption risks. Other countries, like Mexico, see that direct government intervention in the oil industry allows them greater freedom to pursue internal economic and social objectives.

However, access is not the only issue since it is effectively restricted only in Saudi Arabia and Mexico. Even in countries where access to reserves is allowed, there may be important obstacles that could delay or prevent investment by international oil companies.

Degree and ease of access are highly dependent on the institutional framework of the host country, which governs the relationship between IOCs, NOCs, and other bodies that are involved in the oil industry, such as the Ministry of Petroleum, the Ministry of Finance, and in some cases an independent regulatory agency that regulates all commercial activities related to the energy sector. Vagueness concerning the roles of these players and possible overlap in their functions can lead to uncertainties and delays in implementation of investment projects.

It is important for IOCs to fully understand the role of the NOC in a certain country as this role differs from one country to another. Some NOCs have played an operational role and have gained greater confidence in their capacity to develop their own resources. With the support of service companies they are now able to tackle tasks that were not feasible in the past. In many instances their own capabilities have matured. IOCs have not fully recognized this change of circumstances and have not adequately explored new forms of engagement with some of the stronger NOCs.

Tension in the relationship between IOCs and NOCs could also arise from the fact that although both parties are in the same industry, they have different objectives and face different concerns. In other words, there is misalignment in interests. While the IOC is mainly concerned with profitability, share prices, and risk management, the NOC has to deal with government demands, local politics, and to seek ways to access capital and technology where they sometimes face more difficulties than the IOCs. However, many consider that such differences can act as ground for cooperation between the IOC, which is seeking attractive investment opportunities in below-ground resources, and the NOC which in turn seeks above-ground resources, namely technology and managerial skills that can be provided by the IOC.

The terms and conditions demanded by the NOC owners have been hardening over time because the oil market conditions have tightened in recent years. These trends will be further reinforced by growing long-term demand for oil and gas, and by decreasing opportunities open to international oil companies. The IOCs, on the other hand, often insist on a high rate of return for their investments as they are under pressure from shareholders to return their money to them if they cannot deploy it in very profitable ventures.

Under these circumstances, IOCs are sometimes asked to develop alternative business models that better fit the strategic objectives of NOCs' owners. But the large international oil companies find it difficult to adopt the flexibility that is thus demanded. This may open opportunities for service companies to step in and offer to undertake the tasks that IOCs are reluctant to perform; but the service companies themselves are under commercial constraints as they would not want to offend the IOCs which are major customers. The opportunities really exist for other sets of players. These include oil companies from China, Malaysia, Russia, India or Japan for example, which are eager to develop their international investments in search for new sources of supply, and some small companies which enjoy 'niche' advantages. These companies do not generally seek as high a rate of return on their investments as the IOCs, and they are likely to be flexible in negotiating contracts with NOCs and their governments. Like in other periods of the industry's history, marginal players in search of a new identity vis-à-vis the established club of majors might assume the challenge. In many instances, however, these newcomers have to convince host countries that they have the necessary capabilities for the tasks to be undertaken. There is also a political factor as some countries prefer to deal with companies related to Western superpowers than with companies from the third world. Given time, however, the newcomers will inevitably develop expertise and a range of capabilities and be able to pose a credible challenge.

The relationship between IOCs and national governments has been complicated by a phenomenon often referred to, rather misleadingly, as oil nationalism, but which is that of the centrality of the oil sector in the economic and political life of oil countries.

The case of Project Kuwait highlights some of the problems that the quest for access to large upstream resources may involve. Project Kuwait, a $7 billion undertaking initiated in 1997 by the Supreme Petroleum Council (SPC) of Kuwait is part of a general strategy aimed at increasing crude oil production capacity to 4 mb/d by 2020 from the current level of around 2.5 mb/d. The project would allow international oil companies to invest in upstream production in five northern oil fields (Abdali, Bahra, Ratqa, Raudhatain and Sabirya) near the border of Iraq to increase their production from the current level of 650,000 b/d to 900,000 b/d within a 3-year period (EIA, 2005b). Nader Sultan (2003), the previous CEO of KPC, identifies the following main objectives of Project Kuwait: reach a target production level within an agreed time frame, increase the reserve base and recovery factors; reduce the potential increase in production costs, allow for transfer of technology and develop the technical capabilities of the national workforce.

The project has been delayed for a number of years by opposition from the National Assembly which had qualms about the idea of foreign investment in the country's oil sector. Their case invoked the Kuwaiti Constitution, which states in Article 25 that 'no concession for exploitation of either a natural resource or a public service may be granted except by a law and for a limited period'. To avoid any constitutional breach, the KPC has called for the agreements with international oil companies to be structured on 'incentivized buy-back contracts' (IBBC) (EIA, 2005b). Based on this type of contract, international oil companies will not own oil reserves and hence will not be able to book them in their balance sheets. Instead oil companies would be paid a fee per barrel, an incentive fee for increasing reserves, and allowances for their capital investment. This structure allows the government to retain full control of its oil reserves and oil policy.

Project Kuwait is an interesting case because it shows that access of international oil companies to exporting countries with large reserves involves complex political factors that cause delays between planning and implementation, and in some extreme cases prevent foreign investment from taking place. The Project Kuwait bill was finally approved by a parliamentary committee in June 2005 but with amendments limiting its scope to four fields (the Bahra field was excluded). Although this represents a step forward, the implementation of Project Kuwait may still have to wait. First, the bill has to be approved by the National Assembly. Second, KPC needs to complete its negotiations and conclude the legal and financial terms of the contracts with the international oil companies. This is not a straightforward process given the complexity of the IBBCs. Assuming that things run smoothly, Project Kuwait will be in position to start around mid-2006 or sometime later.

3.4   *International Oil Companies' Investment Decisions*

The high liquidity of the international oil companies and the strength of their balance sheets allow them to act decisively in expanding their capital expenditures. However, recent and planned increases in capital expenditures have been quite modest despite the large increases in cash flows. Many international oil companies have used their huge cash reserves to buy back shares and issue generous dividends.

There are many reasons for this behaviour.

First, many take the view that before the 1980s, management loyalty was mainly confined to the corporation and not to shareholders. Furthermore, boards that were supposed to be protecting the rights of shareholders were ineffective in carrying out their duties. This created divergence between the interests of shareholders and those of managers with many corporations subsidizing negative net present value projects and expanding in size beyond the level that would maximize shareholders' wealth. In an influential paper, Jensen (1986) cites the oil industry as an example where agency costs between managers and shareholders were quite high. In the late 1970s and early 1980s, instead of oil companies reducing their exploration and development expenditures in the face of reduced expectations of future oil prices, these companies continued to use their excess cash flows to fund a wave of exploration and development and increase their production. In some instances, oil companies expanded horizontally and diversified into industries in which they had little knowledge about and no experience of running.

The situation has now changed. Nowadays, international oil companies have strongly pursued shareholder short-term value maximization. Maximizing value for shareholders was not to be found in investing capital in exploration and development in low return projects given that they then operated in a low oil price environment. Instead, the focus was mainly in cutting costs and maximizing the returns from existing assets in order to boost returns to shareholders. Thus, rather than investing their abundant cash flows in exploration and developing reserves, many international oil companies opted for returning the excess cash to shareholders through buyback schemes or issuing dividends.

As Stevens (2004) explains:

> In the past, high oil prices would have encouraged ever greater investments in exploration and production, thereby creating a self-adjusting mechanism. High prices increase investments which increase quantity supplied which reduce price. However, in recent years this has failed to materialize and indeed in this decade, according to Deutsche Bank, the major companies have cut their exploration budgets by 27%. The explanation for the potential lack of investment lies in the dominance of value-based management theories as the driving force of financial strategy in the major oil companies ... This problem of funds leaching out of the industry is compounded because returning funds to shareholders is becoming a key source of competition between the major oil companies to keep their shareholders happy...The danger is that the very short-term benefits to share price will be at the expense of future investment in maintaining and developing crude capacity.

International oil companies however cannot remain insensitive to their growing cash mountains indefinitely. High oil prices and aggressive share buyback programmes have a limit under these circumstances and might reinforce calls for windfall taxes which corporations hate. Increasing the capital expenditures budgets may help in fending off the threat of this type of taxes. In any case, growing investment is necessary if the oil industry does not want to face the charge that it is constraining world economic growth. It is also necessary to make up for past underinvestment.[4] The companies have realized that more investment is needed in exploration. In fact, in the past three years, with reserves replacement levels falling and production growth failing to meet corporate targets, they have tried to kick start investment. In 2003, world capital expenditure in exploration and production rose by 5.6 percent to reach $114 billion. In 2004, the growth of E&P expenditure rose to 10 percent to reach $125 billion and in 2005 the world E&P investment is expected to amount to $135 billion. Part of these expenditure increases is due to higher steel prices, rig-day rates and petroleum services costs. Nevertheless, these investment figures are still 20 percent below the 1981 peak of $170 billion in 2004 dollars (IFP, 2004).

Many international oil companies have also engaged in a wave of mergers and acquisitions in an attempt to improve their profitability. An important motive behind the merger strategy was to improve reserve

---

4   When looking at buyback programmes one must be aware that the oil industry is not unique. In many other sectors they have increased in volume. They are part of a more generalized market sentiment and behaviour after the stock market bubble burst in 2000.

production ratios by purchasing other companies' booked reserves. This was viewed as an alternative to investment in exploration and development.

Maximizing shareholder value also meant focusing on independent profit centres: rather than treating upstream and downstream as parts of the same supply chain, investment was diverted away from relatively low profit centres such as refining, as will be discussed later.

Second, in the appraisal of upstream projects a number of inbuilt biases militate against rapid expansion. Price assumptions for testing projects involve one of these biases. Although the oil price hurdle used in the appraisal of upstream projects which was set for a long time at $15 per barrel has been raised to $20 and recently to $25 by many companies, these remain assumptions erring significantly on the low side. It would be interesting to find out which and how many projects that were deferred or rejected would have survived a more reasonable test of, say, $35 per barrel. In today's financial environment, the predilection for short-term market valuations and results poses an important challenge to projects with a 20 to 30 year life. Discounting expected future cash flows in long-term projects does not capture adequately their probable benefits.

Third, for the large international oil companies (but not for the independents) size and scale of investments projects are central to their investment decisions. As Jeroen van der Veer (2005) explains, 'size and scale have always been key characteristics of our industry. But as we respond to huge increases in demand for energy, size and scale are going to be even more of a feature in the future.' By investing in large and high quality assets, international oil companies can achieve substantial efficiencies that help them maximize the return on capital employed. In many instances, profitable projects can be turned down if they do not pass the materiality threshold of a company or if there is not enough potential for repeat investment opportunities so that the oil company can build a critical mass (Zanoyan, 2004).

Fourth, some have argued that growth in production capacity is limited by increasing finding and development costs. There is some evidence, that after a long period of decline in costs from the early 1980s to mid-1990, finding and development costs have begun to climb. Cost increases imply that higher oil prices will be needed to maintain profitability.

Finally, as previously discussed, restriction of access to natural reserves is considered as an important barrier to investment. However, such access will not necessarily ensure that oil companies will achieve the high return that they are after. The problem with large reserves is that the countries that own them know that they are resource rich. These countries know that the production costs are low and that there is no exploration risk, because the reserves have been discovered. They also know that these reserves are often in shallow offshore or simply onshore. For all these reasons, the terms of an upstream contract acceptable to the host country will not always be very attractive to oil companies. Thus large reserves appear to be attractive, but they do not necessarily produce the type of rate of return that an investor will want. It may be more attractive for IOCs to go deep offshore where they may only realize a small margin at the beginning, but where technical progress year after year will substantially increase that margin. In a large reserve country, the margin may remain constant. One may ask then: why do oil companies need to go there? The answer is probably to follow their competitors lest they get criticized by financial analysts. The access problem is not that simple (Mabro, 2003).

3.5   Volatility and Data Problems

Oil prices have often been volatile blurring the distinction between transitory and permanent price movements. As suggested in the literature of irreversible investment under uncertainty the large investment outlays in oil projects and the irreversible nature of these investments have the effect of increasing the value of the option to wait. There is thus a case for delaying the investment until new information about market conditions arrives, especially information about expected global demand and oil supplies from other countries. For the oil industry, the option to wait is very valuable. After all, the decision to wait rather than invest immediately and increase production is more profitable than to invest and increase production in the face of falling global demand. In other words, it is more profitable for all oil investors (national and private companies) to err on the side of potentially under-producing to some extent and for a certain period of time.

The lack of oil supply and demand data causes further volatility and uncertainty to oil markets (Lynch, 2002). Although oil prices have become more transparent over the years, information about crude oil consumption and production has not improved either in quality or timing. On the demand side, data on the consumption of oil, even those for OECD countries, are uncertain, subject to major revisions, and published with a considerable lag. This problem is becoming more acute with the increasing importance of some developing countries such as India and China, both major oil consumers, with even less reliable data. On the supply side, there is uncertainty about actual OPEC production levels. Furthermore, the advent of many small oil producers involves further difficulties in the collection of reliable and timely information. The quality of inventory data is also subject to uncertainties and revisions. Since in tight markets every barrel matters, all these uncertainties about supply, demand and stock levels contribute to the volatility of oil markets by distorting perceptions about any supply/demand gap and increasing the impact of speculation.

The difficulty of forecasting volumes and price movements also contributes to higher uncertainty. Historically, the various international agencies have made poor predictions regarding future demand and supply conditions. In analysing this issue, Horsnell (2004) finds that over the period 2001 and 2002 alone, the various forecasts overestimated global oil supply by 1 mb/d and underestimated global oil demand by 2 mb/d i.e. a combined swing of 3 mb/d for year 2003. This was equivalent 'to finding out, with no notice at all, that the world needed the oil production of another Norway'. In this particular case, the gap was met by OPEC's spare capacity. Poor predictions however can easily work in the opposite direction i.e. agencies can overestimate the quantity of global oil demand and underestimate the global oil supply. In such a case, an investment decision based on these predictions will look very foolish later on. Such flips in oil gap predictions reduce their value in formulating investment decisions. This problem of poor prediction is amplified by the fact that any investment should look not only at incremental demand and its sources but also at maintaining existing capacity determined by the natural decline of oil fields. Horsnell (2004) claims that 'if we are to make major mistakes in assessing the need for more capacity, it is more likely to come from misjudging the rate of depreciation of capital in energy rather than imputing incorrect energy demand growth'.

The IEA (2001) tests the relationship between volatility and investment by using an empirical model in which the annual change in the amount of investment in global oil exploration and production is a function of changes in annual aggregates of daily oil price volatility and annual changes in price levels, both lagged by one year. Their results suggest a strong negative inverse relationship between investment changes and oil price volatility. This means that an increase in oil price volatility induces a decline in investment in exploration and production. Specifically, according to the IEA estimates, a 1 percent increase in volatility causes a 0.11 percent decline in investment. This view receives support from many observers and policy makers. For instance, Ali Naimi the Saudi Petroleum and Mineral Resources Minister has recently argued that 'investors in any industry, and the petroleum industry is no different, crave stability and predictability. As we all know too well, price volatility is not conducive to either stability or predictability. Only speculators benefit from oil price volatility. We therefore need to address the ongoing price volatility that continues to impact oil markets and often delays investment decisions' (Naimi, 2005).

It is important to note that the relationship between volatility and investment is more complex and that it can work in the other direction i.e. from investment to volatility. Specifically, one could argue that it is the lack of investment which caused a decline in spare capacity that has induced greater oil price volatility. Since the IEA's empirical evidence makes no attempt to control for potential endogeneity of volatility, the above estimates are likely to be biased and thus the sizeable impact of volatility on investment should be treated with caution. In fact, Pindyck (2001) finds that although changes in volatility may in principle influence market variables, his empirical evidence suggests that for the petroleum complex these effects are rather modest. He also finds that market variables such as interest rates and exchange rates do very little to explain oil volatility and that the only variable that can explain volatility is its own past values.

3.6   Geopolitical Factors

Unfavourable geopolitical factors can prevent capacity expansion in many oil-producing countries through creating an adverse investment climate. For example, political strife in Iraq, Venezuela, and Nigeria may prevent these countries from undertaking the necessary investment in their oil sectors thus restricting oil supplies. Economic and political sanctions in Libya, Iraq and Iran have hindered investment and deferred the development of these countries' resources. In the Gulf area, military intervention has negatively affected the investment climate.

4.   Underinvestment in Downstream Oil

Investment in downstream activities such as pipelines, refineries and tankers has also lagged behind the growth in global oil demand contributing to the bottlenecks in petroleum product markets and weakening the ability of the oil industry to deal with even minor shocks.

4.1   Refining

The dramatic rise in oil product prices in 2004−2005 highlighted the

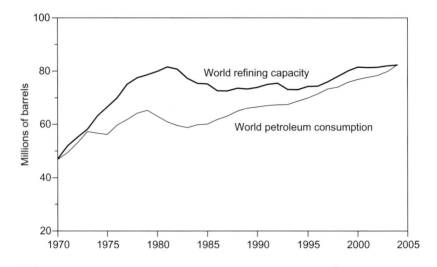

Source: EIA Website

Figure 3: World Refining Spare Capacity

(handwritten annotation: oil peak in 2003-5 due to low refining capacity.)

importance of shortage in refining capacity which has declined from
a peak of 21 mb/d in 1981 to almost zero in 2004 (See Figure 3).
The extent of this shortage is in part reflected by the record levels of
utilization rate (the ratio of barrels input to the refinery to the operating
capacity of the refinery) which rose throughout the 1990s and reached
sometimes the unsustainable level of 95–96 percent in 2004 and 2005.
These are effectively maximum utilization rates given that refineries
should be subject to maintenance turnarounds. The 2005 spike in
oil prices has been blamed by many analysts to lack of refining. This
argument was also made forcefully in a recent speech by the Saudi
Foreign Minister Saud Al-Faisal where he stated that,

> the basic problem of the current energy crisis ... is that the current refiner-
> ies are incapable of meeting demand on oil products, shortage in storage
> capacity and restrictions imposed on the oil industry thus paralyzing it
> from building more refineries... Not a single refinery was built in the
> United States in the last three decades, while the difference in standard
> requirements on oil products from one state to the other inside the United
> States, particularly due to environmental concerns, has greatly deteriorated
> the energy crisis.[5]

---

5   Oil crisis very dangerous, lack of refining capacity to blame, Saudi minister,
    Downloadable from: *http://www.mabico.com/en/news/20050922/government_
    ministries/article47770/*

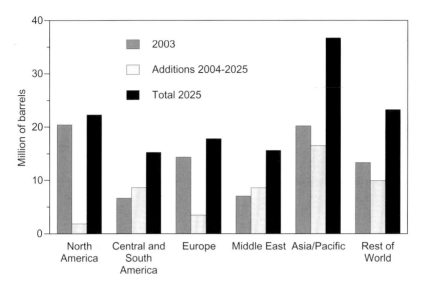

Source: EIA, *Annual Energy Outlook 2005*

Figure 4: Worldwide Refining Capacity by Region, 2003 and 2025

According to the EIA (2005a), in order to meet the growth in international oil demand, worldwide refining capacity has to increase from its current level of 82 mb/d to more than 131 mb/d by 2025 (see Figure 4). OPEC estimates the required additions to primary and secondary refining capacities between 2004 and 2015 at 9.9 mb/d for primary distillation, 15.9 mb/d for desulphurization and 8.6 mb/d for the other types of refining plants.[6] (See Chapter 2) Most of this increase in refining capacity is expected to come from outside North America and Europe where stringent environmental regulations will not permit rapid expansion. Furthermore, building new refining capacity will also have to take into account the increase in global demand for lighter products and provide transportation fuels with reduced lead distillate and residual fuels with lower sulphur levels. The decline in the quality of the marginal barrel towards heavier and sourer crude oil is likely to exert an extra burden and magnify costs of adapting and expanding refining capacity.

Some of the factors affecting investment in upstream have also affected investment in downstream. For instance, excess refining capacity and low historical margins for a period of 20 years curtailed the investment to very low levels. Only mandated product quality improvements

---

6   Vacuum distillation coking, catalytic cracking and reforming, and hydro cracking.

and environmental protection obligations generated investment projects irrespective of returns. Compare the present situation in which the current spike in oil prices is blamed on shortage of refining capacity to that of ten years ago as expressed by an internal Texaco document:

> As observed over the last few years and as projected well into the future, the most critical factor facing the refining industry on the West Coast is the surplus refining capacity, and the surplus gasoline production capacity. The same situation exists for the entire U.S. refining industry. Supply significantly exceeds demand year-round. This results in very poor refinery margins, and very poor refinery financial results. Significant events need to occur to assist in reducing supplies and/or increasing the demand for gasoline.

Excess surplus refining capacity has marked the psyche of oil industry management and delayed or prevented investment despite a radical change of circumstances.[7]

But there are some additional factors specific to downstream which are worth emphasizing.

4.1.1 Independent Profit Centres

In attempting to maximize shareholder value, many international oil companies have shied away from investing in new refining capacity given that the returns on refining are low especially when compared to oil and gas upstream production. Furthermore, the need for additional investment to meet environmental mandates has squeezed gross refinery margins to very low levels. Figure 5 shows that the profit rate in oil and gas production was much higher than refining and marketing in most of the years 1997−2002. Although this is consistent with maximizing shareholder value in the short term, treating each stage of the supply chain as an independent profit centre has decreased the overall flexibility in the products market.

4.1.2   Environmental Restrictions and Government Regulations

Strict environmental regulations have made building refineries more difficult or nearly impossible (IMF, 2005). Refining capacity expansion has also been constrained by uncertainty about the extent and timing of government regulations that refineries must comply with. In the

---

7   Fortunately, the tide had begun to change in the mid-1990s. In the USA, 1.4 million barrels per day of distillation capacity was added between 1996 and 2003. The announcement by ConocoPhillips in spring 2004 of a $3 billion-refinery investment programme and the construction by Marathon of conversion facilities in the Mid-West are clear examples of the direction in which the wind is blowing. Stock prices of pure refiners like Valero are also indicative.

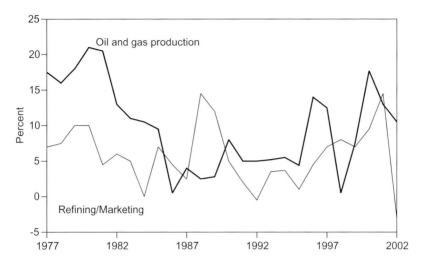

Source: EIA, *Oil Market Basics*

Figure 5: Profit Rates by Industry Segment, 1997−2002

USA for example, refineries confront a complex and time-consuming permitting process involving federal, state and local authorities with often-conflicting priorities. Due to the various regulations, oil companies have found it more profitable to consolidate operations and increase the output at existing refineries than to build new refineries[8] or simply buy other refineries.

4.1.3 Shortage of Capacity for Refining Heavy Crudes
The refining industry has been operating its conversion plants at full effective capacity and has even increased its use of distillation capacity. It has lost much needed flexibility given the changes in the structure of demand for its products, the mandated modification of product specifications and other environmental restrictions, and the changing mix of its crude slate, due to higher incremental volumes of sour and heavier crudes. This has resulted in higher margins, particularly in complex refineries, and much wider light heavy differentials. The Maya-WTS spread tended to be in the 2-dollar range and the Maya-WTI spread in the 3-dollar plus range. In November 2004, this differential was above

8   In its testimony before the Federal Trade Commission, Murphy Oil notes that in 1980 'the average U.S. refinery produced just below 60,000 barrels of petroleum products daily. Last year (2000), the average refinery was capable of producing over 110,000 barrels of gasoline, heating oil, diesel oil and other products every day.'

18 dollars, narrowed to 15 dollars, but given market conditions it is possible for the differential to increase again toward previous peaks.

Investment in refineries is a function of expected mid-cycle margins and demand for products, while investment in conversion units depends on light-heavy differentials.[9] What troubles refiners is the sustainability of high margins and differentials in this highly cyclical industry. Not all are convinced that what we are now facing is the consequence of structural change. Many still think that oil prices and refinery margins will revert to the mean after a short period of adjustment. However, if they fail to react in a timely manner they might lose the opportunities that are being provided by a fundamental disequilibrium.

Another source of uncertainty for any one of the regional refining centres is the strategic reactions of key global industry players, given the global nature of this industry. It is clear that refiners in the USA must expand conversion capacity and other processing units that will enable them to process sour and heavier crudes; European refiners must accommodate a greater share of middle distillate demand and Asian refiners must meet the rapid growth of oil product demand, particularly of automotive fuels. However, many are concerned by the role that Middle East producers might play in refining their own crude supply. They could have a real competitive edge in products that tend to travel far. However, it appears that Middle East planned capacity expansion is mainly geared to domestic and possibly regional demand growth. Saudi Arabia, for example is planning the expansion of two refineries. The one on the West Coast would be directed to meeting domestic demand and the one on the East Coast would export products to Asia.

In short, there are two aspects to the refining problem: an insufficient volume of total capacity and a mismatch between the structure of refining plants and that of the world crude oil slate. The latter means that there is a need to invest in deep conversion and de-sulphurization facilities because a significant proportion of incremental crude oil production is of the heavy/sour varieties. Figure 6 shows the structure of capacity in world refining.

---

9   Also note that causality can run in the opposite direction. Certain investment decisions can cause the differential to widen. For instance, when returns in refining were low for most of the past decade, many refiners were unable to justify significant investment in de-sulphurization. As a result, many refineries could not meet the new regulatory specifications. As recommendations were turned into laws, many refiners chose the easy option of purchasing lower sulphur crude oil feedstock rather than the option of investing in sulphur removal facilities. This has pushed refiners to bid up prices of 'light-sweet' crude oils in order to attain the new environmental standards.

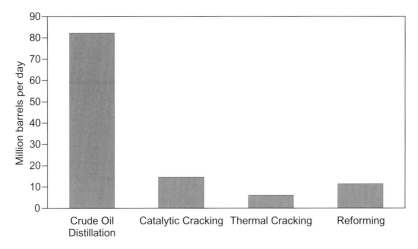

Source: EIA Website

Figure 6: World Crude Oil Refining Capacity by Process as of January 2004

4.2 Oil Tankers

The shortage of tankers is an additional sign of how strong demand and a lack of investment have left the oil industry's infrastructure stretched thin. While oil consumption occurs mainly in the West, oil is produced mainly in the Middle East, former Soviet Union, West Africa, and South America. As a result, significant volumes of oil are traded in international markets. Crude oil and petroleum products can move mainly in two ways: through oil tanker ships and oil pipelines. According to the EIA (2004), about two-thirds of the world's oil trade (both crude oils and refined products) moves by tanker.

Worldwide tanker capacity reached its peak in the late 1970s. During the early 1980s, oil tanker capacity declined but so did the demand for tankers because of a fall in the global demand for oil. This left a sizeable surplus capacity in the oil tanker market. In the mid-1980s, the oil tanker fleet began to grow but at a much slower rate than global oil demand. As a result, surplus capacity in oil tankers began to decline and by 2004, the demand for tankers came close to the available capacity. The shortage in oil tankers has been reflected in soaring freight costs in the spot market in recent years. As can be seen from Figure 7 there has been a large surge in freight costs between 1999 and 2000 and 2002 and 2004 when on most routes the freight cost increased by more than $2 per barrel.

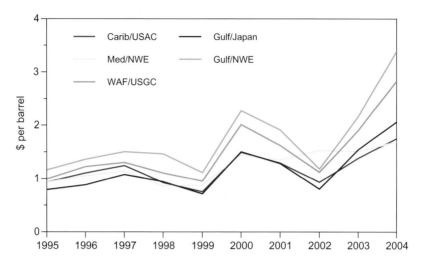

Notes: Carib = Caribbean, USAC = US Atlantic Coast, Med = Mediterranean, NWE = north-west Europe, WAF = West Africa, USGC = US Gulf Coast. Freight costs are based on the weighted averages of spot freight rates in the typical size range.

Source: *OPEC Annual Statistical Bulletin 2004*

Figure 7: Freight Costs in the Spot Market

This more generalized underinvestment problem across the different stages of the oil supply chain has important policy implications. Even if investments in the upstream oil sector materialize and result in an increase in oil production, bottlenecks in refining capacity, pipelines and tankers means that this higher crude oil production will not necessarily translate into higher volumes of petroleum products which consumers want. Since OPEC has little control of the global oil logistics systems, a close coordination of investment plans is required between oil-consuming countries and oil-producing countries to address the bottlenecks in the oil industry in order for increased production to reach ultimate consumers.

5.   Conclusions

As we have seen, the interaction of an array of historical, economical, political and geological factors, and others relating to corporate behaviour, determine the rate of investment in the oil sector. In the past two decades, these forces combined together to produce an environment conducive to low investment rates. The low investment rates both in the upstream and downstream have generated bottlenecks in all parts of the oil supply chain leading to spikes in oil prices and higher oil price volatility. The long gestation periods and lead times of energy investment projects do not have an impact during a short-term cyclical downturn but would affect conditions in the next upturn. Investment undertaken today would only ease supply constraints in the next growth cycle. Because of these investment lags there is not much that OPEC, other oil-producing countries, or the oil industry in general, can do to alter the current market conditions in the next two or three years. Unless geopolitical problems cool down, or high prices and a slowdown of the world economy depress the global demand for oil, the tight conditions that are producing high and volatile oil prices are likely to continue. An action that national oil and international oil companies can take is to send signals to market participants about credible plans to increase investment in both refining and crude oil production. To what extent these signals can alter expectations in the futures markets remains to be seen.

The main impact of underinvestment in the oil sector is felt on the erosion of spare capacity. The challenges facing the oil industry nowadays are not only to invest to meet expected growth in global demand for oil, but more importantly to invest in new spare capacity that can provide the global oil system with the flexibility it needs to counter supply shocks. A production system as big and complex as world petroleum cannot function smoothly without the cushion of a significant volume of spare capacity. The main question is: who is going to bear the costs of investing in spare capacity? The international oil companies are not willing to undertake such a role. In fact, investing in spare capacity contradicts the principle of shareholder value maximization. They will have to be compelled by governments to hold a certain amount of spare capacity in their refineries as happened after 1973 when many OECD countries imposed mandatory oil inventory holdings on private oil companies. On the other hand, most NOCs – due to many constraints discussed above – may not be willing or even able to invest in new spare capacity. Saudi Arabia is a major exception as its declared policy is to maintain a volume of spare capacity of 2–3 million barrels per day to enable it to stabilize the market when demand suddenly increases as in 2004 or when supplies are disrupted by technical incidents or political events.

The spare capacity that emerged in the past was not the result of an optimal investment decision. It was the expansion of non-OPEC production in the 1970s and 80s and the contraction and low growth of demand during the 80s and 90s that generated excess capacity in OPEC member countries. These specific circumstances were critical in defining the oil price regime that prevailed until recently and imposed the role played by OPEC, and Saudi Arabia in particular, of managing excess capacity in order to protect price levels and revenue streams. The exceptional successes of the 1970s in expanding non-OPEC production cannot be replicated. Extraordinary discoveries in Alaska, the North Sea and Mexico were followed by the build-up of Soviet production that reached a peak in the late eighties. Later developments in the Caspian, West Africa and Brazil, among others, have been of a smaller aggregate size and would not have been able to compensate for oil production decline in OECD and other countries in the absence of the Russian recovery of 2000–2004.

One of the most important items that should be put on the agenda of the oil-producing/oil-consuming countries dialogue is the spare capacity issue. How should the financing burden of holding the necessary volumes be equitably shared between all the parties involved? The implicit assumption, mentioned before, that the role of non-OPEC is to supply the world with as much of the incremental oil demand that obtains every year and that of OPEC is to provide the capacity cushion has become totally irrelevant today. We have entered a different world. Failure to address new issues simply means more price spikes, more price volatility, and wilder cycles in many years to come

References

Arnott, R. (2004), 'Oil Depletion or Depleted Policies', Oxford Institute for Energy Studies, *http://www.oxfordenergy.org/presentations/Oil* Depletion

Energy Information Administration (EIA) (2004), 'World Oil Transit Choke-points', USA: EIA. *http://www.eia.doe.gov/emeu/cabs/choke.html*

Energy Information Administration (2005a), *Annual Energy Outlook 2005*, USA: EIA.

Energy Information Administration (2005b), *Country Analysis Brief. http://www.eia.doe.gov/emeu/cabs/kuwait.html*

Horsnell, P. (2004) 'Energy Investments and Impediments', Paper presented at the International Energy Forum, The Netherlands.

Institut Français du Pétrole (IFP) (2004), *Panorama 2005: Exploration-Production Activity and Market*, Paris: IFP.

International Energy Agency (IEA) (2001), *Energy Price Volatility: trends and consequences*, Paris: International Energy Agency.

International Energy Agency (2003), *World Energy Outlook 2003*, Paris: International Energy Agency.

International Monetary Fund (IMF) (2005), *Oil Market Developments and Issues*, Washington: International Monetary Fund.

Jensen, M. C. (1986), 'Agency Costs of Free Cash Flow, Corporate Finance, and Takeovers,' *American Economic Review*, 76(2):323−2.

Lynch, M. (2002), 'Causes of Oil Price Volatility', Background Paper for the 8th International Energy Forum, Osaka, Japan.

Mabro, R. (2003), 'Setting the Scene', *OPEC Review*, XXVII(3):187−90.

Myers, S. C. (1977), 'Determinants of corporate borrowing', *Journal of Financial Economics*, 5:147–75.

Myers, S. C. and N.S. Majlouf (1984), 'Corporate Financing and Investment Decisions When Firms Have Information That Investors Do Not Have', *Journal of Financial Economics*, 13:187−221.

Naimi, A. (2005), 'Shaping the Energy Future: The Role of Saudi Arabia', *Middle East Economic Survey*, XLVIII(40).

Pindyck, R. (2001), 'Volatility and Commodity Price Dynamics', MIT Centre for Energy and Environmental Policy Research Working Paper.

Stevens, P. (2004), 'The Future Price Of Crude Oil', *Middle East Economic Survey*, XLVII(37).

Sultan, N.H. (2003), 'The Challenges of Opening up the Upstream to International Investors: A Kuwaiti Perspective', Paper presented to the Oil and Money 2003 Conference, London.

Van der Veer, J. (2005), 'What is the International Oil Company of the Future Going to Look Like?' Shell Website, downloadable from: *http://www.shell. com/static/media-en/downloads/speeches/jvdv_oilsummb.pdf*

Zanoyan, V. (2004), 'The Oil Investment Climate', *Middle East Economic Survey*, XLVII(26).

CHAPTER 5

GLOBAL PETROLEUM RESERVES, RESOURCES AND FORECASTS

Thomas S. Ahlbrandt

1. Introduction

Under most energy projections, fossil fuels will remain the dominant source of energy for the remainder of this century (Salvador, 2005). Consequently, an understanding of the distribution, quantities, and availability of fossil fuel resources will continue to be important to policy makers. Petroleum resources are periodically reassessed, not just because new data become available and better geologic models are developed, but also because many non-geologic factors determine which part of the crustal abundance of petroleum will be economic and acceptable over some foreseeable future. Misunderstanding of reserve and resource terminology and the nature of resources commonly leads to strong disagreements over the size of remaining reserves and resources (Figures 1, 2). Because our concepts of petroleum reserves and resources and their measurements and estimates are changing especially with a greater emphasis now being placed on uncertainties and risk (probabilistic rather than deterministic estimates) and on recent significant reserve revisions associated with these terms, the primary focus of this paper is on the definitions and methodologies applicable to reserve and resource studies and assessments in the forecasting of oil and natural gas supply.

Detailed global petroleum information was published in a recent world assessment of petroleum resources conducted by the U.S. Geological Survey (USGS) in their World Petroleum Assessment 2000 (US-Geological Survey, 2000). Although that report was released in 2000, the cut-off date for the assessment was January 1, 1996. Throughout this report, it will be referred to as 'USGS 2000 assessment' or in some cases shortened to 'USGS 2000'. The foundation of the USGS 2000 assessment was geologic insights gained from studying a large number of the total petroleum systems of the world (Ahlbrandt et al., 2000, 2005). That assessment provides estimates of the quantities

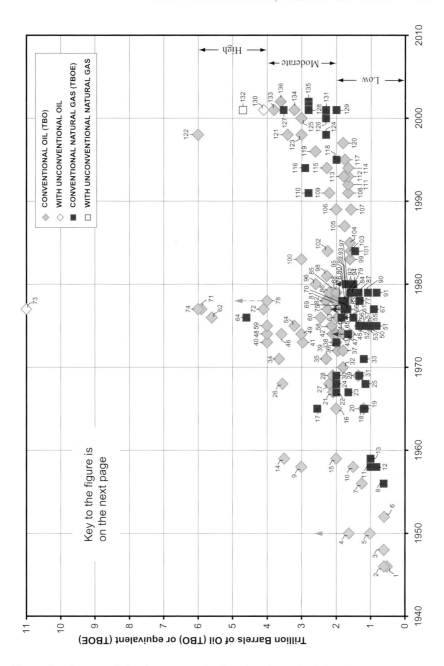

Note: Captions to all the figures may be found at the end of the chapter

Figure 1: Comparison of World Oil and Natural Gas Resource Endowment Estimates.

Key to Figure 1

1.   Duce, 1946
2.   Pouge, 1946
3.   Weeks, 1948
4.   Levorsen, 1950
5.   Weeks, 1950
6.   Pratt, 1952
7.   Hubbert, 1956
8.   MacKinney, 1956*
9.   Weeks (H), 1958
10. Weeks (L), 1958
11. Weeks (H), 1958*
12. Weeks (L), 1958*
13. Weeks, 1959*
14. Weeks (H), 1959
15. Weeks (L), 1959
16. Hendricks, 1965
17. Hendricks, 1965*
18. MacKinney, 1965*
19. Weeks, 1965*
20. Weeks, 1965
21. Ryman, 1967
22. Ryman, 1967*
23. SHELL, 1967*
24. MacKinney, 1968*
25. Weeks, 1968*
26. Weeks (H), 1968
27. Weeks (L), 1968
28. Hubbert (H), 1969
29. Hubbert (L), 1969
30. Hubbert (H), 1969*
31. Hubbert (L), 1969*
32. Moody, 1970
33. Weeks, 1971*
34. Weeks (H), 1971
35. Weeks (L), 1971
36. Bauquis, 1972
37. Warman, 1972
38. Hubbert, 1973
39. Hubbert, 1973*
40. Odell, 1973
41. Schweinfurth, 1973
42. Hubbert (H), 1974
43. Hubbert (L), 1974
44. Kirkby, Adams (H), 1974
45. Kirkby, Adams (L), 1974
46. Parent, Linden, 1974
47. Parent, Linden (and up), 1974*
48. Parent, Linden (H), 1974
49. Parent, Linden (L), 1974

50. Adams and Kirkby (H), 1975*
51. Adams and Kirkby (L), 1975*
52. MacKay, North (H), 1975
53. MacKay, North (L), 1975
54. Moody, Esser (H), 1975
55. Moody, Esser (L), 1975
56. Moody, Geiger, 1975*
57. Nat.   Academy of Science, 1975*
58. Nat.   Academy of Science, 1975
59. Odell and Rosing, 1975
60. Barthel, BGR, 1976
61. Barthel, BGR, 1976*
62. Grossling (H), 1976
63. Grossling (L), 1976
64. Grossling (H), 1976*
65. Grossling (L), 1976*
66. Folinsbee, 1977
67. International Gas Union, 1977*
68. Klemme, 1977
69. Parent, Linden (H), 1977*
70. Parent, Linden (L), 1977*
71. Seidl, IIASA (H), 1977
72. Seidl, IIASA (L), 1977
73. Styrikovich, 1977
74. Styrikovich, 1977
75. World Energy Conference, 1977
76. Desprairies (H), 1978*
77. Desprairies (L), 1978*
78. IFP (4 estimates >4 TBO), 1978
79. Klemme, 1978
80. McCormick, AGA, 1978*
81. Moody, 1978
82. Nehring (H), 1978
83. Nehring (L), 1978
84. Bois, 1979*
85. Halbouty, 1979
86. Meyerhoff, 1979
87. Meyerhoff, 1979*
88. Nehring (H), 1979
89. Nehring (L), 1979
90. Nehring (H), 1979*
91. Nehring (L), 1979*
92. De Bruyne, 1980
93. Parent, Linden (H), 1980*

94. Parent, Linden (L), 1980*
95. Schubert, 1980*
96. World Energy Conference, 1980
97. World Energy Conference, 1980*
98. Halbouty, 1981
99. Masters, 1983
100. Odell and Rosing, 1983
101. Masters, 1984*
102. Masters (H), 1984
103. Masters (L), 1984
104. Martin, 1985
105. Masters, 1987
106. Bookout, 1989
107. Campbell, 1989
108. Campbell, 1991
109. Masters, 1991
110. Masters, 1991*
111. Campbell, 1992
112. Campbell, 1993
113. Laherrere, 1993
114. Campbell, 1994
115. Masters, 1994
116. Masters, 1994*
117. Campbell, 1995
118. Riva, 1995*
119. MacKenzie, 1996
120. Campbell, 1997
121. BP, 1998
122. Odell (H), 1998
123. Odell (L), 1998
124. Schollnberger, 1998*
125. USGS 2000
126. USGS 2000*
127. CEDIGAZ (H) 2001*
128. CEDIGAZ (L) 2001*
129. Deffeyes 2001*
130. SHELL 2001
131. SHELL 2001*
132. SHELL 2001*
133. SHELL (H) 2001
134. SHELL (L) 2001
135. BGR 2002*
136. Edwards 2002

* - gas estimates
(H) - high estimate
(L) - low estimate

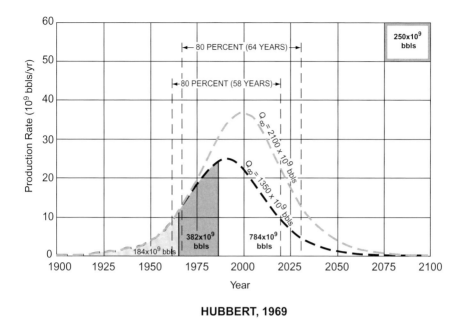

HUBBERT, 1969

Figure 2:   Ultimate World Oil Supply Estimates – both a low scenario and high
scenario based on the Hubbert prediction model

of *conventional* technically recoverable oil, gas, and natural gas liquids
(NGL) outside the United States that have the potential to be added
to reserves (Figure 1). *Conventional* accumulations are associated with
structural or stratigraphic traps, commonly bounded by a down-dip
water contact, and are affected by the buoyancy of petroleum in water.
In contrast, *unconventional (continuous)* accumulations are really extensive
reservoirs of petroleum not necessarily related to conventional structural
or stratigraphic traps. They lack well-defined down-dip petroleum to
water contacts, and are not dependent on petroleum buoyancy in water
(Ahlbrandt, 2004; Klett, 2004). In this paper, petroleum is a collective
term for crude oil, natural gas, and natural gas liquids.

Many groups, such as the International Energy Agency (2000, 2001,
2002, 2003, 2004), the Energy Information Administration of the
Department of Energy (2001, 2002), and the Organization of Petroleum Exporting Countries (2004) use the USGS 2000 estimates as the
baseline for world petroleum resources. Similarly numerous individuals
have utilized these estimates in forecasting future petroleum potential
or peak oil and gas scenarios including Cavallo (2002, 2004), Edwards
(2001, 2002a,b), Greene (2003), Green (2004), and Salvador (2005).

The USGS 2000 assessment is also the basis for estimates of reserve growth and undiscovered resources for this report. Those estimates, for oil (Figure 3) and natural gas (Figure 4), exclusive of the USA, as well as consumption projections demonstrate the petroleum endowment distribution for these resources and their concentration. A detailed analysis by regions, provinces, total petroleum systems and their assessment units is provided in Ahlbrandt et al. (2000, 2005), Ahlbrandt (2002, 2004), and Charpentier (2004).

However, the USGS does not make projections of peak oil or future oil and natural gas production as shown by the early estimates of Hubbert (1969) (Figure 2). Reserves and cumulative production data are derived from the IHS Energy database (Petroconsultants, 1996; IHS Energy Group, 2001) and for Canada from Nehring and Associates (NRG Associates, Inc., 1995). Recently, calibration of the results of the assessment, based on data as of January 1996, suggests that the USGS 2000 assessment is reasonable in light of petroleum discovery and reserve growth data from January 1996 through December, 2003 (Klett et al., 2005).

The four critical components of assessment of conventional petroleum resources are (1) cumulative production (how much has been used, which is useful for historical perspective); (2) remaining reserves (how much has been found but not yet produced); (3) reserve growth (how large will the fields be when they are fully developed); and (4) undiscovered resources (how much additional petroleum may be found in as yet undiscovered fields). Each component is a subject for debate and the confusing nomenclature, as well as a long history of use and misuse of terms, commonly cloud discussions. This paper will explore the various components of an assessment and the assumptions behind them.

2.   Reserve and Resource Definitions

Because (1) petroleum reserves and resource terminology have long been misused or misunderstood, and (2) policymakers increasingly focus on the significance of petroleum reserves and resources, it is important to more clearly define these terms.

Reserves are that portion of an identified resource from which a usable mineral and energy commodity can be economically and legally extracted at the time of determination (U.S. Geological Survey and U.S. Bureau of Mines, 1976). By this resource/reserve scheme, the definition of reserves may be updated to: an estimate within specified

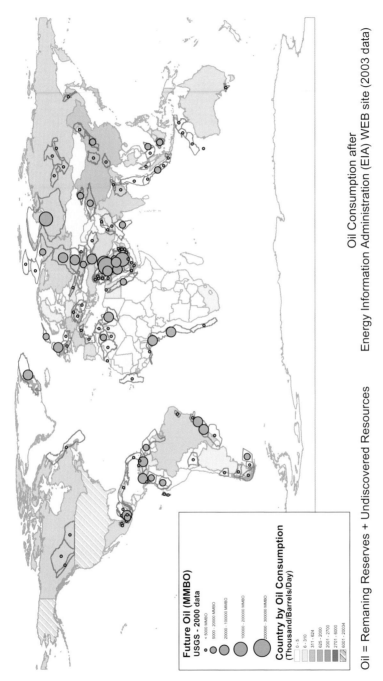

Figure 3: Future Oil Resources and Oil Consumption (assessed provinces excluding USA)

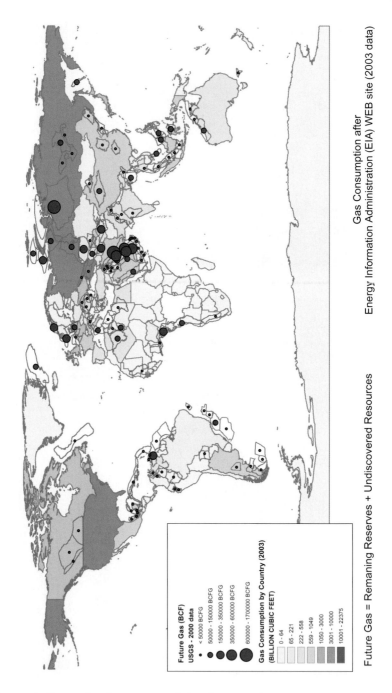

Figure 4: Future Gas Resources and Gas Consumption (assessed provinces excluding USA)

accuracy limits of the valuable metal or mineral content of known deposits that may be produced under current economic conditions and with present technology (Schantz and Ellis, 1983; American Geologic Institute, 1997). The USGS and U.S. Bureau of Mines do not distinguish between extractable and recoverable reserves and include only recoverable materials (American Geological Institute, 1997).

Resources are defined as a concentration of naturally occurring solid, liquid, or gaseous materials in or on the Earth's crust in such form that economic extraction of a commodity is currently or potentially feasible (U.S. Geological Survey and U.S. Bureau of Mines, 1976).

In view of varying definitions of reserves and resources, some background is needed to understand the current debates about the size and significance of reserves and resources. The USGS has used the classification by McKelvey (McKelvey, 1972) as shown in Figure 5C, and it is useful for framing a discussion of reserves and resources as this or derivative classifications are widely used in the USA and Europe (for example see American Petroleum Institute (API, 1995 a, b).

The greatest misunderstanding surrounds the term 'proved (or proven) reserves'. However, as pointed out by the American Petroleum Institute (1995a) 'Proved reserves have no bearing whatsoever on future potential.' Although, widely cited as such in reserve-to-production ratios (R/P), the estimates of proved reserves have always been viewed as minimum estimates related to inventories rather than future supplies (American Petroleum Institute, 1995a).

Proved reserves are of critical importance for U S financial investments and regulatory responsibilities, because these are the reserves of greatest significance when the Securities and Exchange Commission (SEC) formally adopted them into the tax structure of the USA in 1975. The critical linkage between financial and accounting considerations and reserve definitions requires an understanding of the definitions given in SEC Regulation 4-10 (Brock et al., 1996; their Appendix 1). Some argue that these definitions are inherently conservative or restrictive thereby causing initial conservative estimates of reserves (for example Campbell, 1997; Schuyler, 1999), but they are understandably conservative given the history of speculation and over zealous promotion of petroleum that occurred in the early days of the industry (Yergin, 1991).

Definitions of proved oil and gas reserves including two sub-categories — proved developed reserves and proved undeveloped reserves — as given in SEC Regulation 4-10, Section 210.4-10 and as generally used by industry, are as follows:

Proved oil and gas reserves are the estimated quantities of crude oil,

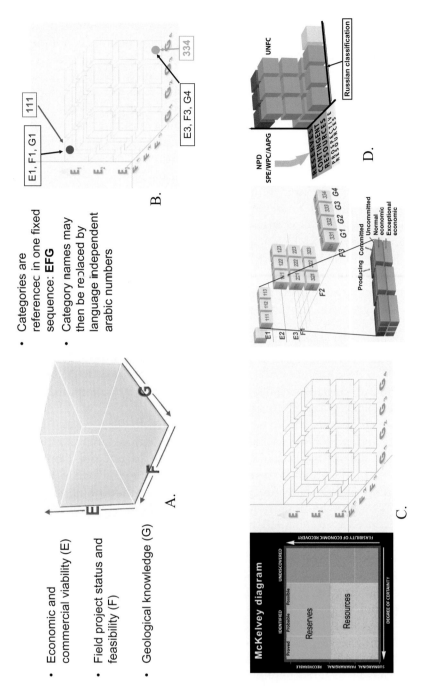

Figure 5: The United Nations Framework Classification (UNFC) System for Petroleum Resources

natural gas, and natural gas liquids that geological and engineering data demonstrate with reasonable certainty to be recoverable in future years from known reservoirs under existing economic and operating conditions (that is, prices and costs as of the date the estimate is made). The regulation goes on to define proved reserves as those that can be economically produced on the basis of either actual production or conclusive formation testing; such reserves would include only those delineated by drilling and defined by gas-oil and oil-water contacts, and (or) are in immediately adjoining portions not yet drilled, but which can be reasonably judged to be economically productive on the basis of available geological and engineering data. In the absence of information on fluid contacts, the lowest known structural occurrence of hydrocarbons controls the lower proved limit of the reservoir. Proved reserves do not include: (1) oil that may become available from known reservoirs but is classified separately as 'indicated additional reserves'; (2) crude oil, natural gas, and natural gas liquids, the recovery of which is subject to reasonable doubt because of uncertainty as to geology, reservoir characteristics or economic factors; (3) crude oil, natural gas, and natural gas liquids that may occur in undrilled prospects, and (4) crude oil, natural gas, and natural gas liquids that may be recovered from oil shale, coal, gilsonite, and other such sources.

Proved developed oil and gas reserves are those that can be expected to be recovered from existing wells with existing equipment and operating methods. Additional oil and gas expected to be obtained through the application of fluid injection or other improved recovery techniques for supplementing the natural forces and mechanisms of primary recovery should be included as 'proved developed reserves' only after testing by a pilot project or after the operation of an installed program has confirmed through production response that increased recovery will be achieved.

Proved undeveloped reserves are those that are expected to be recovered from new wells on undrilled acreage or from existing wells where a relatively major expenditure is required for recompletion. Reserves on undrilled acreage are limited to those drilling units offsetting productive units that are reasonably certain of production when drilled. Proved reserves for other undrilled units can be claimed only where it can be demonstrated with certainty that there is continuity of production from the existing productive formation. Under no circumstances should estimates for proved undeveloped reserves be attributable to any acreage for which an application of fluid injection or other improved recovery technique is contemplated, unless such techniques have been proved effective by actual tests in the area and in the same reservoir.

In 1997, the World Petroleum Congress (WPC) and the Society of Petroleum Engineers (SPE) jointly published petroleum reserve definitions that added the element of probability to the deterministic definitions then in common use. The WPC/SPE definitions build on the SEC definitions by including probabilistic estimates; these definitions are given by Seba (1998) as follows:

Reserves: Those quantities of petroleum that are anticipated to be commercially recovered from known accumulations from a given date forward.

Proved Reserves are those quantities of petroleum, which, by analysis of geological and engineering data, can be estimated with reasonable certainty to be commercially recoverable, from a given date forward, from known reservoirs and under current economic conditions, operating methods, and government regulations. Such reserves can be categorized as developed or undeveloped. If deterministic methods are used, the term 'reasonable certainty' is intended to express a high degree of confidence that the quantities will be recovered. If probabilistic methods are used, there should be at least a 90 percent probability that the quantities actually recovered will equal or exceed the estimate.

Unproved Reserves are reserves that are based on geologic and (or) engineering data similar to the data used in estimates of proved reserves; but technical, contractual, economic, or regulatory uncertainties preclude such reserves being classified as proved. Unproved reserves may be further divided into probable reserves and possible reserves.

Probable Reserves are those unproved reserves which analysis of geological and engineering data suggests are more likely than not to be recoverable. In this context, when probabilistic methods are used, there should be at least a 50 percent probability that the quantities actually recovered will equal or exceed the sum of estimated proved plus probable reserves.

Possible Reserves are those unproved reserves which analysis of geological and engineering data suggests are less likely to be recoverable than probable reserves. In this context, when probabilistic methods are used, there should be at least a 10 percent probability that the quantities actually recovered will equal or exceed the sum of estimated proved plus probable plus possible reserves.

Much of the discussion in this chapter relates to reserves, in that three of the four components of the petroleum endowment are related to reserves (the fourth component involves undiscovered resources). Reserves identified in the upper left corner of Figure 5C, include, in the USGS (McKelvey) system (McKelvey, 1972), measured reserves − those having greater economic feasibility and greater geologic certainty than indicated or inferred reserves. Undiscovered resources have even greater uncertainty than the previous categories (U.S. Geological Survey, 1995). Many industry groups use the terms proved, probable, possible and speculative (Potential Gas Committee, 1999) that roughly correlate to categories of measured, indicated, inferred, and undiscovered in the McKelvey diagram, respectively (Figure 5C). Proved, probable and possible reserves are commonly called P1, P2, P3, respectively, and considerable analysis has been applied to relations among these reserve categories to estimate ultimate recoverable resources (Grace et al., 1993; International Energy Agency, 1998; Schuyler, 1999, Ahlbrandt, 2001). The WPC/SPE effort in 1998 included a deterministic estimate as well as a probabilistic estimate for proved (P1, 90 percent probability), probable (P2, 50 percent probability) and possible (P3, 10 percent probability). The inclusion of a probability associated with reserve estimates reflects the increasing awareness that reserve estimates change through time and reflect renewed investment, application of new technologies, and development of new reservoirs within a field, all of which can alter the view of proved, probable, and possible reserves.

Recognizing that resources are also a critical component of future petroleum supply, definitions for petroleum resources were included in a document prepared in the year 2000 by the World Petroleum Congress (WPC), the Society of Petroleum Engineers (SPE) and the American Association of Petroleum Geologists (AAPG). In addition to reserves, the WPC/SPE/AAPG system provided categories for petroleum resources that were either contingent (discovered but subcommercial) or prospective (undiscovered resources) (Figure 5D). This classification system (WPC/SPE/AAPG) has been extended to a new international classification system recently adopted by the United Nations that is also compatible with many classification systems (Figures 5 C, D).

Recent international financial concerns related to accurate reporting of petroleum reserves and resources have underscored the need for an internationally recognized petroleum reserve and resource terminology. A committee of the United Nations Economic Commission for Europe (UNECE) prepared an international, harmonized, petroleum classification system that was adopted by the United Nations in July, 2004. This classification system is known as the United Nations Framework Classification (UNFC) (Ahlbrandt et al., 2004). The UNFC is a three-dimensional system whereby reserves and resources can be categorized with respect to three sets of criteria or axes. These axes are the economic and commercial viability axis (E axis), the field projects status and feasibility axis (F axis) and the level of geologic knowledge (G axis; Figure 5A). These categories can then be fixed into a sequence (EFG), and replaced by Arabic numbers; for example E1, F1, G1= 111, Figure 5B). The UNFC system harmonizes both 'western' (US/Europe) petroleum classification systems such as that used by the USGS (the McKelvey diagram) and many other groups that fall on the E and G axes (Figure 5C). 'Eastern' (Russia/Former Soviet Union) classification systems fall on the F and G axes (Figure 5D). The most widely utilized petroleum classification system in the west, the WPC/SPE/AAPG system, as well as the Norwegian Petroleum Directorate System (NPD) are compatible with the UNFC system (Figure 5D). Thus, an international petroleum classification system is now in place to (1) address concerns about international and national resources management and industry management of business processes to achieve efficiency in exploration and production, and (2) provide an appropriate basis for documenting the value of reserves and resources in financial statements (Ahlbrandt et al., 2004).

3.   Reserve Growth (Reserve Appreciation or Field Growth)

The problem of reporting reserves at several levels of certainty or probability (P) (for example, P90 for proved, P50 for probable, and P10 for possible) involves the recognition that reserve estimates are not deterministic or fixed; however, the changes in reserve volumes must in some way be recognized to better appraise the ultimate recoverable volume of oil or natural gas in a well, field, basin, province, region, or the world. Reserve (or field) growth refers to the increases in estimated sizes of fields that typically occur through time as oil and gas fields are developed and produced. This growth of reserves is typically reported as the recoverable volumes of oil and natural gas, which are some fraction of the original amount of oil and gas in place in a field based on the recovery factor for the individual accumulation.

It has long been known that the initial reserve estimates for many fields tend to be conservative and that the ultimate recoverable reserves are commonly several times larger than early estimates. However, much of this revision does occur relatively early in the life of a field, as development drilling proceeds to determine its potential. On average, there is a substantial upward revision, but there are also cases of reserve decline. Reserve growth is a measure of average growth of all fields in an area (Klett and Schmoker, 2003); in the USA, for example, reserve growth has added 85 percent of reserves in the last 15 years, much of it due to additions in existing fields rather than to new field discoveries (U.S. Geological Survey, 1995; Gautier et al., 1996; Schmoker and Attanasi, 1996, 1997; Verma, 2000).

3.1 United States Reserve Growth

In the USA the pioneering discussion of reserve (field) growth by Arrington (1960) led to many subsequent scientific investigations into the subject (Attanasi and Root, 1994; Attanasi and Schmoker, 1997; Attanasi et al., 1999; Schmoker and Attanasi, 1996, 1997, Klett, 2003). Reserve growth in the USA reflects at least two aspects: (1) initial conservative reporting of 'proved reserves'; and (2) application of new technology in exploration (such as 3-D and 4-D seismic data) and development practices to better define reservoirs, new production technology (smart well, dynamic monitoring systems, enhanced oil and gas recovery efforts), and new drilling technology (horizontal wells, new fracturing and completion techniques).

The first aspect of US reserve growth relates to the reporting of 'proved reserves'. Reserve (field) growth obviously would not occur if the initial estimates of a well or field were proved ultimately to be correct. From a tax and financial accounting perspective in the USA, only proved reserves are considered real assets, and penalties for failing to appropriately state proved reserves accurately are severe at all levels, as outlined by Brock et al. (1996).

There are several incentives to be conservative in reporting proven reserves, as there are (1) a variety of taxes levied against these reserves, including ad valorem, severance, conservation, and production taxes; and (2) several tax-related ramifications − among them, Income Tax Credits, deferred tax liability, depreciation, depletions allowances, at-risk rules, and passive loss rules. All of these are dependent upon the initially declared proved reserves. Compliance with regulations governing reserve reporting is thus imperative if a company is to survive financially in the USA (Brock et al., 1996). As previously discussed, proved reserve estimates are intended to provide minimum inventory estimates and *do not* represent future resources (American Petroleum Institute, 1995a). Thus, it is readily understandable why the US proved reserve estimates are initially conservative. The consequence is that as a field ages, there is increased investment in new wells, as well as improved infrastructure or production capabilities, that commonly add new reserves. The phenomenon of reserve growth thus becomes increasingly significant in those fields where new technology and investment are undertaken.

The growing significance of reserve growth for the USA can be shown by comparing the two most recent national assessments of the lower 48 states and offshore areas conducted by the USGS. In 1989, the USGS estimated 21.2 billion barrels of oil (BBO) reserve growth.

By 1995, the estimate was increased to 60 BBO despite six years of continuing production (U. S. Geological Survey, 1995, Gautier et al., 1996). The reserve growth number in the 1995 estimate for oil was more than twice as large as the mean undiscovered resource estimate of 23.5 BBO. Natural gas reserve growth in 1989 for the lower 48 and State waters was estimated by the USGS to be 92.7 trillion cubic feet (TCFG) or 15.5 billion barrels of oil equivalent (BBOE). By 1995, the reserve growth estimate increased to 322 TCFG or 53.7 BBOE, which was larger than the mean estimate for undiscovered natural gas of 258.7 TCFG or 43.1 BBOE.

The second aspect of reserve growth in the USA, and by far the most important, involves technological advances. For example, a significant component of overall reserve growth is enhanced oil recovery (Verma, 2000). Primary recovery including water flooding generally recovers only about one-third of the original oil in place (OOIP). New techniques, which utilize gases such as CO_2, nitrogen, or chemicals to enhance recovery, may result in the recovery of 50 percent or more of the OOIP. In heavy oil areas, recoveries of OOIP in excess of 90 percent are to be expected in some places (Verma, 2000). Thus, the estimated future potential additions to reserves in the USA for both oil and gas from reserve growth are far greater than they were from new field discoveries as has been demonstrated in recent studies (for example, Attanasi and Root, 1994; Schmoker and Attanasi, 1996, 1997).

3.2   World Reserve Growth

An important difference between reserve growth studies in the USA and the world is that the US database for these analyses report proved reserve estimates only, whereas reserve databases for the world such as I.H.S. Energy Group Inc. report 'known' reserve estimates, which includes proved reserves plus a portion (P50) of probable reserves. On one hand, this difference would suggest that world reserve growth may be less than US reserve growth algorithms because more of the total reserves are included. On the other hand, the world's oil and gas fields are on average about 20 years younger than US fields (Klett and Schmoker, 2003) and would likely experience greater reserve growth than US fields because they are younger and much of the reserve additions occur soon after discovery and diminish with time as noted for both the CGF and AFC methods.

Reserve growth in oil and gas fields worldwide, as in the USA, has long been known. Odell (1973), for example, documented what he called reserve appreciation in the world's oil fields, and provided an algorithm for its volumetric contribution to world oil reserves. The USGS 2000 assessment estimated that reserve growth additions would be comparable to new field discoveries, and applied a US algorithm at the global level (Ahlbrandt et al., 2000; Schmoker and Klett, 2000; Klett, 2003; Klett et al., 2005). In fact, reserve growth in the form of added reserves in existing fields for the last several decades in all parts of the world has actually been at least three times more important volumetrically than new field discoveries (Klett et al., 2005).

3.3   *Measuring Reserve Growth and Recent Reserve Growth Insights*

Reserve growth studies commonly look at the phenomena in either of two ways. One approach is to compare cumulative reserve estimates for a field, commonly on an annual basis, with the original recoverable reserve estimates from the date of discovery. This gives a multiplier of the original reserve estimate and is generally in the 4 to 9 times range. When treated in this manner, the original estimates at best commonly represent only about 25 percent of the ultimate amount of recoverable petroleum, and at worst possibly only 10 percent. Because much of this reserve growth occurs in the first few years of a field's development, a cumulative plot will show convex-upward growth curves, known as Cumulative Growth Factor (CGF) becoming tangential to a maximum volume through time (Attanasi and Root, 1994; Schmoker and Klett, 2000).

A second approach to reserve growth is essentially the opposite type of analysis, in that reserves are computed annually in a frequency plot. As the fields become older, the annual increment of reserve growth decreases to the point that for many US fields (Klett, 2003), as well as for fields worldwide (Odell, 1973) the annual reserve growth is only about 1 percent. These are convex-downward plots known as Annual Fractional Changes (AFC), becoming tangential to smaller volume additions through time to some point when there are no more reserve additions (Klett and Gautier, 2005). Under those circumstances, production will eventually cease, and the field will be either abandoned or dead (field mortality). However, with increasing commodity prices or new technology, a previously abandoned field may be rejuvenated so it is difficult to declare when a field has been permanently abandoned. Some US fields, particularly heavy oil fields, have shown significant growth for more than 100 years (Schmoker and Attanasi, 1996, 1997; Verma, 2000, 2005). It should be noted that reserve growth estimates related to offshore fields are generally more conservative than for onshore fields due to a more complete initial field development necessitated by the extensive infrastructure planning inherent in offshore activities (Minerals Management Service, 1996; Klett, 2003).

Although both of the approaches described above demonstrate the same phenomena – that is, most reserve growth occurring early in the history of a field, commonly in the first five to seven years – algorithms have been developed reflecting this early growth followed by the diminishing quantities of reserve growth through time (Klett, 2003; Verma, 2005). The USA, with some 80 percent of all wells drilled in the world, can be used as a global indicator for a mature oil and gas area, in that its fields are older, by about twenty years, than non-U.S.fields (Ahlbrandt and Klett, 2005).

Reserve growth algorithms for natural gas, when viewed either from the convex-upward cumulative growth factor (CGF) or the convex-downward annual fractional change (ACF), are larger than those estimated for oil (Root and Mast, 1993; Root et al., 1997; Schmoker and Attanasi, 1996, 1997; Klett, 2003; Klett and Gautier, 2003; 2005) owing to several factors. Because it was not considered to be a valuable resource until the latter half of the twentieth century, much of the natural gas discovered in the late 1800s and early 1900s was flared, particularly if associated with oil production. In fact, natural gas is much less utilized today worldwide compared to oil –11 percent of the world's gas endowment has been used compared to 23.5 percent of the world's oil endowment (Figures 6, 7, 8). Thus, the true size of a gas field was not commonly determined early in its productive history, if at all, and much of the gas was not measured nor put into storage or pipelines, but vented to the atmosphere. Those fields that did contain sufficient gas for economic viability during these early years have grown extensively, as documented recently by Klett (2003) and Klett et al. (2005) (Figures 9, 10). As shown in Figures 9 and 10, reserve growth for the world's natural gas fields added more natural gas reserves than new field discoveries and, volumetrically, more reserve growth globally has occurred for natural gas than for oil in the seven-year period from 1996 to 2003.

Potential additions to reserves from reserve growth estimates were made only on a worldwide basis for the USGS 2000 assessment (Table 1), but have subsequently been allocated to the eight regions individually (Table 2) (Ahlbrandt, 2004; Charpentier, 2004; Ahlbrandt et al., 2005). Prior to the USGS 2000 assessment, reserve growth for the world has not been estimated by the USGS, but estimates were made by others (Schollnberger, 1998; Edwards, 1997, 2001, 2002a, b; International Energy Agency, 1998, 2000, 2001). Schollnberger (1998) estimated field growth for the world at 500 BBO, and the International Energy Agency

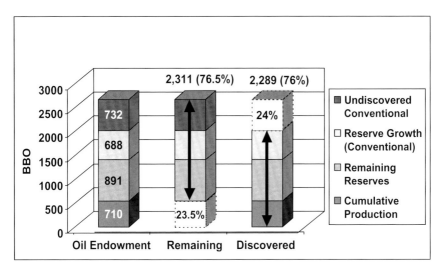

Figure 6: World Conventional Oil Endowment

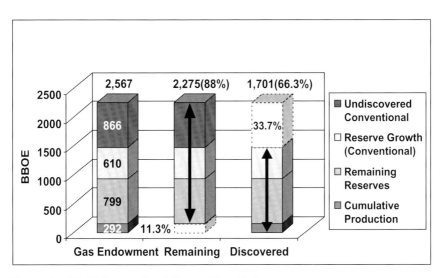

Figure 7: World Conventional Natural Gas Endowment

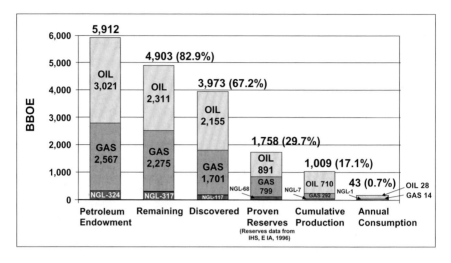

Figure 8: World Conventional Petroleum Endowment

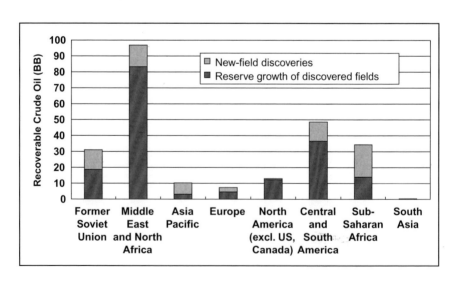

Figure 9: Additions to Crude Oil Reserves from January 1996 to December 2003 by region

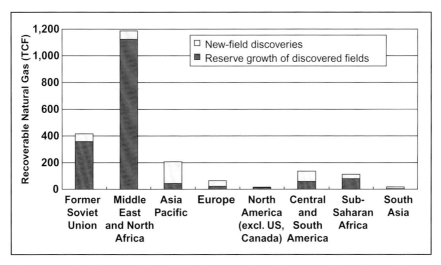

Figure 10: Additions to Natural Gas Reserves from January 1996 to December 2003 by region

(1998) estimate was between 418 and 1,136 BBO. The USGS 2000 assessment was for a 30-year forecast span (1995–2025), in contrast to previous USGS assessments that presented estimates of ultimate petroleum resources without a time constraint. The latter portion of even a 30-year time frame is difficult to predict because of continuous technological progress and a rapidly evolving world economy. Additions to the world reserves from reserve growth are at least three times greater than those from new field discoveries for the period January 1, 1996 to December 31, 2003 for oil (Figure 9) and natural gas (Figure 10) using the I.H.S. Energy Group Inc. database (Klett et al., 2005). Detailed reserve growth studies outside the United States have recently been completed for the Volga–Ural province (Verma et al., 2000) and the West Siberian Basin (Verma and Ulmishek, 2003) in Russia, and for the North Sea (Klett and Gautier, 2003, 2005), the Middle East (Verma et al., 2004), and Canada (Beliveau, 2003; Verma and Henry, 2004).

Some conclusions from these recent reserve growth studies are:

1. The US-based reserve growth algorithm used at the world level in the USGS 2000 global assessment fits the observed reserve growth volume additions (Klett, 2005).
2. Additions to world reserves from reserve growth for the period 1996 to 2003, the first seven years since the cut-off date of the USGS 2000 assessment (January 1, 1996), have contributed at least three times the resource volumes compared to new field discoveries in that

Table1:   World Level Summary of Petroleum Estimates for Undiscovered Conventional Petroleum and Reserve Growth for Oil, Gas, and Natural Gas Liquids (NGL)

[BBOE, billions of barrels of oil equivalent. Six thousand cubic feet of gas equals one barrel of oil equivalent. F95 represents a 95 percent chance of at least the amount tabulated. Other fractiles are defined similarly. Production and reserves normalized to 1/1/96. Shading indicates not applicable]

| | Oil | | | | Gas | | | | BBOE | NGL | | | |
| | Billion Barrels | | | | Trillion Cubic Feet | | | | | Billion Barrels | | | |
	F95	F50	F5	Mean	F95	F50	F5	Mean	Mean	F95	F50	F5	Mean
World (excluding United States)													
Undiscovered conventional	334	607	1,107	649	2,299	4,333	8,174	4,669	778	95	189	378	207
Reserve growth (conventional)	192	612	1,031	612	1,049	3,305	5,543	3,305	551	13	42	71	42
Remaining reserves*				859				4,621	770				68
Cumulative production*				539				898	150				7
Total				2,659				13,493	2,249				324
United States													
Undiscovered conventional**	66		104	83	393		698	527	88		Combined with oil		
Reserve growth (conventional)**				76				355	59		Combined with oil		
Remaining reserves				32				172	29		Combined with oil		
Cumulative production				171				854	142		Combined with oil		
Total				362				1,908	318				
World Total (including United States)				3,021				15,401	2,567				

*World reserve and cumulative production data reflect only those parts of the world actually assessed and are from Petroconsultants (1996) and NRG Associates (1995).

**U.S. data from Gautier and others (1996) and Minerals Management Service (1996).

Table 2: World Oil Resources and NGL Resources by Region (in million barrels)

Region*	Cumulative Production	Remaining Reserves	Allocated Reserve Growth	Undiscovered Resources	Endowment
OIL					
1 Former Soviet Union	112,106	152,173	104,537	115,985	484,801
2 Middle East and North Africa	218,644	528,620	363,141	229,882	1,340,287
3 Asia Pacific	47,722	39,980	27,465	29,780	144,947
4 Europe	31,235	27,572	18,941	22,292	100,040
5 North America (incl. U.S.)**	208,832	59,143	95,341	153,491	516,807
6 Central and South America	64,096	50,272	34,535	105,106	254,009
7 Sub-Saharan Africa and Antarctica	22,437	26,049	17,895	71,512	137,893
8 South Asia	5,131	7,071	4,858	3,580	20,640
Total (incl. U.S.)	**710,203**	**890,880**	**666,712**	**731,628**	**2,999,423**
NGL					
1 Former Soviet Union	1,801	14,753	9,071	54,806	80,431
2 Middle East and North Africa	1,413	35,376	21,750	81,747	140,286
3 Asia Pacific	1,022	4,906	3,016	15,379	24,323
4 Europe	649	6,400	3,935	13,667	24,651
5 North America (incl. U.S.)**	1,980	1,462	899	7,853	12,194
6 Central and South America	486	2,016	1,239	20,196	23,937
7 Sub-Saharan Africa and Antarctica	10	2,932	1,803	10,766	15,511
8 South Asia	24	467	287	2,604	3,382
Total (incl. U.S.)	**7,385**	**68,312**	**42,000**	**207,018**	**324,715**

All estimates reported here are mean estimates. Oil and NGL resource totals from U.S. Geological Survey, cumulative oil production and remaining reserves from IHS Energy as of 1/96. Allocated mean reserve growth from U.S. Geological Survey estimates, allocated (for non-U.S.) proportional to mean volumes of remaining reserves. All volumes in MMBO or MMBNGL.

* USGS Region and Code.   ** For U.S. portion of North America, some oil and gas estimates are combined.

time period (Figures 9, 10; Klett and others, 2005). The implication is that the reserve growth algorithm used by the USGS for the 2000 assessment may in fact be conservative.

3. Reserve growth is demonstrable in both 'western' (U.S./European) and 'eastern' (Russia/Former Soviet Union) regions, which are represented by very different petroleum industries (Verma et. al, 2000; Verma and Ulmishek, 2003), thus is in part independent of financial and reporting systems.

4. Reserve growth can also be demonstrated in both individual fields and pools as evidenced by studies of Canadian (Saskatchewan) pool data (Beliveau, 2003; Verma and Henry, 2004).

5. There are multiple, at least eight, US reserve growth algorithms that can be utilized at a regional or subregional level in certain areas of the world (Klett, 2003; Klett and Gautier, 2003, 2005; Verma and Ulmishek, 2003; Verma, 2005).

6 Offshore US algorithms are reasonable for offshore global provinces (Klett and Gautier, 2003, 2005).

7. In 'eastern' (Russia/Former Soviet Union) areas and in offshore areas, reserve growth algorithms using year of first production rather than year of field discovery seems to better predict observed reserve growth (Verma et. al., 2000; Verma and Ulmishek, 2003; Klett and Gautier, 2003, 2005).

4.   Global Reserve and Resource Estimates

There are disagreements and differing perspectives on petroleum reserve and resource supplies (Williams, 2003). Global resource estimates have ranged from conservative or pessimistic (*low estimates*), commonly referred to in economic terms as Malthusian, to optimistic (*high estimates*), the proponents of which are referred to as Cornucopians (McCabe, 1998, O'Neil, 2001); an intermediate position *(moderate estimates)* which includes a group referred to in economic terms as Incrementalists (O'Neil, 2001). To facilitate discussion, (1) *low estimates* are defined here as those estimating ultimate recoverable endowment of a petroleum commodity (oil or natural gas) as less than or equal to 2 trillion barrels of oil equivalent (TBOE, in some literature GBOE is used). (2) *moderate estimates* are those between 2 and 4 TBOE, and (3) *high estimates* are those greater than 4 TBOE (see Figure 1).

Further complicating dialogue on petroleum resources are comparisons that either include, or selectively exclude conventional and unconventional resources; the latter are commonly less economic to produce but potentially have much greater volume than conventional resources (McCabe, 1998). Differing methodologies, varying levels of effort associated with estimates, vested interests, differing vintages of data, and in-place versus recoverable volumes are all complicating factors in analysing ultimate petroleum resources. For example, Ahlbrandt and Klett (2005) demonstrate that for a given set of reserve data in the Neuquen Basin in Argentina, estimates of undiscovered conventional resources have varied by an order of magnitude, depending entirely upon the assessment methodology chosen. Many discussions are limited to a fairly narrow perspective, either from one cultural or commodity perspective − that is, the 'western' (US/European) view of resources or the 'eastern' (Russia/Former Soviet Union) view − or are limited to a discussion of oil with only a passing reference to natural gas or natural gas liquids, which, on an energy equivalent basis are as important as oil. Although this chapter focuses on presently defined conventional resources; it should be recognized that some resources now classified as unconventional may possibly be reclassified for future resource assessments. For example, the Athabasca tar sands in Canada are now considered to be conventional rather than unconventional oil reserves (Radler, 2002)

4.1 Early Twentieth-century Estimates

A historical perspective of the evolution of resource estimates is given by reviewing some of the earliest estimates of US and world oil resources by the USGS. In 1909, the USGS estimated that between 10 and 24.5 BBO would ultimately be found in the USA, but by 1915 that estimate was reduced to 9 BBO. US resource estimates gradually increased until the First World War, but pessimism soon again dominated. For example, (1) in 1918, Gilbert and Pogue of the Smithsonian Institute argued imminent exhaustion of US oil reserves stating 'there is no hope that new fields, unaccounted in our inventory, may be discovered of sufficient magnitude to modify seriously the estimate… (World War I) has merely brought into the immediate present an issue underway and scheduled to arrive in the course of a few years'; and (2) in 1919, David White of the USGS predicted exhaustion of US oil resources in the early 1920s. In sharp contrast to the foregoing, the USA has now produced some 171 BBO, and current estimates are for an oil endowment of 362 BBO which includes cumulative production, remaining reserves, reserve growth, and undiscovered resources (Table 1; U.S. Geological Survey, 1995, 2000).

The conservative bias in early US resource estimates led to

clarifications of terminology used in reporting these estimates. In 1922, the term 'reserves' was introduced by the USGS and the American Association of Petroleum Geologists (AAPG) to describe remaining oil resources (American Petroleum Institute, 1995a). The 1915 USGS/AAPG estimate of 9 BBO was divided into 5 billion barrels of 'oil in sight' and 4 billion barrels as 'prospective and possible'. The former category was judged as 'reasonably reliable' and the latter 'absolutely speculative and hazardous?' The effects of these early conservative estimates caused the greatest effect on the petroleum industry and its relations to government from then to the present. For example, The Federal Oil Conservation Board was established in 1924 to study the government's responsibility and enlist the full cooperation of representatives of the oil industry to safeguard the national security through oil conservation (American Petroleum Institute, 1995a). This fear for energy security stemmed from the fact that 80 percent of the oil used in the First World War had been domestically supplied, and the fear was that domestic resources were nearing depletion.

Similar conservatism dominated world oil supply estimates. In 1920, David White of the USGS estimated ultimate world oil resources to be 43 BBO and he predicted that an imminent peak in US oil production would be reached by 1922. Thus, the range of estimates for world oil resources have ranged from White's 0.043 TBOE, to estimates as large as 15 TBOE from Russian investigators (Modelesky et al. 1985) with most older estimates generally in the 2 TBOE or less range, and more modern estimates essentially greater than 2 TBOE (Figure 1).

4.2   *Low Estimates*

For about the last 40 years, numerous proponents of the *low estimate* category have settled on an ultimate oil resource of about 2 TBOE or less (Figure 1). Particularly strong proponents of the *low estimate* scenario include Weeks (1958), Hubbert (1967), (low estimate, Figure 2), Warman (1972), Campbell (1989); Campbell and Laherrère (1998), Deffeyes (2001) and Goodstein (2004). Some within this group perceive an imminent period of difficult economic times or even an end to civilization based upon perceptions of a worldwide oil supply crisis and the resulting conflicts from competition for these valuable resources (for example, Bartlett, 1978, 1994, 1996; Campbell, 1989, 1997; Youngquist, 1997; Duncan and Youngquist, 1999; Duncan, 2000; Deffeyes, 2001; Goodstein, 2004; Simmons, 2005). This conservative bias has been so pervasive in petroleum resource estimation, particularly early estimates, that the API concluded: 'This (credibility of resource estimates) led to a general recognition of a conservative bias to such estimates, and to attempts to qualify those estimates sufficiently to reduce this bias' (American Petroleum Institute, 1995a, p. 9).

4.3   Moderate Estimates

Moderate estimates are those between 2 and 4 TBOE (Figure 1). The USGS 2000 recoverable oil endowment mean estimates of 3 TBOE are in this moderate estimate range as are a number of more recent estimates (Figure 1). Some estimates cross the *low* and *moderate* categories. For example, Hubbert (1967) gave two estimates of ultimate recoverable world oil – one low (1.3 TBO) and one moderate (2.1 TBO) (Figure 2). The 1994 USGS petroleum estimates (Masters et al., 1994) were virtually identical at the mean, within 6 percent, of the USGS 2000 estimates (Ahlbrandt et al., 2000, 2005), although different approaches were used in the two assessments. The USGS 2000 endowment is larger, however, because it includes the reserve growth component that was not included in previous USGS world assessments. Until recently, relatively few studies have included reserve growth in world oil and natural gas analyses. Examples of global level assessments that do include reserve growth are Odell (1973 and subsequent publications), Schollnberger (1998), U.S. Geological Survey (2000), Edwards (2001, 2002a,b), Cavallo (2002, 2004), Greene (2003), Green (2004), Salvador (2005) as well as many detailed regional studies discussed later (for example, Attanasi and Root, 1994; Verma et al., 2000, 2004; Klett, 2003, Klett and Schmoker, 2003; Stark, 2003; Klett and Gautier, 2003, 2005; Verma and Ulmishek, 2003, Verma and Henry, 2004). The role of new technology in increasing reserves in existing fields over time is sufficiently important to have warranted its detailed discussion in the previous section.

In economic discussions, some *moderate estimates* fall within a group referred to as Incrementalists. This view basically holds that oil should remain available in response to oil supply and demand and investment requirements; in essence, an eventual tightening of supplies will prompt higher prices, which leads to incremental and orderly development and substitution of other energy sources. These changes occur over a few decades or longer time frames driven by economic, energy efficiency or societal needs (Adelman, 1993; Nakicenovic, 1993; Lynch, 1994, 1998; Adelman and Lynch, 1997; Ausebel, 1998, 2000, 2001, Ausebel and Marchetti, 2001).

4.4   High Estimates

High estimates are those greater than 4 TBOE (Figure 1). They are the least numerous and, in economic discussions, proponents are sometimes referred to as optimists or Cornucopians (McCabe, 1998; O'Neil, 2001). There are also a few geoscientists who are considered Cornucopians and argue that there is abundant petroleum generation from vast sources deep within the crust of the earth resulting in a far greater resource than most earth scientists envision (Porfir'yev, 1974; Gold, 1999) or that there are vast as yet untapped energy resources (Huber and Mills, 2005).

'Western' (US/Europe) estimates are generally more conservative than some 'eastern' (Russia/Former Soviet Union) estimates (Odell, 1983). Some examples in excess of 4 TBOE are given by the Institut Français du Pétrole (in Odell, 1983) and recent estimates by major international companies such as Shell (2001), are shown in Figure 1. The most recent oil endowment estimate by ExxonMobil (2005) is between 6 to 8 TBO (in place) for conventional oil, over 4 TBO (in place) for extra heavy oil and oil sands (unconventional), and 3 TBO from oil shale (unconventional). Some eastern (Russian) estimates are as large as 11 TBOE for world ultimate oil resources (6 TBO of conventional oil and 5 TBO of unconventional oil) although a competing Soviet view estimated as much as 15 TBO (Styrikovich, 1977; Modlesky et al., 1985; Odell and Rosing, 1985; this highest estimate is not shown in Figure 1).

4.5   Evolution of Estimates

In general, ultimate petroleum estimates have increased over time (Figure 1), and there have been some relatively sharp increases, such as in the mid-1970s and early 2000s (Figure 1) driven by several factors associated with new exploration, reserve revisions (reserve or field growth) and application of new technology particularly in new frontiers. Understandably, a limited knowledge base can cause a pessimistic view of future resources, but a *moderate estimate* is favoured because reserve volumes, for example of consumed oil (~1.0 TBO), when added to remaining reserves (~1.2 TBO) currently exceed the *low estimate* of 2 TBO. Several sources (for example, IHS Energy Group Inc., 2001; the *Oil and Gas Journal* as reported by Radler, 2002; and Organization of the Petroleum Exporting Countries, 2004,) document world oil reserves at greater than 1 TBO, the latest being 1.2 TBO as reported in the *Oil and Gas Journal*; or 1.38 TBO if Canadian tar sands are included

(Radler, 2001). Furthermore, significant amounts of new oil and reserve growth of oil continue to be added to the estimates, as indicated by the reserve volumes that have been added by new field discoveries and reserve growth in existing fields in the last seven years (Klett et al., 2005). The USGS 2000 estimates of 3 TBO are well within the range of moderate estimates (Figure 1) and are suitable for more detailed analysis as they are the reference values used by the International Energy Agency (2000, 2001, 2002, 2003, 2004), the Energy Information Administration of the U.S. Department of Energy (EIA 2001, 2002) and many analysts (for example, Edwards, 2002b; Cavallo, 2002, 2004; Greene 2003; Green 2004, Salvador, 2005). The USGS 2000 estimates are used as a reference for conventional oil resources in this report.

5.   Projections of Oil and Gas Supply

Components of oil and gas supply or the energy endowment that have been discussed in earlier sections form the basis for an examination of various projections of future energy supply. Many older projections have not predicted the future well (Lynch, 1994; 1998; Demming, 2001; Cavallo, 2004, Charpentier, 2005), having been made largely for conventional petroleum resources with far fewer estimates made for unconventional (continuous) accumulations. Conventional petroleum resources will be considered initially for comparison with recent estimates and models that use such estimates to predict future petroleum supply. Using the USGS 2000 world's potential oil endowment for provinces outside the USA when added to the most recent estimates for the USA (U. S. Geological Survey, 1995, Minerals Management Service, 1996) as a reference, gives an oil endowment of 3,021 BBO at the mean (Table 1, Figure 6). Of this assessed quantity 710 BBO (23.5 percent) of oil resource has been consumed, 891 BBO (29.5 percent) are remaining reserves, 688 BBO (22.8 percent) are as potential reserve growth, and 732 BBO (24.2 percent) remains undiscovered as of January 1, 1996 (mean values, Table 1, Figures 6, 8). Because uncertainty is inherent in any estimate, a range of probabilities is also given for petroleum resource estimates in Table 1 − conservative (F95), median (F50), or optimistic (F5).

The USGS 2000 natural gas endowment estimate of 2.57 TBOE is both volumetrically less than the oil endowment and much less utilized relative to the oil endowment (Figures 7, 8). Eighty-eight percent of the natural gas endowment remains, and one-third remains undiscovered (Figure 7) compared to only one-quarter of the estimated undiscovered oil (Figure 6). In the USA, there is an approximate energy equivalence in terms of volume for natural gas and oil, but there is 0.5 TBOE less of natural gas relative to the oil endowment (2.5 vs. 3.0 TBOE, respectively) in the world (Ahlbrandt et al., 2005). This difference probably will be equalized by future development of gas-prone provinces in the Arctic (Ahlbrandt, 2002; Ahlbrandt et al., 2000, 2005)

Table 1 incorporates resource estimates that include recent US assessments made by Gautier et al. (1996) and the Minerals Management Service (1996) to provide a comprehensive world analysis. Within the 128 assessed provinces outside the United States, 159 total petroleum systems (TPS) containing 274 assessment units (AU) were defined and of these 149 TPSs and 274 AUs were assessed. Combining the estimates from these provinces with the US estimates (done by different groups as noted earlier), the endowment of recoverable oil, which includes cumulative production, remaining reserves, reserve growth, and undiscovered resources is estimated at about 3 TBO as of January 1, 1996 (Table 1; Figure 6). The natural gas endowment is estimated at 2.6 TBOE (Table 1; Figure 7). Of the combined oil and gas endowments of about 5.6 TBOE, the world has consumed about 1 TBOE, or 18 percent, leaving about 82 percent yet to be found and (or) utilized based on data as of January 1, 1996, the cut-off date of the USGS 2000 assessment (Ahlbrandt, 2002; Figure 8). Oil reserves are currently 1.1 TBO; world consumption is about 0.028 TBO per year. Natural gas reserves are about 0.8 TBOE; world consumption is about 0.014 TBOE. Thus, without any additional discoveries of oil, gas or natural gas liquids, about 2 TBOE of proved petroleum reserves remain.

As of the cut-off date of the USGS 2000 assessment (January 1, 1996), potential additions to reserves from reserve growth were estimated to be nearly as large as the estimated undiscovered resource volumes. The additional remaining reserves and the estimates for reserve growth from these known fields imply that 76 percent of the world's conventional oil endowment (cumulative production, remaining reserves, reserve growth, and undiscovered resources at the mean, Figure 7) and 67 percent of the world's conventional gas endowment have already been discovered in the assessed areas (Figure 8). For these areas, 23.5 percent of the world's total conventional oil endowment and 11.3 percent of the world's total conventional gas endowment had been produced by the end of 1995 (Figures 7, 8). As previously mentioned, reserve growth was actually three times greater than new field discoveries in the first seven years following the date of the USGS 2000 assessment (Klett et al., 2005).

By means of comparison, the overall petroleum endowment volumes estimated in the USGS 2000 assessment at the mean (cumulative production, remaining resources, field growth and undiscovered resources) is 5.91 TBOE (Figure 8). This estimate is close to an estimate of 5.79 TBOE (produced, proven, field growth, and undiscovered resources) made by Schollnberger (1998). For oil and natural gas endowments, the USGS 2000 mean estimates for world resources, when combined with estimates for the USA (U.S. Geological Survey, 1995; Minerals Management Service, 1996) are comparable to many others as shown in Figure 1.

5.1   Projections Using Low Estimates

A group of supply forecasters who fall into the *low estimates* group (also referred to as Malthusians, Catastrophists, or Trendologists) can be traced back to the predictions of M. King Hubbert (McCabe, 1998). Hubbert who worked for Shell Oil Company and the USGS, made several predictions based upon what he called a logistic curve analysis (as shown in Figure 2) of US peak oil production. After several earlier attempts (Hubbert 1956, 1959), he correctly predicted a peak in the early 1970s (Hubbert, 1967) based upon extrapolating production rates, but underestimated the ultimate volume of oil to be produced. That production (or logistic) curve for the USA was hand-derived from data at the time of his analysis (Hubbert, 1967) as were those he also constructed for the world (Hubbert, 1967, 1969, 1973, 1974; Figure 3). Hubbert's predictions of US natural gas supply and of world oil peak and volumes have also been shown to be incorrect for the world as well (Lynch, 1994; McCabe, 1998; Demming, 2001; Cavallo, 2004; Charpentier, 2005), inasmuch as he predicted peak oil production would be reached prior to 2000 in one scenario and by 2000 in another scenario (Figure 2).

There are several striking aspects of the Hubbert methodology that have been adopted by proponents such as Campbell (1989, 1997); and Deffeyes (2001); Campbell, for example predicted the world's oil production to peak in 1989 (Figure 11). Hubbert (1959) recognized that if there were 'retardation effects' a multimodal rather than a unimodal approach should be used for predicting production. Although Campbell and Laherrère (1998) published a production scenario using a Hubbert model, Laherrère (2000) later recognized the multimodal nature of oil production and explained the strengths and weaknesses of a Hubbert analysis, as did Duncan and Youngquist (1999) in this *low estimate* group. Deffeyes (2001) basically relied on a symmetrical distribution for his analysis, little different than Hubbert's original efforts. McCabe (1998),

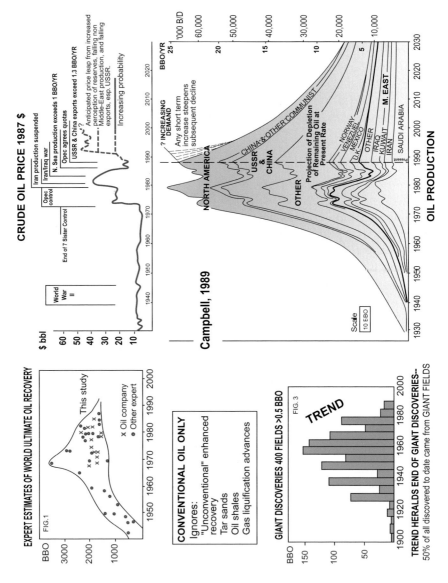

Figure 11: A View from a Low Estimate Proponent (Campbell, 1989)

Cavallo (2004), and Charpentier (2005) outlined the problems associated with using the Hubbert model for world oil supplies; their analyses demonstrate that the model is inappropriate for predicting ultimate oil reserves and that it should be considered an econometric model.

5.2   Projections Using Moderate Estimates

Examples using *moderate estimates* of the USGS 2000 assessment of recoverable oil are shown in Figures 12 and 13. However, there are other
examples in the economic world including Adelman (1993), Lynch,
(1994), Adelman and Lynch (1997), and Ausabel (1998, 2000, 2001).
The estimated oil volumes presented by the USGS have been used by
agencies such as the Energy Information Administration (EIA) of the
U.S. Department of Energy to predict 12 different peaking scenarios
for global oil production. Of the different USGS estimates (F95, mean,
F5) and demand scenarios (as much as 3 percent annual increase), the
EIA favours 2036 as the peak production year (Energy Information
Administration, 2001, Figure 12). Other groups at the Department of
Energy predict oil peaking in the period 2015–2020 in non-OPEC
nations, based on the same USGS data (Cavallo, 2002). Edwards (1997,
2001, 2002a,b) also used USGS estimates, combined with other data,
to predict world oil peaking from 2020 to 2035 depending on differing
assumptions. Another example is the model by Green (2004) showing
the gradual substitution of energy fuels shifting from oil to natural gas
to renewables and unconventional resources (Figure 13).

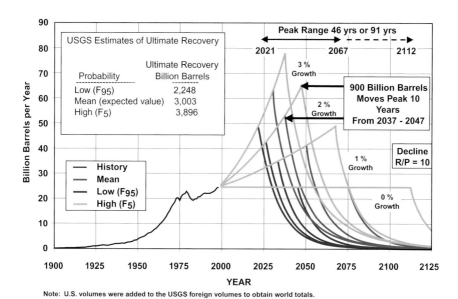

Figure 12: Twelve Oil Production Scenarios

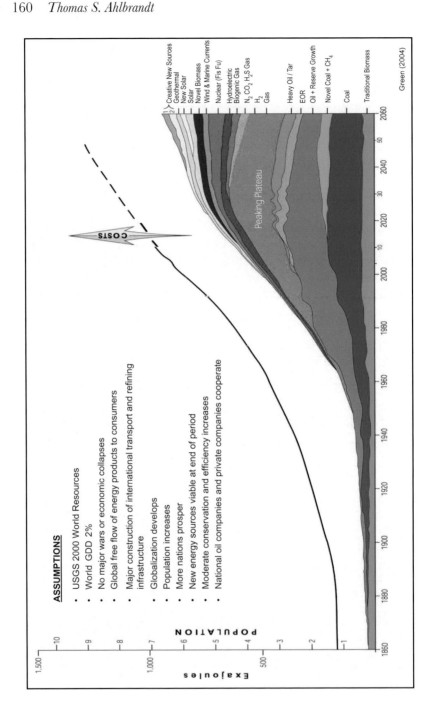

Figure 13: A View of a Moderate Estimate Proponent (Green, 2004)

5.3   Projections using High Estimates

An example of a high estimate scenario (or Cornucopian model) is by Odell (1998; as shown in Figure 14). Interestingly, Odell (1998) used a 3 TBO volume for the conventional oil endowment, which is comparable to the USGS 2000 estimate, but then added another 3 TBO of unconventional oil resources (such as heavy oils and tar sands) thereby extending the oil resource considerably. However, even larger oil estimates such as the 11 TBO by Stryikovich (1997) which includes 6 TBO of conventional oil and 5 TBO of heavy oil, make the above analysis seem conservative. Clearly, there are oil resources that can be derived from other sources such as oil shale estimated to contain 3 TBO (ExxonMobil, 2005) and the estimated 175 BBO reported by the *Oil and Gas Journal* (Radler, 2002) from the Athabasca tar sands of Canada. Adding these amounts would raise world oil reserves to more than 1.2 TBO. These reserves are more than five times the world's oil reserves at the end of the Second World War (1945) and 36 percent larger than the effective date of January 1, 1996 used in the USGS (2000) assessment. Clearly much oil is being found and fields are now growing beyond the supposed 2 TBO barrier (Figures 1, 10, 11).

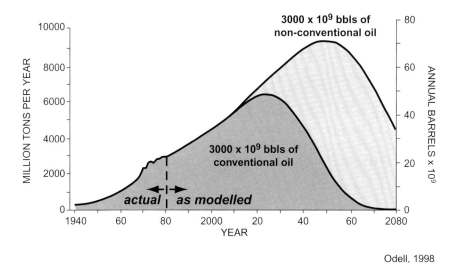

Odell, 1998

Figure 14:   A Prospective Depletion Curve for the World's Conventional and Unconventional Oil to 2080, from a High Estimate Proponent (Odell, 1998)

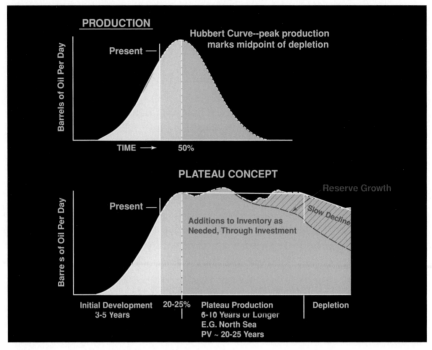

Figure 15: Views of Conventional Oil Production

5.4   Peaks or Plateaus?

A moderate estimate view of a plateau of production with additional petroleum reserves being added as needed is shown by Green (2004; Figures 13, 15). Following the concept that reserves at any given time are more a measure of needed inventory and have stayed relatively constant versus production for a long period of time as shown by American Petroleum Institute (1995) and McCabe (1998), then additional reserves and capital are added at a point when it is financially prudent, and capital is withdrawn when it becomes too costly to add further reserves in a given area. The plateau concept of field production is well established in petroleum economics (Seba, 1998), and current world oil production suggests this is a reasonable concept. The plateau is also seen in the EIA models in Figure 12, with a rather precipitous drop off once maximum production is encountered.

6.   The Interplay of Energy Sources

Although debates about the relative remaining amounts of petroleum continue, it is clear that a transition to energy sources other than oil will occur. In contrast to oil (Figure 6), only 11 percent of the natural gas endowment of the world has been used and fully one-third is yet to be discovered (Figures 7, 8). With the development of technologies to convert natural gas to liquid petroleum (GTL), a large resource will become available, helping to offset oil production concerns. In the first half of the twenty-first century, petroleum resources will increasingly be developed offshore where more than 50 percent of future petroleum is expected to be found. Beyond that, the next great frontier for un-discovered petroleum resources will likely be the Arctic regions (above latitude 65° N,) where about 25 percent of the undiscovered potential is indicated in USGS 2000 assessment and also reported by Ahlbrandt (2002) and Ahlbrandt et al. (2000, 2005). Both offshore and Arctic areas are environmentally sensitive and considerable technologic and international cooperation will be required to access these resources.

However, the most viable source of energy, particularly for electrical generation in the second half of the twenty-first century will probably be nuclear. Coal resources remain abundant and are viable throughout the century dependent upon environmental regulations affecting this vast resource. Unconventional oil resources such as the heavy oils of Orinoco or the tar sands of Canada mentioned previously are large (estimates of recoverable resources in the hundreds of billions of barrels of oil for each) and represent potentially viable sources of petroleum, although they will be expensive to develop. Supplies of unconventional (continuous) natural gas from basin-centre gas and coal-bed meth-ane are significant and largely undeveloped throughout the world. With all these undiscovered and unassessed volumes, combined with the significant amount of reserves currently proven or estimated, the world's supply of petroleum should be adequate for several decades − with significant caveats. The rate of production of these resources is dependent upon several factors, including investments in exploration and development, political and social conditions, and demand levels, making it difficult to forecast rates of production.

The recent scenario authored by Arthur Green (2004), retired Chief Scientist of ExxonMobil, incorporated all of the elements discussed above and utilized the moderate estimates of the USGS 2000 assess-ment. He incorporated the concepts of reserve growth, a plateau, and transitional energy supplies, and placed them in a time framework as shown in Figure 13. Salvador (2005) made similar projections, incorporating reserve growth, and energy substitution, and making forecasts for the twenty-first century.

Regardless of the scenario used, fossil fuels, particularly petroleum will remain the critical energy resource throughout at least the first half of this century, and serve as the pivotal resources for alternative sources such as hydrogen. Methodologies designed to more accurately determine and (or) estimate the size and distribution of various categories of resources are continuing to be developed to provide a much more comprehensive picture of the petroleum endowment of the world. Such information is critical to the utilization of the finite petroleum resources of the world. Increasingly the world will rely on petroleum resources that were once viewed as unconventional such as tar sands, heavy oils, basin-centre (continuous) oil and natural gas systems, and shale gas, thus requiring an increasing need to focus attention on their characterization, evaluation, and utilization.

Acknowledgments

I thank Dr. Peter McCabe, Dr. Mitchell Henry and Dr. Richard Keefer of the U.S. Geological Survey for their helpful reviews, and my colleagues on the USGS World Energy Project for their discussions and insights on resource investigations.

References

Adams, T.D., and M.A. Kirkby (1975), 'Estimate of world gas reserves', *in* Proceedings of the 9th World Petroleum Congress, Tokyo, May, Chichester: John Wiley & Sons, no. 9, v. 3, pp. 3−9.

Adelman, M.A. (1993), *The Economics of Petroleum Supply*, Cambridge, Massachusetts: Massachusetts Institute of Technology Press, 576 pp.

Adelman, M.A., and M.C. Lynch (1997), 'Fixed view of resource limits creates undue pessimism', *Oil and Gas Journal*, April 7, 1997, pp. 56−60.

Ahlbrandt, T.S. (2001), 'Reserves and Resources', in *MacMillan Encyclopedia of Energy*, J. Zumerchik, ed., New York: MacMillan Reference, v. 3, pp. 1007−1.

Ahlbrandt, T.S. (2002), 'Future petroleum energy resources of the world', *International Geology Review*, v. 44, pp. 1092−104.

Ahlbrandt, T.S. (2004), 'Oil and Natural Gas Liquids: global magnitude and distribution', *Encyclopedia of Energy*, Elsevier, v.4, pp.569−79.

Ahlbrandt, T.S., Blaise, J.R., Blystad, Per, Kelter, Dietmar, Gabrielyants, Gregorio, Heiberg, Sigurd, Marinez, Anibal, Ross, J.G., Slavov, Slav, Subelj, Andrei,

Young, E.D. (2004), 'Updated United Nations Framework Classification for reserves and resources of extractive industries', Society of Petroleum Engineers Paper 90839, 7 p.

Ahlbrandt, T.S., Charpentier, R.R., Klett, T.R., Schmoker, J.W. and Schenk, C.J. (2000), Analysis of Assessment Results, Chapter AR in USGS World Petroleum Assessment 2000—Description and Results, U. S. Geological Survey Digital Data Series DDS-60, 323 p. (available at *http://energy.cr.usgs.gov/oilgas/wep*).

Ahlbrandt, T.S., Charpentier, R.R., Klett, T.R., Schmoker, J.W., Schenk, C.J., and Ulmishek, G.F. (2005), *Global Resource Estimates from Total Petroleum Systems*, American Association of Petroleum Geologists Memoir 86, 324 p.

Ahlbrandt, T.S. and T.R. Klett (2005), 'Comparison of methods used to estimate conventional undiscovered petroleum resources: world examples', *Natural Resources Research*, v. 14, no.3, pp.185–208.

American Geological Institute (1997), *Glossary of Geology*, Julia A. Jackson, ed., 4th Edition, Alexandria, Virginia, 769 p.

American Petroleum Institute (1995), *Are we running out of oil?* American Petroleum Institute Discussion Paper #081, E. Porter, ed., 56 p.

American Petroleum Institute (1995b), *Standard Definitions for Petroleum Statistics*, American Petroleum Institute, 5th edition, 62 p.

Arrington, J.R. (1960), 'Predicting the size of crude reserves is key to evaluating exploration programs', *Oil & Gas Journal*, v. 58, no. 9, pp. 130–34.

Attanasi, E. D., and D. H. Root (1994), 'The enigma of oil and gas field growth', *American Association of Petroleum Geologists Bulletin*, v. 78, pp. 321–32.

Attanasi, E.D., and J.W. Schmoker (1997), 'Long-term implications of new United States gas estimates', *Nonrenewable Resources*, March, v. 6, no. 1, pp. 53–62.

Attanasi, E.D., Mast, R.F., and Root, D.H. (1999), 'Oil and gas field growth projections: wishful thinking or reality?' *Oil and Gas Journal*, v.97, no. 4, pp.79–81.

Ausubel, J.H. (1998), 'Resources and environment in the 21st century: seeing past the phantoms', *World Energy Council Journal*, July, pp. 8–16.

Ausubel, J.H. (2000), 'Where is energy going?' *Industrial Physicist*, v.6, pp.16–19.

Ausubel, J.H. (2001), 'Some ways to lessen worries about climate change', *The Electricity Journal*, v. 14, pp. 24–33.

Ausubel, J.H. and C. Marchetti (2001), 'The evolution of transport', *The Industrial Physicist*, v. 7, pp. 20–24.

Barthel, F., P. Kehrer, J. Koch, F.K. Miscus, D. Weigel, 1976, Die kunftige Entwicklung der Energienachfrage und deren Deckung: in Perspektiven bis zum Jahr 2000. Section 3, Das Angebot von Energie-Rohstoffen; Hannover, Federal Republic of Germany: Bundesanstalt f r Geowissenschaften und Rohshtoffe, pp. 89–189.

Bartlett, A.A. (1978), 'Forgotten fundamentals of the energy crisis', *American Journal of Physics*, v. 46, pp. 876–88.

Bartlett, A.A. (1994), 'Reflections on sustainability, population growth, and the environment', *Population and Environment*, v. 16, pp. 5–35.

Bartlett, A.A. (1996), 'The exponential function, XI: The new Flat Earth Society', *The Physics Teacher*, v. 34, pp. 342–43.

Bauquis, P.R., R. Brasseur, J. Masseron, 1972, Les réserves de pétrole et les perspectives de production a moyen et long terme. Petroleum reserves and production possibilities in the medium and long range, *Revue de l'Institut Français du Pétrole et Annales des Combustibles Liquides*, v. 27, no. 4, pp. 631–58.

Beleveau, Dennis (2003), 'Reserves growth: enigma, expectation or fact?' Society of Petroleum Engineers Annual Technical Conference and Exhibition, Denver, Colorado, SPE 84144 paper, 10p.

BGR, 2002, (Bundesministerium fur Wirtschaft and Arbeit or Federal Institute for Geosciences and Natural Resources, Hannover, Germany), 'Reserves, resources and availability of energy resources', Technical Report Number 15, E. Schwiezbart'sche Verlagsbuchhandlung (Nagele u. Obermiller), Johannes strasse 70174 Stuttgart, Germany 41 p. (Note: this is a BWMA-Documentation ISSN 0342-9288 published by the Federal Ministry of Economics and Labor.

Bois, C., P. Bouche, R. Pelet (1979), 'Histoire Géologique et repartition des reserves dans le monde', Report no. 27.542, Paris, Institut Français du Pétrole.

Bookout, J.F. (1989), 'Two centuries of fossil fuel energy', *Episodes*, v. 12, no. 4, pp. 257–62.

Brock, H.R., Jennings, D.R., and Jones, D.M. (1996), *Petroleum accounting: principles, procedures and issues*, PriceWaterhouseCoopers, 4th Ed., Professional Development Institute, University of North Texas, 794 p. plus appendices.

British Petroleum (1998), *BP Statistical Review of World Energy*, London, United Kingdom (Also available at *http://www.bp.com/centres/energy*).

British Petroleum (2001), *BP Statistical Review of World Energy* June 2002. London, United Kingdom [Also available at *http://www.bp.com/centres/energy.*]

Campbell, C.J. (1989), 'Oil price leap in the early nineties', *Noroil*, v. 17, no. 12, pp. 35–8.

Campbell C.J. (1991), *The Golden Century of Oil 1950-2050: the depletion of a resource*, Dordrecht, Netherlands: Kluwer Academic Publishers, 345p.

Campbell, C.J. (1992), 'The depletion of oil', *Marine & Petrol. Geol.*, v. 9, Dec. 1992, pp. 666-71.

Campbell, C.J. (1993), 'The Depletion of the world's oil', *Petrole et technique*, No. 383, Paris, France, pp. 5–12.

Campbell, C.J. (1994), 'The imminent end of cheap oil-based energy', *SunWorld* 18/4, pp. 17–19.

Campbell, C.J. (1995), 'The next oil price shock: the world's remaining oil and its depletion', *Energy Exploration and Exploitation*, 13/1, pp. 19–46.

Campbell, C.J. (1997), *The Coming Oil Crisis*, Essex England: Multi-Science Publications Co, 210 p.

Campbell, C.J., and Laherèrre, J.H. (1998), 'The end of cheap oil', *Scientific American*, v. 278, pp. 78–83.

Cavallo, A.J. (2002), 'Predicting the peak in world oil production', *Natural*

Resources Research, v. 11, no. 3, pp. 187–95.

Cavallo, A.J. (2004), 'Hubbert's petroleum production model: an evaluation and implications for world oil production forecasts', *Natural Resources Research*, v. 13, no.4, pp. 211–21.

CEDIGAZ, 2001, 'Natural gas in the world: 2001 survey', Paris: Institut Français du Pétrole. (Cited in IEA, 2002).

Charpentier, R.R. (2004), 'Global distribution of natural gas resources', *Encyclopedia of Energy*, v. 4, Elsevier, Netherlands, pp. 249–56.

Charpentier, R.R. (2005), 'Estimating undiscovered resources and reserve growth: contrasting approaches', *in* Dore, A.G., and Vining, B.A., eds., Petroleum Geology: North-West Europe and Global Perspectives—Proceedings of the 6th Petroleum Geology Conference, Geological Society of London, v. 1, pp. 3–9.

De Bruyne, M.D. (1980), 'Financing problems in the oil industry', *in* Proceedings of the 10th World Petroleum Congress, v. 1, pp.51–4.

Deffeyes, K.S. (2001), *Hubbert's Peak: the impending world oil shortage*, Princeton, New Jersey: Princeton University Press, 208 p.

Demming, D. (2001), 'Are we running out?', *in* M.W. Downey, J.C. Threet, and W.A. Morgan, eds., *Petroleum provinces of the twenty-first century*, American Association of Petroleum Geologists Memoir 74, pp. 45–56.

Desprairies, P., W.T. McCormick Jr., L.W. Fish, R.B. Kalisch, T.J. Wander, 1978, 'Worldwide petroleum supply limits', *in* World Energy Resources 1985-2020. Oil and gas resources, Worlds Energy Conference, Guildford, United Kingdom, IPC Science and Technology Press, New York, New York, United States (USA), ISBN 0902852957.

Duce, J.T. (1946), 'Post-war oil supply areas', *The Petroleum Times*, (London), April 13.

Duncan, R.C. (2000), 'The peak of world oil production and the road to the Olduvai Gorge', Pardee Symposium, Geological Society of America Summit, Reno, Nevada, November 13, 2000, 6 p.

Duncan, R.C., and W. Youngquist (1999), 'Encircling the peak of world oil production', *Natural Resources Research*, v. 8, pp. 219–32.

Edwards, J.D. (1997), 'Crude oil and alternative energy production forecasts for the twenty first century: the end of the hydrocarbon era', *American Association of Petroleum Geologists Bulletin*, v. 81, pp. 1291–1305.

Edwards, J.S. (2001), 'Twenty-first century energy: decline of fossil fuel, increase of renewable nonpolluting energy sources', *in* Downey, M.W., Threet, J.C., and Morgan, W.A , eds., *Petroleum provinces of the twenty-first century*, American Association of Petroleum Geologists Memoir 74, pp. 21–34.

Edwards, J.S. (2002a), 'Twenty-first century energy: transition from fossil fuels to renewable, nonpolluting energy sources', *in* Gerhard, L.C., P.P. Leahy, and V.J. Yannacone, eds., *Sustainability of Energy and Water through the 21st Century*, published jointly by the Kansas Geological Survey and the American Association of Petroleum Geologists and the AAPG Division of Environmental Geosciences, Proceedings of the Arbor Day Farm Conference, pp. 37–48.

Edwards, J.S. (2002b), 'Twenty-first century energy transition fromhFolinsbee, R. (1977), 'World view from Alph to Zipf', *Geological Society of America*, v. 88, pp. 897−907

Gautier, D. L., Dolton, G. L., Takahashi, K. I., and Varnes, K. L., eds., (1996, 1995), National assessment of United States oil and gas Resources - results, methodology, and supporting data: U.S. Geological Survey Digital Data Series DDS-30.

Gautier, D. L., G.L. Dolton, and E.D. Attanasi (1998), 1995 National oil and gas assessment and onshore Federal Lands: U.S. Geological Survey Open-File Report 95-75-N; 64 p.

Gold, T. (1999), *The Deep Hot Biosphere: the myth of fossil fuels,* Springer Verlag, New York: Copernicus Press, 235 p.

Goodstein, David (2004), *Out of Gas: the end of the age of oil,* W.W. Norton and Company, 144 p.

Grace, J.D., R.H. Caldwell, and D.I. Heather (1993), 'Comparative reserves definitions: U.S.A., Europe, and the Former Soviet Union', *Journal of Petroleum Technology,* Society of Petroleum Engineers, v. 45, no. 9, pp. 866−72.

Green, A.R. (2004), 'Global energy − the next decade and beyond', American Association of Petroleum Geologists 2004-2005 Distinguished Lecture (available at *www.aapg.org*).

Greene, D.L., J.L. Hopson, and J. Li, 2003, 'Running out of and into oil: analyzing global oil depletion and transition through 2050', Oak Ridge National Laboratory, TM-2003/259, October, 2003, 124 p.

Grossling, B.F. (1976), *Window on Oil: A Survey of World Petroleum Resources,* London: Financial Times, Ltd., 140 p.

Halbouty, M.T., and J.D. Moody (1979), 'World ultimate reserves of crude oil', *in* Proceedings of the 10th World Petroleum Congress, Bucharest, Hungary, v. 2, pp. 291−301.

Halbouty, M.T. (1981), 'Prospectos futuros de exploración y las reservas petrolíferas mundiales', Future exploration prospects and the world petroleum reserves: *Boletín de la Asociación Mexicana de Geólogos Petroleros,* v. 33, no. 2, pp. 3−22.

Hendricks T.A. (1965), 'Resources of oil, gas, and natural-gas liquids in the United States and the world', U. S. Geological Survey Circular 522, 20 p.

Hubbert, M.K. (1956), 'Nuclear energy and the fossil fuels'; Amer. Petrol. Inst. Drilling & Production Practice. Proceedings Spring Meeting, San Antonio, Texas. pp. 7−25.

Hubbert, M.K. (1959), 'Techniques of prediction with application to the petroleum industry', Publication 204, Shell Development Company, Short Course Notes for presentation at the 44th Annual Meeting of the American Association of Petroleum Geologists, Dallas, 1959, 32 p.

Hubbert, M.K. (1967), 'Degree of advancement of petroleum exploration in United States', *American Association of Petroleum Geologists Bulletin,* v. 51, no. 11, pp. 2207−27.

Hubbert, M.K. (1969), 'Energy resources', *in* Resources and man, Committee on Resources and man, National Academy of Science—National Research

Council, San Francisco, California,W.H. Freeman, pp. 157–242.

Hubbert, M.K. (1973), 'Survey of world energy resources', *Canadian Mining and Metallurgy Bulletin* 66, 735, Montreal, Canada, pp. 37–53.

Hubbert, M.K. (1974), U.S. 'Energy resources: A review as of 1972: part 1. A National Fuels and Energy Policy Study', U.S. Congress, Senate Committee on Interior and Insular Affairs, Washington, D.C., U.S. Government Printing Office, serial no. 93-40.

Huber, P.W., and M.P. Mills (2005), *The Bottomless Well: the twilight of fuel, the virtue of waste, and why we will never run out of energy,* Basic Books, 256 p.

IHS Energy Group, 2001, International petroleum exploration and production database. [Database available from IHS Energy Group, 15 Inverness Way East, Englewood, Colorado, 80112, U.S.A.].

Institut Français du Pétrole (IFP) (1978), p. 24. (Cited in Odell, 1983).

International Energy Agency (1998), *World Energy Outlook*, Paris, France: International Energy Agency/OECD, 2000 Edition, 475 p.

International Energy Agency (2000), *World Energy Outlook*, Paris, France: International Energy Agency/OECD, 2000 Edition, 457 p.

International Energy Agency (2001), *World Energy Outlook—2001 Insights*, Paris, France: International Energy Agency/OECD, 2001 Edition, 421 p.

International Energy Agency (2002), *World Energy Outlook*, Paris, France: International Energy Agency/OECD, 2002 Edition, 530 p.

International Energy Agency (2003), *World Energy Outlook—2003 investment outlook, 2003 insights,* Paris, France: International Energy Agency/OECD, 2001 Edition, 511 p.

International Energy Agency (2004), *World Energy Outlook*, Paris, France: International Energy Agency/OECD, 2002 Edition, 570 p.

International Gas Union (1977), Discussed in 'Survey of energy resources. Hydrocarbons: Part A2 by E. Schubert': 1980 11th World Energy Conference, Munich, Germany, pp. 75–169.

Kirkby, M.A., and T.D. Adams (1974), 'The search for oil around the world up to 1999', *Petroleum Times*, London, United Kingdom, November 1, 1974, pp. 25–9.

Klemme, H. D. (1977), 'World oil and gas reserves from analysis of giant fields and petroleum basins (provinces)', *in* Meyer, R. F., ed., *The Future Supply of Nature-made Petroleum and Gas*, Technical Reports, Pergamon Press, New York, New York, United States (USA), pp. 217–60.

Klemme, H. D. (1978), 'Worldwide petroleum exploration and prospects', Landwehr, M. L., ed., New ideas, new methods, new developments, *Exploration and Economics of the Petroleum Industry*, v.16, pp.39–101.

Klett, T.R. (2003), Graphical comparison of reserve-growth models for conventional oil and gas accumulations: U.S. Geological Survey E-Bulletin 2172, 98 p., Appendices [CD-ROM] (available at http://energy.cr.usgs.gov/oilgas/noga).

Klett, T.R. (2004), 'Oil and natural gas resource assessment: Classification and terminology', *Elsevier Encyclopedia of Energy*, v. 4, Elsevier, Netherlands, pp.595–605.

Klett, T.R. (2005), 'United States Geological Survey's Reserve-Growth Models and Their Implementation', *Natural Resources Research*, v. 14, no. 3, pp. 247–62.

Klett, T.R. and D.L. Gautier (2003), Characteristics of Reserve Growth in Oil Fields of the North Sea Graben Area [abs.]: 65th EAGE Conference and Exhibition, European Association of Geoscientists and Engineers, Stavanger, Norway, June 2–5, 2003, 4 p.

Klett, T.R. and D.L. Gautier (2005), 'Reserve growth in oil fields of the North Sea', *Petroleum Geoscience*, v. 11, pp. 179–90.

Klett, T.R., D.L. Gautier, and T.S. Ahlbrandt (2005), 'An evaluation of the U.S. Geological survey World Petroleum Assessment 2000', *American Association of Petroleum Geologists Bulletin*, v.89, no. 8, pp. 1033–42.

Klett, T.R., and J.W. Schmoker (2003), 'Reserve growth in the world's giant oil fields', *in* Halbouty, M.T., ed, *Giant oil and gas fields of the decade 1990–1999*, American Association of Petroleum Geologists Memoir 78, pp. 107–22.

Laherrère J.H. (1993), 'Le pétrole, une ressource sure, des réserves incertaines', *Pétrole et Techniques*, v. 383, pp. 13–19.

Laherrère, J.H. (2000), 'Learn strengths, weaknesses to understand Hubbert Curve', *Oil & Gas Journal*, pp. 63–4, 66–8, 70, 72, 74–6.

Levorsen A.I. (1950), 'Estimates of undiscovered petroleum reserves', *in* Proceedings of the United Nations Scientific Conference on the Conservation and Utilization of Resources, v. 1, Plenary meetings, United Nations, pp. 107–10.

Lynch, M.C. (1994), 'Bias and theoretical error in long-term oil market forecasting', *in* J.R. Mooney, ed., *Advances in the Economics of Energy and Natural Resources*, Cambridge, Massachusetts: JAI Press, pp. 53–87.

Lynch, M.C. (1998), *Crying wolf: warnings about oil supply*, Cambridge, Massachusetts: Massachusetts Institute of Technology, 28 p.

MacKay, I.H., and F.K. North (1975), 'Undiscovered oil reserves', *in* Haun, J.D. ed., Proceedings of AAPG Conference on Methods of Estimating the Volume of Undiscovered Oil and Gas Resources, Stanford University, Palo Alto, California, August 21–23, 1974, American Association of Petroleum Geologists Studies in Geology no. 1, p.. 76–86.

Martin, A.J. (1985), 'Prediction of strategic reserves in prospect for the world oil industry', *in* Niblock, T. and R. Lawless, eds., University of Durham, United Kingdom, pp. 16–39.

Masters, C.D. (1987), 'Global oil assessments and the search for non-OPEC oil', *OPEC Review*, Summer 1987, pp. 153–69.

Masters, C.D., E.D. Attanasi, and D.H. Root (1994), 'World petroleum assessment and analysis', Proceedings of the 14th World Petroleum Congress: London, United Kingdom: John Wiley and Sons, pp. 529–41.

Masters, C.D., D.H. Root, and W.D. Dietzman (1983), 'Distribution and quantitative assessment of world crude oil reserves and resources', U. S. Geological Survey Open File Report OF 83-0728, 23 p.

Masters, C.D., D.H. Root, and W.D. Dietzman (1984), 'Distribution and quantitative assessment of world crude oil reserves and resources', Proceedings of the 11th World Petroleum Congress, Chichester, United Kingdom: John Wiley and Sons, v. 2, pp. 229−37.

Masters, C.D., D.H. Root, and E.D. Attanasi (1991), 'World Resources of crude oil and natural gas', Proceedings of the 13th World Petroleum Congress, Chichester, United Kingdom: John Wiley and Sons, pp. 51−64.

McCabe, P.J. (1998), 'Energy resources—cornucopia or empty barrel', *American Association of Petroleum Geologists Bulletin*, v. 82, pp. 2110−34.

McCormick, W.T. Jr., L.W. Fish, R.B. Kalisch, and T.J. Wander (1978), 'The future for world natural gas supply: World Energy Resources 1985−2020. Oil and gas resources', The full reports to the conservation commission of the Worlds Energy Conference. Guildford, United Kingdom: IPC Science and Technology Press for the World Energy Conference, pp. 49−56.

McKelvey, V.E. (1972), 'Mineral resource estimates and public policy', *American Scientist*, v. 60, no. 1, pp. 32−40.

Meyerhoff, A. (1979), 'Proved and ultimate reserves of natural gas and natural gas liquids in the world: World reserves of oil and gas', *in* Proceedings of the 10th World Petroleum Congress, Bucharest, Hungary, v. 2, pp. 303−43.

Minerals Management Service (1996), 'An assessment of the undiscovered hydrocarbon potential of the Nation's Outer Continental Shelf', Minerals Management Service OCS Report MMS 96-0034, 40 p.

Modelesky, M.S., G.S. Gurevich, and E.M. Kartukov (1985), 'World cheap crude oil and natural gas resources', Available in an unpublished form by Academician Styrikovich at the June 1985 meeting of the International Institute for Applied Systems Analysis Energy Workshop in Laxenburg, Austria (cited in Odell, 1985).

Moody, J.D. (1970), 'Petroleum demand of future decades', *American Association of Petroleum Geologists Bulletin*, v. 54, pp. 2239−45.

Moody, J.D. (1978), 'The world hydrocarbon resource base and related problems', *in* Philip, G.M., and K.L. Williams, eds., *Australia's Mineral Resources Assessment and Potential*, University of Sydney Earth Resources Foundation Occasional Publication 1, pp. 63−9.

Moody, J.D., and R.W. Esser (1975), 'An estimate of the world's recoverable crude oil resource', *in* Proceedings of the 9th World Petroleum Congress, Tokyo, Japan, May 1975, Applied Science Publ. Ltd. London, United Kingdom, v. 3, pp. 11−20.

Moody, J.D., and R.E. Geiger (1975), 'Petroleum resources: How much oil and where?', *Technology Review*, Massachusetts Institute of Technology, 77, April 1975, Cambridge, Massachusetts, pp. 38−45.

Nakicenovic, N. (1993), 'The methane age and beyond', *in* Howell, D., ed., *The Future of Energy Gases*, U.S. Geological Survey Professional Paper 1570, pp. 661−76.

National Academy of Sciences (1975), Reported in 'Survey of energy resources. Hydrocarbons: Part A2 by E. Schubert', 1980 11th World Energy Conference, Munich, Germany, pp. 75−169.

Nehring, R. (1978), 'Giant oil fields and world oil resources', CIA report R-2284-CIA, Santa Monica, California, Rand Corporation, 162 p.

Nehring, R. (1979), 'The outlook for conventional petroleum resources', Rand Corporation Paper P-6413, Santa Monica, California, 21 p.

NRG Associates, Inc. (1995), The significant oil and gas fields of Canada database:, NRG Associates, Inc. Database available from NRG Associates, Inc., P.O. Box 1655, Colorado Springs, Colorado 80901 U.S.A..

Odell, P.R. (1973), 'The future of oil: a rejoinder', *Geographical Journal*, v. 139, pp. 436–54.

Odell, P.R. (1979), 'World energy in the 1980's: the significance of non-OPEC oil supplies', *Scottish Journal of Political Economy*, v. 26, pp. 215–31.

Odell, P.R. (1992), 'Global and regional energy supplies: recent fictions and fallacies revisited', *Energy Policy*, v. 20, no. 4, pp. 285–96.

Odell, P.R. (1998), 'Oil and gas reserves: retrospect and prospect', *Energy Exploration and Exploitation*, v. 16, pp. 117–24.

Odell, P.R., and K.E. Rosing (1975), 'Estimating world oil discoveries up to 1999: The question and method', *Petroleum Times*, London, February 1975, pp. 26–9.

Odell, P.R., and K.E. Rosing, (1983), *The Future of Oil, World Oil Resources and Use*, London, United Kingdom: Nichols Publishing, 224 p.

Odell, P.R., and Rosing, K.E. (1985), 'World oil resources: east-west differ on estimates', *Petroleum Economist*, v.LII, pp. 329–31.

O'Neil, W. (2001), 'Oil as a strategic factor: the supply of oil in the first half of the 21st century, and its strategic implication for the U.S.', Center for Naval Analyses, Issue Concept Paper, 98 p.

Organization of Petroleum Exporting Countries (2004), *Oil Outlook to 2025*, 51 p.

Parent, J.D. (1980), 'A Survey of United States and Total World Production, Proved Reserves and Remaining Recoverable resources of Fossil Fuels and Uranium as of December 31, 1978', Chicago, Illinois Institute of Gas Technology, 140 p.

Parent, J.D, and H.R. Linden (1974), 'A study of potential world crude oil supplies', *Energy World*, London, United Kingdom, January 1974, pp. 3–9.

Parent, J.D, and H.R. Linden (1974), 'Analysis of world energy supplies', *in* Proceedings of the 9th World Energy Conference, Detroit, Michigan,September 1974, ch. 1.2, pp. 25–9.

Parent, J.D, and H.R. Linden (1977), 'A survey of United States and total world production, proved reserves and remaining recoverable resources of fossil fuels and uranium as of December 31, 1975', Chicago, Illinois: Institute of Gas Technology, 50 p.

Petroconsultants (1996), Petroleum exploration and production database: Houston, Texas, Petroconsultants, Inc. [Database available from Petroconsultants, Inc., P.O. Box 740619, Houston, Texas, 77274-0619 U.S.A.]

Pogue, J.E. (1946), 'Oil in the world', *Yale Review*, New Haven Connecticut: Yale University, v. 35, pp. 623–32.

Porfir'yev, V.B. (1974), 'Inorganic origin of petroleum', *American Association of Petroleum Geologists Bulletin*, v. 58, pp. 3–33.

Potential Gas Committee (1999), 'Potential supply of natural gas in the United

States. A report of the Potential Gas Committee', Potential Supply of Natural Gas, 1998: Colorado School of Mines, Golden, Colorado, 195 p.

Pratt, W.E. (1952), 'Toward a philosophy of oil finding', *American Association of Petroleum Geologists Bulletin*, v. 36, no. 12, pp. 2231−6.

Radler, M. (2002), 'Worldwide reserves increase as production holds steady', *Oil and Gas Journal*, v.99, no.52, pp.125−8.

Rempel, Hilmar (2002), 'Availability of energy resources: non-renewable energy sources, conventional crude oil and NGL, conventional natural gas', 17th World Petroleum Congress, Rio de Janeiro, Brazil, Abstracts with Program, p. 293.

Riva, J.P. (1995), 'World production after year 2000: business as usual or crises?', Congressional Research Service, Library of Congress, Rept. for Congress, August 18, 1995, 20 p.

Root, D.H., and R.F. Mast (1993), 'Future growth in known oil and gas fields', *American Association of Petroleum Geologists Bulletin*, v. 77, no. 3, pp. 479−84.

Root, D.H., E.D. Attanasi, R.F. Mast, and D.L. Gautier (1995), 'Estimates of inferred reserves for the 1995 National Oil and Gas Resources Assessment', U.S. Geological Survey Open-File Report 95-75-L, 29 p.

Ryman, W.P. (1967), Discussed by Hubbert M.K, 'Energy resources': Resources and Man, Committee on Resources and Man, National Academy of Sciences–National Research Council, San Francisco, California, W.H. Freeman and Co., 1968, pp. 157-242.

Salvador, A. (2005), *Energy: a historical perspective and 21st century forecast*, American Association of Petroleum Geologists Studies in Geology #54, 208 p.

Schanz, J. J., Jr., and J.G. Ellis (1983), 'Assessing the mineral potential of the Federal Public lands', Congressional Research Service.

Schollnberger, W.E. (1998), 'Projections of the world's hydrocarbon resources and reserve depletion in the 21st century', *Houston Geological Society Bulletin*, Houston, Texas, November, 1998, pp. 31−7.

Schmoker, J.W., and E.D. Attanasi (1996), 'The importance of reserve growth to the Nation's supply of natural gas', U.S. Geological Survey Fact Sheet FS-202-96, 2 p. (available at http://energy.cr.usgs.gov/oilgas/noga).

Schmoker, J.W., and E.D. Attanasi (1997), 'Reserve growth important to U.S. gas supply', *Oil and Gas Journal*, v. 95, no. 4 , January 27, pp. 95−6.

Schmoker, J.W., and T.R. Klett (2000), 'Estimating potential reserve growth of known (discovered) fields', A component of the USGS World Petroleum Assessment 2000-- Chapter RG, U.S.Geological Survey World Petroleum Assessment 2000-Description and Results: U.S. Geological Survey Digital Data Series DDS-60, CD-ROM, 19 p. (available at *http://energy.cr.usgs.gov/ oilgas/wep*).

Schubert, E. (1980), 'Survey of energy resources. Hydrocarbons', Part A2, 11th World Energy Conference, Munich, Germany, pp. 75−169.

Schuyler, John (1999), 'Probabilistic reserves definitions, practices need further refinement', *Oil and Gas Journal*, May 31, 1999, pp. 64−7.

Schweinfurth, S.P. (1973), 'Crude oil and natural gas liquids resources of the United States and the world', Special reports, U. S. Federal Power Commission, Washington, D. C., pp. 55–72.

Seba, R.D. (1998), *Economics of Worldwide Petroleum Production*, Tulsa, Oklahoma: Oil and Gas Consultants International, Inc. Publishing, 2nd Edition, 576 p.

Seidl, R. (1977), 'Oil: The picture is changing', Options, International Institute for Applied Systems Analysis, Laxenburg, Winter, p. 4.

Shell (1967), Reported in 'Survey of energy resources. Hydrocarbons: Part A2 by E. Schubert', 1980, 11th World Energy Conference, Munich, Germany, pp. 75–169.

Shell (2001), Energy needs, choices and possibilities scenarios to 2050, *http:// www.shell.com/files/media-en/scenarios.pdf*

Simmons, M.R. (2005), *Twilight in the Desert: the coming Saudi oil shock and the world economy*, John Wiley and Sons, 448 p.

Stark, P.R. (2003), 'Report finds most new oil around the world comes from old fields', First Break (news feature), v. 21, p. 11-13 (also available from IHS Energy 2003 World Petroleum Trends).

Styrikovich, M.A. (1977), 'The long range energy perspective', *Natural Resources Forum*, v. 1, pp. 252–3.

U.S. Geological Survey and U.S. Bureau of Mines (1976), 'Principles of the mineral resource classification system of the U.S. Bureau of Mines and U.S. Geological Survey', *U.S. Geological Survey Bulletin* 1450-A, 5 p.

U.S. Geological Survey (1995), 'National assessment of United States oil and gas resources', U.S. Geological Survey Circular 1118, 20 p.

U.S. Geological Survey (2000), U.S. Geological Survey World Petroleum Assessment 2000 – Description and Results by U.S. Geological Survey World Energy Assessment Team, U.S. Geological Survey Digital Data Series DDS-60, a four CD-ROM set. (available at http://energy.cr.usgs.gov/oilgas/wep).

Verma, M.K. (2000), 'The significance of field growth and the role of enhanced recovery', U.S. Geological Survey Fact Sheet FS-115-00, 4 p.

Verma, M.K. (2005), 'A new reserve growth model for the United States oil and gas fields', *Natural Resources Research*, v. 14, no. 2, pp. 77–89.

Verma, M.K., T.S. Ahlbrandt, and M. Al-Gailani (2004), 'Petroleum reserves and undiscovered resources in the total petroleum systems of Iraq: reserve growth and production implications', *GeoArabia*, v. 9, no. 3, pp. 51–74.

Verma, M.K., and M.E. Henry (2004), 'Historical and potential reserve growth in oil and gas pools in Saskatchewan', *in* Summary of Investigations 2004, v. 1, Saskatchewan Geological Survey, Saskatchewan Industry Resources, Miscellaneous Report 2004-4.1, CD ROM, Paper A-1, 20 p.

Verma, M.K., and G.F. Ulmishek (2003), 'Reserve growth in oil fields of West Siberiaan Basin, Russia', *Natural Resources Research*, v. 12, no. 2, pp. 105–19.

Verma, M.K., G.F. Ulmishek, and A.P. Gilberstein (2000), 'Oil and gas reserve growth—a model for the Volga-Ural province, Russia', Society of Petroleum Engineers Publication 62616 presented at the 2000 SPE/AAPG Western Regional Meeting, Long Beach, California, June 19–23, 2000, 7 p.

Warman, H.R. (1972), 'The future of oil', *Geographical Journal*, v. 138, pp.

287–97.

Weeks L.G. (1948), 'Highlights on 1947 developments in foreign petroleum fields', *American Association of Petroleum Geologists Bulletin*, v. 32, pp. 1093–160.

Weeks, L.G. (1950), Discussion of 'Estimates of undiscovered petroleum reserves by A.I. Levorsen', Proceedings of the United Nations Scientific Conference on the Conservation and Utilization of Resources, New York, New York, v. 1, pp. 107–10.

Weeks, L.G. (1958), 'Fuel reserves of the future', *American Association of Petroleum Geologists Bulletin*, v. 42, no. 2, pp. 431–41.

Weeks, L.G. (1959), 'Where will energy come from in 2059?' *The Petroleum Engineer*, Dallas, Texas, August 1959, p. A-24-A-31.

Weeks, L.G. (1965), 'World offshore petroleum resources', *American Association of Petroleum Geologists Bulletin*, v. 49, no. 10, pp. 1680–93.

Weeks, L.G. (1968), 'The gas, oil, and sulfur potential of the sea', *Ocean Industry*, Houston, Texas, June 1968, pp. 43–51.

Weeks, L.G. (1971), 'Marine geology and petroleum resources', *in* Proceedings of the 8th World Petroleum Congress, Moscow, Russia, v. 2, pp. 99–106.

Williams, Bob (2003), Debate over peak-oil issue boiling over, with major implications for industry, society (the first in a series of six special reports on Future Energy Supply), *Oil and Gas Journal*, v. 101, no. 7, July 14, pp. 18–29.

World Petroleum Congress and Society of Petroleum Engineers (1997), Reserve definitions (reported in Seba, 1998).

Yergin, D. (1991), *The Prize*, New York: Simon and Shuster, 884 p.

Youngquist, Walter (1997), *GeoDestinies – the inevitable control of Earth resources over nations and individuals*, Portland, Oregon: National Book Company, 500 p.

Captions

Table 1: Summary of Estimated Volumes of Conventional Oil, Gas, and Natural Gas Liquids (NGL) making up the World Endowment of these Commodities. Included by commodity are amounts of cumulative production, remaining reserves, estimated volumes of reserve growth, and estimated undiscovered amounts. Data represent 128 provinces exclusive of the United States as reported in U.S. Geological Survey (2000); United States estimates are those of U.S. Geological Survey (1995) and Minerals Management Service (1996).

Table 2: World Oil Resources and NGL Resources by Region.

Figure 1: Oil estimates are in trillion barrels of oil (TBO), natural gas estimates are in trillion barrels of oil equivalent (TBOE). Modified from Ahlbrandt et al. (2005).

Figure 2: Both a low scenario and high scenario based on the Hubbert prediction model (from Hubbert, 1969).

Figure 3: Future Oil (remaining reserves plus USGS 2000 mean estimate of undiscovered resources) in the assessed provinces (non-USA). Evaluated by the U.S. Geological Survey on the basis of data to January 1, 1996. Oil consumption information from the Energy Information Administration 2003 website. MBO, thousand barrels of oil; MMBO, million barrels of oil.

Figure 4: Future Natural Gas (remaining reserves plus USGS 2000 mean estimate of undiscovered resources) in the assessed provinces (non-USA). Evaluated by the U.S. Geological Survey on the basis of data to January 1, 1996. Natural gas consumption information from the Energy Information Administration 2003 website. BCFG, billion cubic feet of gas.

Figure 5: A. The economic (E), field status and feasibility (F), and geologic knowledge (G) axes of the system. B. Codification of the UNFC system relative to the three axes, for example E1, F1, G1=111 as illustrated. C. The UNFC system as it relates to western classification systems such as the McKelvey diagram (McKelvey, 1972) used by the U.S. Geological Survey and others. The McKelvey diagram projects to the E and G axes of the UNFC. D. The UNFC classes may be grouped or subdivided as required. The system is comparable to the Norwegian Petroleum Directorate (NPD), Society of Petroleum Engineers/World Petroleum Congress/American Association of Petroleum Geologists (SPE/WPC/AAPG) and Russian classifications as shown. (modified from Ahlbrandt et al., 2004).

Figure 6: Estimates include both the 128 world provinces assessed by the U.S. Geological Survey (2000) and the U.S. provinces assessed by the U.S. Geological Survey (1995) and Minerals Management Service (1996) as of January 1, 1996. Total oil endowment includes the four components shown on the right side of the diagram (from Ahlbrandt, 2002).

Figure 7: Estimates include both the 128 world provinces assessed by the U.S. Geological Survey (2000) and the U.S. provinces assessed by the U.S. Geological Survey (1995) and Minerals Management Service (1996) as of January 1, 1996. Total oil endowment includes the four components shown on the right side of the diagram (from Ahlbrandt, 2002).

Figure 8: As assessed from data representing 128 provinces exclusive of the United States as reported in USGS (2000); United States estimates are those of U.S. Geological Survey (1995) and Minerals Management Service (1996). Components of this endowment include cumulative production, remaining reserves, and mean estimates of reserve growth, and undiscovered resources for oil, gas, and natural gas liquids. The second bar from the left (remaining reserves) is the total endowment less cumulative production; the third bar (discovered) subtracts out the undiscovered resource estimates, and the fourth bar (proven reserves) is the endowment less cumulative production, reserve growth and undiscovered resources. The fifth bar (cumulative production) reflects production as of January 1, 1996, and annual consumption shown in the sixth bar represents recent consumption worldwide as reported in Energy

Information Administration (2001) (from Ahlbrandt, 2002; data from IHS Energy Group, 2001, and Energy Information Administration, 2001)

Figure 9: As identified by the USGS 2000 assessment. The greatest amount of reserve growth occurred in the Middle East and North Africa, whereas the greatest contribution from new-field discoveries occurred in sub-Saharan Africa (from Klett et al., 2005).

Figure 10: As identified by the USGS 2000 assessment. The greatest amount of reserve growth occurred in the Middle East and North Africa, whereas the greatest contribution from new-field discoveries occurred in the Asia Pacific region. (From Klett et al., 2005).

Figure 11: Demonstrating that the peak of world oil production should have occurred in 1989, and illustrating the attendant consequences. World oil production (2004) is about 28 billion barrels of oil per year.

Figure 12: Under Various Growth and U.S. Geological Survey 2000 estimate Levels, which are in the moderate estimate group. Ultimate recovery volumes include both U.S. and non-U.S.oil resources. R/P, reserve to production ratio. F95 (the 95th fractile) represents a 95 percent chance of the occurrence of at least the amount tabulated; other fractiles are defined similarly (from the Energy Information Administration, 2001).

Figure 13: This energy scenario demonstrates the interacting variables in forecasting the future. For example, uncertain reserve figures, the plateau concept of oil resources, reserve growth and substitution of energy resources are demonstrated in this scenario.

Figure 14: This scenario suggests comparable conventional and unconventional oil volumes.

Figure 15: Views of conventional oil production either as a symmetrical peak when half of the production is reached, or an undulating plateau of production where reserves are added either through new discoveries or through reserve growth until such point as continued production is no longer economically viable. Production then falls off of the plateau and declines.

CHAPTER 6

TECHNOLOGIES TO EXTEND OIL PRODUCTION

Andrew Gould

1. Introduction

Technology is key to the continuing success of the oil exploration and production industry. Advances in technology have enabled the energy sector to provide for today's energy needs, but further advancement will be necessary to meet future needs by improving exploration success and increasing reservoir recovery.

The exploration and production (E&P) industry relies on seismic, drilling, logging, completion, stimulation, testing, modelling and monitoring methods that were not widely available a decade ago; some were not available at all. These techniques facilitate exploration successes to find new resources, or production operations to extend the commercial lives of wells and fields.

Much of the upstream petroleum industry today is focused squarely on mature fields, because about 70% of oil production comes from fields more than 30 years old. Improving recovery from current reservoirs by just 1% would add almost 12,000 million barrels of oil reserves.[1]

The International Energy Agency (IEA) predicts that continued increases in world oil production will be needed in the future to meet rising demand for energy (Figure 1). IEA figures indicate a rapid decline in production from existing capacity, although the decline rate may be less in some areas, such as Saudi Arabia and other Middle Eastern countries. Additional production must come from further expansion of

---

Q is a mark of WesternGeco. FlowScan Imager, PeriScope 15 and RapidSeal are marks of Schlumberger. Casing Drilling is a mark of Tesco Corporation.

[1] This value is based on a 2004 world reserve value of nearly 1,200 thousand million barrels of oil as cited in 'Putting Energy in the Spotlight,' *BP Statistical Review of World Energy* (June 2005): 4.

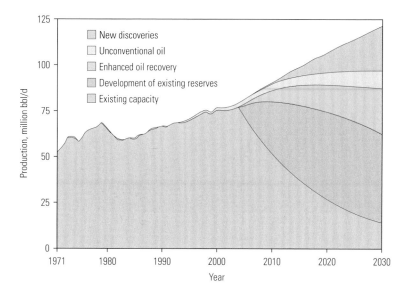

Figure 1: Prediction of World Oil Production

Existing capacity will experience a normal decline during the next 25 years. New discoveries and unconventional oil will provide some new reserves, but development of existing reserves and enhanced oil recovery will provide the majority of the production. (Figure used with permission, © IEA: World Energy Outlook 2004.)

existing proven reserves, enhanced oil recovery, unconventional sources of oil and new discoveries.

This is achievable. For example, Statoil increased the ultimate recovery factor in Statfjord field from 49.4% in 1986 to 65.6% in 2000, with an expectation of achieving 68% recovery. Over the same time period, the company's Gullfaks field recovery factor increased from 46.5% to 54.5%, with a future expectation of capturing 62% of the original reserves in place. Statoil credits effective resource management and technology application for these improvements (Reinertsen, 2003).

This chapter describes some important technologies to locate, access and recover oil and gas. It concludes with sections about carbonate reservoirs and technologies of the future.

2.   Highlights of Recent Advances

Global demand for oil has increased to a point that the excess supply cushion has been considerably reduced. The E&P industry has to steadily add production to meet the growing demand and to replace what is lost because of natural decline in producing fields. In response, Saudi Arabia announced that it would increase its oil production capacity from 11 to 12.5 million bbl/d by 2009, with the potential for a later increase to 15 million bbl/d.[2] At the time of this announcement in 2005, the country produced 9.5 million bbl/d. The Saudi petroleum minister cited application of technology as an important aspect of the strategy.

To the extent that technology can reduce costs, its application extends the life of a field, makes smaller fields economical and can even enable the redevelopment of fields that have already been abandoned. Application of new technology has resulted in an overall cost reduction and efficiency improvement. The high incremental cost of additional productive capacity in many parts of the world means that investment in exploration and development is highly sensitive to small changes in the price of oil.

One of the most remarkable achievements of the past two decades has been the ability of the industry to provide technologies that have reduced finding and development costs in a period of sustained low commodity prices. Finding and development costs fell by as much as two-thirds between 1981 and 2003 in constant dollar terms.[3] Much of this reduction was due to investment in new technologies that have enabled the industry to do a better job of exploring for, defining and developing reservoirs. Three-dimensional (3D) seismic data and extended-reach and horizontal drilling are the most obvious examples of technologies that have greatly impacted the cost of finding and accessing reserves.

Technology has not only helped reduce exploration costs, it has also accelerated recovery through more efficient reservoir drainage. An important question is: how can new or existing technology help manage production decline in mature fields in a manner that maximizes net-present value without jeopardizing ultimate recovery? This is a different challenge from those experienced during the exploration and initial development phases of a field. It involves, for example, management of technical options that were chosen many years before and that might limit what can be done today.

---

2   'Saudi Minister Addresses International Conference,' The Saudi Arabia Information Resource, May 24, 2005, available at *www.saudinf.com/main/y8196.htm* (accessed August 22, 2005).

3   Finding and development costs have been rising slowly over the past few years.

A fully detailed discussion of all technologies involved in locating, accessing and recovering reserves is beyond the scope of this chapter and would itself require an entire book. Instead, the remainder of this chapter will highlight some important technologies, starting with the first of these three areas, locating reserves.

3.   Locating Reserves: The Advantage of Looking Around

The success of 3D seismic technology is well known in the industry. Although the concept had been in existence for many years, advances in seismic acquisition and data processing that enabled 3D surveys occurred only in the 1980s. Expansion from single seismic lines to an integrated cube of seismic data allowed geophysicists to examine a two-dimensional (2D) slice cut at any angle through a volume. Complex structures could now be investigated with improved visualization techniques. This technology increased exploration and development activity around the world.

Three-dimensional seismic data offered a new way to reduce finding and producing costs, and seismic acquisition grew rapidly in the 1990s until most offshore activity was 3D rather than 2D. This increase in activity is even more impressive when the cost-saving effect of acquiring and processing seismic data more quickly is factored in — some estimate that such improvements increased overall efficiency in seismic activities ten-fold during this period.

Seismic data are used to find resources in both new and existing provinces. Many of the new provinces, such as those found in ultradeep water and in arctic regions, are difficult to access. These areas present environmental challenges for equipment and personnel, and require vigilance to avoid damage to delicate ecosystems, particularly in the arctic. In areas of existing production, infrastructure that was put in place to exploit large fields becomes underutilized when those fields go into decline. To extend the life of that infrastructure, small, previously uneconomical satellite fields may become economical to develop by tying them back to existing infrastructure.

As exploration targets, such as satellite fields, become smaller, the quality of seismic data becomes more critical to resolve ambiguous features. The recent introduction of high-fidelity Q single-sensor seismic acquisition and processing methodology has greatly improved seismic resolution. For example, Kuwait Oil Company (KOC) reported positive results from a pilot Q seismic study over the Minagish field (Rached, El-Emam & Anderson, 2005). The 24-km^2 area surveyed had an extraordinary trace density equivalent to more than 16 million single-sensor traces per km^2. Both vertical and lateral resolutions were improved compared with the resolutions of a traditional survey acquired in 1996 (Figure 2).

This high-fidelity system comprises several enabling technologies. The receiver signals are not grouped before processing; each signal is captured individually. Compared with a traditional system, the number of traces captured is vastly increased. Special processing applied to this enormous amount of information corrects for surface effects and heterogeneities. Despite the tremendous increase in the amount of information processed, supplier WesternGeco delivered the Minagish data to KOC six days after completing survey acquisition. In contrast with the weeks, months or even years that seismic processing consumed in the past, this rapid delivery means that decisions concerning drilling or development can be made soon after acquisition.

Single-sensor recording has also proved successful offshore. In a marine environment, seismic receivers are towed behind a vessel on

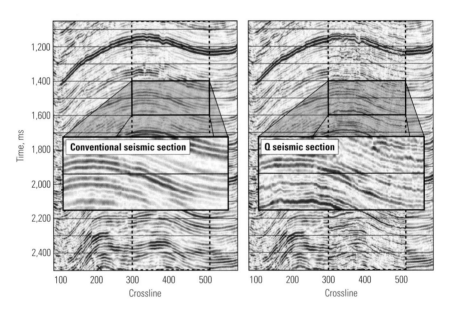

Figure 2: Improvement in Seismic Resolution

A conventional seismic section of Minagish field, Kuwait, was obtained in 1996 (*left*). A Q seismic section, obtained in 2004, is interleaved with the 1996 survey, with the Q data between the vertical dashed lines and the older data outside the lines (*right*). The insets highlight the Minagish formation at about 1,500 ms. (Figure used with permission, © WesternGeco, all rights reserved.)

long streamers. Crosscurrents in the ocean can move the streamers out of alignment, complicating data interpretation. Steering devices for the new Q system placed along the streamers have largely eliminated this drift, called *feathering*. The fins on these devices can move the streamer up, down, left or right, to maintain a precise trajectory within a close tolerance. For example, when the Q system was used in the Norne field in Norway to repeat an earlier conventional survey, the amount of feathering, or departure from the desired course, was less than 0.5° about 75% of the time (Eiken, 2003). This precision becomes essential when a 3D survey is repeated for a time-lapse seismic study of a reservoir, as with the Norne field.

Electromagnetic sensing data can be complementary to seismic data. This technology relies on the same physical property that resistivity-based wireline logging has used for decades: hydrocarbons have significantly higher resistivity than the brines contained within a rock. Offshore, the most common form this technology takes is controlled source electromagnetic measurements (CSEM). An array of receivers is placed on the seafloor, and a low-frequency dipole source is towed above the array. The receivers sense the resistivity of underlying formations.

CSEM does not have the resolution of seismic measurements, so it does not indicate structure. However, once structures are detected through a seismic survey, a CSEM survey can indicate which of them do not contain hydrocarbons. Using these technologies together has the potential to increase exploration success once CSEM technology fully matures.

Seismic or CSEM technologies also can locate bypassed reserves in existing fields. Crosswell surveys, such as the electromagnetic survey performed by Petroleum Development Oman (PDO) to monitor the progress of a waterflood in the Fahud field, provide a more detailed look at a reservoir than is possible with surface detection alone.[4] The objective of such surveys is to determine the structural complexity of the interwell area and to locate bypassed hydrocarbon.

The interwell region is significant because a typical 35% recovery factor means that 65% of the original reserves in place will not be produced − a significant amount of oil. In many cases, locating the remaining producible reserves requires a rigorous evaluation of existing data and a focused programme to gather new data to refine a reservoir model. Until recently, obtaining detailed petrophysical information in formations behind casing in a well was virtually impossible. After

---

4   'Rising to the Challenge,' Petroleum Development Oman Annual Report to His Majesty Sultan Qaboos bin Said, Sultan of Oman, 2004: 14.

many years of development, Schlumberger introduced a new set of behind-casing logging tools.

These tools, introduced starting in 1999, evaluate bypassed pay near existing wells. Such reserves remain for many reasons. It is possible that a well was originally drilled to access other target formations, or certain zones were not drained due to geological compartmentalization or variable permeability. Other causes may include failure to run a full suite of logs when the well was drilled, or the presence of low-resistivity pay zones.

A behind-casing resistivity tool provided guidance for improving productivity in a well in Libya. Oil production from a sandstone formation had decreased gradually, while water production increased significantly (Tchambaz and Belhadj, 2003). The tool identified an 8.5-m rise in the oil/water contact and additional oil above the existing perforations. The water zone was shut off and new perforations accessed the oil zone. In addition to increasing oil production eightfold, the operation decreased water production in the well from 90% of the total fluid volume to 0%.

As this example shows, behind-casing evaluations not only can locate bypassed hydrocarbon reserves, but also may help operators remediate wells by identifying zones that produce unwanted water. Both activities extend the productive life of a well.

The seismic or electromagnetic survey methods described earlier typically locate hydrocarbon accumulations that have not yet been drilled. Before discovered resources can be considered recoverable reserves, an economical plan to access and recover these reserves must be developed.

4. Accessing Reserves: Designer Wells and Novel Completions

Well trajectories have evolved from vertical and near-vertical boreholes to exotic designs that can follow almost any trajectory mapped in a seismic cube. In the Middle East, drilling strategies began to shift from conventional horizontal wells to extended-reach and multilateral wells with customized trajectories during the early 2000s. The primary motivating factors for using these increasingly complex maximum reservoir-contact (MRC) wellbores are to improve well productivity, reduce the number of wells and surface facilities needed to put the wells in production, reduce costs of drilling and production, and to use well spacing more efficiently (Saleri et al., 2004; Nughaimish et al., 2004).

Initially, Middle East operators used horizontal wellbores to reduce or delay water breakthrough, or coning, in reservoirs with a strong bottom waterdrive. Later, high-angle directional, horizontal and extended-reach drilling technologies helped achieve commercial production rates while minimizing early gas coning in reservoirs with a large overlying gas cap and a weak waterdrive from an underlying aquifer.

More recently, PDO began using longer high-angle and horizontal wellbores to develop oil reserves trapped in the thin, low-permeability layers of laminated reservoirs in northern Oman. New logging-while-drilling (LWD) technologies are essential to geosteering operations that maintain a horizontal well path through the highest quality reservoir rock (Hiebert et al., 2005).

These LWD technologies and the MRC techniques initiated by Saudi Aramco also are helping Saudi Arabian Texaco and Kuwait Oil Company recover unswept oil from mature fields in the Neutral Zone between Kuwait and Saudi Arabia. Bypassed reserves in the thin, low-permeability and more heterogeneous upper zones of these reservoirs are being exploited more effectively by drilling and completing new horizontal wells and horizontal sidetracks from existing wellbores (Jha et al., 2005).

Complicated well trajectories are possible as a result of combining several innovations. Novel downhole motor technology makes directional drilling practical. State-of-the-art measurements-while-drilling (MWD) and LWD tools and their immediate interpretation provide the information necessary to steer the wellbore within strata. These tools were improved over the past few decades to reach their current level of operations.

Introduction of the downhole positive displacement motor (PDM) in the 1960s facilitated drilling without full-string pipe rotation. These systems use mud flowing through a turbine or a rotor-stator power section to generate torque downhole. Steerable motors with a fixed bend angle, or bent housing, allow simultaneous control of borehole azimuth and inclination angle, which subsequently resulted in better directional control and routine construction of high-angle wellbores, horizontal borehole sections in the 1980s and, eventually, extended-reach wells.

In the 1990s, rotary steerable systems (RSS) helped operators set new records in extended-reach drilling. This technology facilitates directional control and steering of the bit while continuously rotating the entire drillstring. Steering is accomplished in a unit behind the bit by activating three pistons, separated on the circumference by 120°, in the proper sequence to force the bit in the correct direction.

In the Gulf of Suez, Belayim Petroleum Company (Petrobel), a

joint venture between Eni and Egyptian General Petroleum Corporation (EGPC), is using advanced rotary steerable drilling technology to
provide sufficient rotation rates to drill hard anhydrite stringers in the
mature Belayim Marine field. Interbedded sands and shales present
well-construction challenges, including severe mud losses and differential sticking. To overcome these drilling difficulties, Petrobel decided
against using a conventional mud motor in favour of a powered rotary
steerable system. This saved more than ten days of rig time in the
directional Belayim 113M-86 well. The downhole power resulted in a
drilling rate that was 47% higher than the best previous performance
in the field, and saved at least five rig-days per well in three other
directional wells.

In addition to reduced drilling costs, advanced drilling technology
provides some less obvious advantages, such as decreasing casing wear,
decreasing the time that drilling mud can invade a formation before
formation evaluation, decreasing the time during which a borehole can
degrade, improving zonal isolation and well security, and ultimately,
constructing wells faster to produce oil and gas sooner.

Another group of enabling technologies for complex well trajectories,
MWD tools, was developed to determine wellbore orientation. Sending
that information to surface required high-speed telemetry to transmit
data between a bottomhole assembly and the surface. As more MWD,
and later LWD, measurements became available, the capability of
geosteering to a desired target improved, and telemetry bandwidth had
to increase to keep pace with the demand for information and data.

Within the past few years, drilling targets have become increasingly
better defined. LWD logs provide petrophysical data that locate the
bit within specific strata. Placing a measurement collar nearer the bit
decreases the time lag between finding out where the bit has just gone
and drilling ahead. New LWD technology is now helping drillers see
where the bit is about to go. This helps access hydrocarbon assets in
locations that previously were difficult or costly to produce.

For example, in Oman, PDO wanted to drill into a thin rim of attic
oil in a Shuaiba carbonate reservoir. Veering off course upward into
the mechanically unstable Nahr Umr shale would likely complicate
well construction and completion or even jeopardize the borehole
itself (Daveridge et al., 2005). On the other hand, steering just 3–5 m
beneath the top of the Shuaiba reservoir would place the wellbore in
a low-permeability zone.

Previous attempts to place a wellbore just below the Shuaiba-Nahr
Umr boundary relied on conventional LWD tools. With their shallow
depth of investigation, these older tools provided little advance warning

that the wellbore was about to cross the boundary between the high-resistivity reservoir and the low-resistivity shale. This frequently resulted in unintentional exits from the reservoir, requiring a turn to steer the trajectory back down to the Shuaiba.

PDO turned to the new PeriScope 15 directional, deep imaging while drilling service. This tool propagates electromagnetic signals and uses a unique array of transmitters and receivers to determine the direction of bed boundaries and water zones up to 5 m away from the tool. Real-time measurements from this tool determined that the low-permeability zone lay just 2.5 to 3 m below the top of the Shuaiba reservoir. Guided by PeriScope 15 measurements, PDO drilled the well horizontally for 1,300 m, averaging 1.2 m beneath the Nahr Umr interface. By placing the well so close to the top of the formation, PDO added reserves of attic oil that would otherwise be inaccessible. With 100% of its horizontal section placed in the upper zone of the reservoir, this well has produced oil at significantly higher rates than the field average. This directional technology is advancing, and deeper-reading measurements are coming in the future.

In mature areas, one objective of advanced technology is to control the cost of access to reserves, such as through reentry drilling, multilateral wells and coiled tubing drilling. Saudi Aramco uses coiled tubing drilling of multilateral wellbores to maximize reservoir contact while minimizing the cost per foot of producing a wellbore (Saleri et al., 2004). Multilateral junction technology has developed to keep up with the trend (Figure 3).

Advances in modelling and executing fracture jobs have improved the ability to contact reservoirs. Tip-screenout fracturing techniques create more conductive flow channels. Advances such as coiled tubing fracturing have decreased cost and facilitated selective stimulation of individual zones. Similarly, using viscoelastic surfactants diverts acid from the most permeable or water-bearing zones to stimulate less permeable areas. Perforating using dynamic underbalanced conditions cleans out the perforations more effectively than earlier methods, and so improves well productivity.

Novel ways to drill and complete wells are also being developed. Between 1999 and 2004, more than 300 vertical intervals and about a dozen inclined sections were drilled with casing, mainly in Texas, USA. Casing Drilling operations minimize pipe trips, thereby reducing incidents of borehole collapse from swabbing and surging. Drilling with casing also decreases the chance of an unintentional sidetrack and minimizes wear inside previously installed surface or intermediate casing strings.

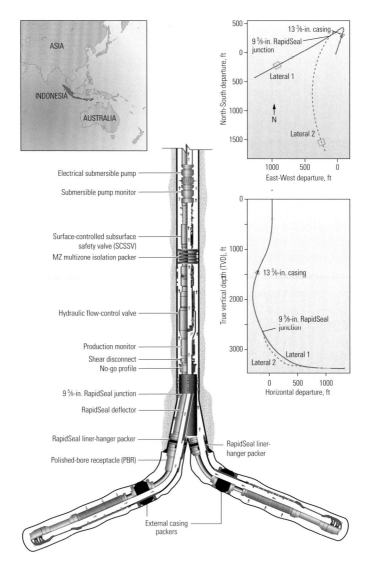

Figure 3: The World's First Intelligent Level 6 Multilateral Junction, in the Java Sea, Indonesia

This Level 6 RapidSeal multilateral completion system junction has a cased and cemented main wellbore and uncemented lateral liners. Hydraulic and pressure integrity are provided by the primary casing at the lateral liner intersection without additional completion equipment inside the main wellbore. In this completion, hydraulic control valves and a production monitor measure pressure, temperature and flow rate for each well branch, making this the first intelligent completion of the Level 6 type. (Figure used with permission, © Schlumberger, Ltd., all rights reserved.)

When a wellbore drilled with casing reaches total depth, the casing is already in place, eliminating the need to pull drillpipe and then install permanent casing. This technique can be more efficient than conventional techniques and has cut drilling time by a nominal 10% to 35% per well in some cases (Shepard et al., 2001).

Expandable tubulars and mechanical sand-exclusion screens represent another innovation that decreases well-construction cost. Originally considered a contingency method for overcoming problems while drilling, expandable tubulars are now used as the planned deployment in situations such as difficult deep-drilling environments in the deepwater Gulf of Mexico. Expandable pipe is run into the borehole and mechanically expanded by several percent in diameter. Fewer step changes in borehole size mean the bottomhole completion has fewer restrictions and more productive capacity, or viewed alternatively, the borehole diameter at surface is smaller. In addition to cost-savings arising from smaller diameter completions near surface, environmental impact is reduced, with disposal of fewer cuttings and less fluids. Similarly, expandable screens provide greater flow area within the tubulars.

These drilling and completion technologies are being applied in newly discovered and existing reserves around the world. However, most of the easily obtainable oil and gas reserves have been found and either have been or are being produced. Much of what remains is costly to access − either because the reserves are in remote locations or because they are in small accumulations.

Activity has been increasing in deep and ultradeep waters, in ultradeep reservoirs, in high-temperature, high-pressure reservoirs and in arctic areas. As the water depth has increased, offshore platforms evolved from fixed platforms to tension-leg platforms and then to spars. There is also a drive to reduce cost. A Norwegian Ministry of Petroleum and Energy study showed that the cost of four fixed platforms developed during the 1980s was double that of four floating production solutions developed between 1997 and 2000, when normalized to daily production (IEA, 2005). Innovative subsea completions are becoming more common for developing satellite fields that are tied back to an existing platform, or that serve as the base for a riser to a floating production facility. Without existing production facilities, these satellite fields might otherwise be uneconomical.

Downhole equipment has been made more rugged to endure higher temperatures and higher pressures in these harsher environments. At the other end of the temperature extreme, durable surface equipment also has been designed for harsh arctic conditions.

Many of today's offshore facilities must be designed before producing wells are drilled. The metallurgy of tubulars and process facilities must be compatible with produced fluids, so sophisticated fluid-sampling programmes must be employed on exploration wells. Fluid samples are used to determine the quantities of gas, oil and water that the facilities may be expected to handle; the physical properties of the formation fluid, such as bubblepoint or dewpoint; the potential for hydrate, asphaltene or wax formation; and the amount of harmful or corrosive fluids such as carbon dioxide or sulphur compounds and scale-forming compounds that could disable a completion. These analyses are important for oil as well as gas and gas-condensate fields.

Almost all the new technologies described here are equally applicable to oil or gas reservoirs. However, the growing importance of natural gas raises particular issues of technology development. The largest remaining gas reserves lie in Russia and the Middle East, where the infrastructure for gas delivery is still sparse. Some countries are now actively producing their gas resources for local use, reserving their oil for export. Other countries are treating gas as another marketable commodity, and are expanding infrastructure to export it. Countries like Qatar and Trinidad have advanced along this path and their gas exports far exceed those of oil.

Traditional methods for gas delivery are by pipeline and liquefied natural gas (LNG) tanker. The transport capabilities of both methods are increasing, with more pipelines being laid in developing areas and more LNG facilities being built. Japan is the biggest consumer of LNG, with most of its supplies coming from the Asia Pacific area, although it also imports a substantial amount from the Middle East.

Other methods of gas transport have not achieved the economy of these two methods. Compressing natural gas is less capital intensive than liquefying it, but the containers are larger, so transport cost is greater. Its niche may lie in shipment of small quantities of gas over short distances. Gas can also be formed into gas hydrate and shipped as frozen pellets, but this technology has not been proved.

Converting gas to liquid products is also feasible. In a gas-to-liquids (GTL) plant, the gas is converted to methanol or dimethyl ether. Several such plants around the world are in operation, and several are planned for Qatar, one of the world leaders in GTL production.

5. Recovering Reserves: Building a Digital Oil Field

Much engineering and design work goes into new fields to meet the global need for additional reserves. However, mature fields are known resources in known locations, and similar engineering efforts in older fields can also yield significant production gains.

From where does a company obtain the information needed to manage a field? The best information comes from the field itself – from historic records of productivity and pressures, from permanent sensors in wells and at surface and from information-gathering activities such as production logging and, as discussed earlier, logging behind casing. Where possible, integration of this information with seismic data taken at surface or with cross-wellbore tomography can provide a link between the near-well information and the distant reaches of the reservoir.

Mature fields already have many wells, so it is critical to optimize production and design cost-effective workovers in existing wells. A detailed field study in a mature area can distinguish good candidate wells from bad, and may provide guidance for converting more wells into good or excellent producers or for decreasing water cut, often with minimal investment. Then, a skillfully planned workover programme can boost production without the cost of drilling new wells. Monitoring can also optimize scheduling of workovers and maintenance.

Logging tools have progressed from measurement of resistivity and porosity to current methods for evaluating permeability, lithology, rock texture and bedding through millimetre-scale imaging. Elemental analysis, obtained using radioactive sources or sourceless radiation generators, can be used to determine formation lithology. Nuclear magnetic resonance (NMR) logs provide information about fluid type and saturation, and pore-size distributions. In addition, NMR correlations are used to estimate permeability. Permeability can also be measured in situ, using a formation test tool. Suction from one test probe placed against the borehole wall generates a response in several others.

An example of this technology to improve understanding of reservoir behaviour is one operator's use of a formation dynamics tester to determine permeability heterogeneity in a well in the United Arab Emirates. Horizontal permeability was determined from the pressure response of a probe at the same level as the source probe, and vertical permeability used probes 0.7 and 4.4 m above the source probe. A subsequent borehole image log clearly indicated that the location of the probes spanned two stylolite zones (Figure 4).

Even relative permeability has been measured using wireline tools. Pressure and fluid sampling data obtained with a formation tester set in the transition zone of a Middle East Cretaceous carbonate formation were used together with a single-well reservoir simulator to deduce the in-situ relative permeability curves. The simulator provided a fractional flow curve over the tested interval (Zeybek et al., 2004).

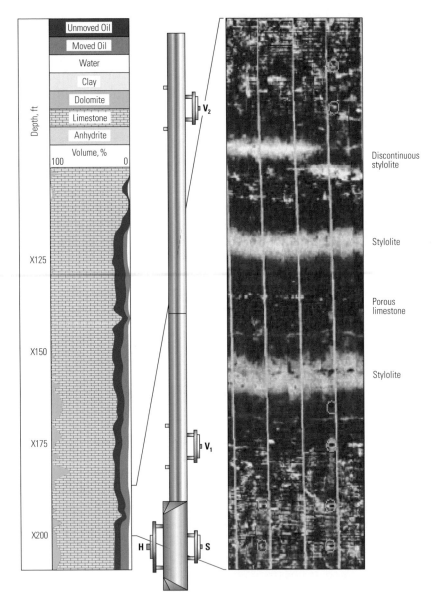

Figure 4: Volumetric Analysis and the Four-probe Formation Dynamics Tester

Volumetric analysis (*left*) and the four-probe formation dynamics tester (*middle*) set across a stylolitic interval in a United Arab Emirates well. The locations of the probes, labeled H, S, V_1 and V_2, match the imprints (circled in green) on a formation microimage log that was run after the permeability survey (*right*). The tool was set twice in this location, and both tests overlapped stylolites. (Figure used with permission, © Schlumberger, Ltd., all rights reserved.)

Formation testers have long been used to determine reservoir pressure in addition to fluid sampling and formation testing. New tools for both MWD and wireline applications are designed specifically to obtain pressure measurements quickly (Pop et al., 2005; Manin, Jacobsen and Cordera, 2005).

The locations of gas, oil and water influx can be determined by production logging, including temperature and spinner surveys and phase sensors. However, requirements for logging deviated wells differ from those for vertical wells. A combination tool with centralized spinner and widely separated phase sensors provides appropriate data in a near-vertical well with stabilized flow, but is inadequate for the complex flow patterns found in deviated wells.[5]

The FloScan Imager production logging system for horizontal and deviated wells has five small spinners and six sets of phase sensors spaced across the well diameter. The tool length is only 3.4 m, providing a significant improvement in measurement spacing compared to the 30- to 40-m length of the combination tool used in vertical wells. This new tool has revealed complex flow regimes in several wells, such as significant water circulation downhole that was not suspected (Baldauff et al., 2004). Logging tools can now be transported into a long, deviated well by wireline-conveyed tractors, a conveyance technique that has advanced in recent years.

On the surface or on the seabed, gas, oil and water phase measurements are possible using new portable, multiphase, periodic, well-testing equipment. These novel flowmeters can be mounted permanently at the wellhead, or can be mounted for short-term use. By continuously measuring the flow of each phase, the devices can reveal transient behaviour that is missed completely in a separator test. These devices allow more frequent testing that helps optimize production.

Reservoir modelling or simulation is an important part of modern reservoir management. Today, models can range from simple single-well models to very large, very complex, full-field models. Complex well configurations, such as multilateral wells, can be modelled. Large models that would have been unthinkable a few years ago are now run routinely. Ultimately, these models are used to make decisions that increase recovery and efficiency. In the near future, efforts such as the joint advanced-simulator project, called Intersect, between Chevron and Schlumberger will make it possible to model more accurately large and complex reservoirs in a fraction of the time it takes now.

---

5   'Complex Flows in Nonvertical Wells Pose Logging-Tool Challenges,' *Journal of Petroleum Technology* 54, no. 4 (April 2004), 26–27.

Compositional modelling is an advance over the older black-oil models. It is necessary when the compositions of the oil or gas phase change significantly with time, such as miscible flooding or gas recycling in a gas-condensate field.

Reservoir models are based on well information from logs and on geologic models for the interwell volumes. Going from one scale to another can be problematic. Well logs measure on the scale of inches, while a seismic data cube measures on a scale of metres. A simulation cell typically has a smaller vertical dimension — a few feet to tens of feet — than the seismic cube but larger lateral dimensions — hundreds of feet.

Through integration of asset teams, production geophysicists often work closely with reservoir engineers to generate or update a reservoir model. Software tools and workflow processes are improving the ability to move between scales and between sources of data. Decisions, such as picks for reservoir tops, can be carried over from one dataset to another. Total Exploration and Production, Balikpapan, used software workflow tools to prepare a 3D model of a field in Indonesia from all available geophysical and geological information. After testing scenarios in this model, the team used the 3D information to predict log responses along a horizontal well path. This information was used to geosteer the well. A great advantage was feeding the information from geosteering back into the model immediately to obtain updated geosteering information for the next wellbore segment (Le Turdu et al., 2004). This allowed the company to be proactive in anticipating events rather than reactive.

Integration of activities also includes the link between the reservoir and production facilities. Companies are taking a systemic approach to modelling flow from the reservoir, into the tubulars and pipelines, and onward to processing facilities. Gas-delivery contracts were the original driving force behind these models to ensure that the contracted amount could be delivered. Now, models are being used to design upgrades and optimize use of facilities to extend their useful lives.

The utility of time-lapse, or four-dimensional (4D), seismic surveys rests on the difference in seismic attributes caused by changes in fluid content or porosity. Surveys taken over an area at two different times, but in the same manner, will exhibit differences that may be discovered by subtracting the two surveys. The differences may be caused, for example, by water movement as oil and gas are produced, or by formation compaction.

Companies operating in the North Sea have been particularly active in using time-lapse seismic monitoring. They have more than recovered the cost of repeat surveys by locating bypassed oil and improving recovery. In Norne field, offshore Norway, Statoil commissioned a series of high-fidelity Q time-lapse surveys (Aronsen et al., 2004). Because of the rapid turnaround in processing − within 11 days of completion of the monitor survey − Statoil had time to adjust horizontal-well drilling plans to avoid a water zone. Processing times for this type of survey have recently improved, and results are now delivered within a few days.

BP installed permanent seabed sensors over Valhall field, also off-shore Norway, and has acquired repeat seismic surveys roughly every four months for more than a year (Barkved et al., 2005). These frequent snapshots of the field help indicate depleted zones and faults defining reservoir compartments. Chalk in the Valhall field compacts with depletion, and compaction effects appear in the time-lapse seismic studies. The surveys clearly show drainage around producing wells, including changes associated with perforated zones along horizontal wells (Figure 5) (Barkved, 2003). The studies indicated water from a waste-injection well entered not only the overburden, as planned, but also the reservoir. The company moved a well target to reduce risk from this injected waste.

Mature fields are often pressure depleted and produce with a significant water cut, making artificial lift necessary. Improving lifting costs by 1% could save the industry as much as US$ 1,000 million annually. Lower lifting costs would also allow operators to keep fields producing longer and increase ultimate recovery. Great potential exists for optimizing artificially lifted wells to significantly improve operations and increase production. Field analysis can identify underperforming artificially lifted wells that show the most promise for artificial lift optimization.

The versatility and reliability of permanent downhole gauges have improved during the past decade. New sensors, such as distributed temperature sensors using fiber-optic cables, are available. These permanently installed temperature sensors can measure temperature at intervals of every metre along the length of an installed fiber-optic cable, to indicate the location of fluid influx into a wellbore, show the inflow pattern changes with time, measure the potential for behind-casing crossflow, and even give early warning of problems with downhole valves (Walker, McAllister and Brown, 2004).

Downhole flow-control valves make it possible to choke back or shut off individual multilaterals or perforated zones. Saudi Aramco has begun installing hydraulic flow-control valves in multilateral maximum reservoir-contact wells to provide a degree of independent control over each lateral (Al-Dossary et al., 2005). Each of these valves has several

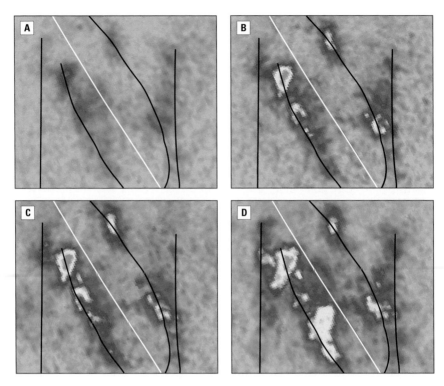

Figure 5: Time-lapse Seismic Difference Maps

Five horizontal wells traverse this area of Valhall field, offshore Norway. The baseline survey for all four images was taken in October 2003. The time-lapse periods were to (A) March 2004, (B) June 2004, (C) September 2004 and (D) March 2005. The colour indicates compaction due to depletion in the reservoir, with grey signifying noise and increasing compaction denoted by orange to yellow and then green. Compaction is a major source of reservoir energy in this field. The wellpath running diagonally across the area (white) is an old well that is no longer producing. It created a compacted, lower pressure zone delaying further compaction, which is evident in images (B) and (C). Time-lapse sequences like this help BP understand compaction in the field and locate new wells in undrained areas. (Figure used with permission, © BP, thanks to Olav Barkved.)

choke settings that can be changed through a hydraulic control line to surface, increasing the value of continual monitoring of wells.

Many terms have been used to describe fully instrumented and monitored fields, including iField, eField, digital oil field, and smart or intelligent oil field. Fundamentally, these terms refer to a field with monitoring devices downhole and at surface, a data-gathering system, software with sufficient intelligence to indicate problems, and control devices to act on the information obtained. Elements of such a system are in place at different locations, but there has been no large-scale implementation of real-time data delivery in an intelligent oil field. Companies are still assessing the trade-off between the cost of permanently installed monitoring systems and increased productivity from wells. Many companies are, however, implementing operation support centres, which is an important step toward development of an intelligent oil field.

6.   Specific Technologies for Carbonate Rocks

Carbonate formations contain about two-thirds of the world's oil and gas reserves. They range from some of the most prolific producers to some of the most difficult to produce. These formations tend to be more complex and heterogeneous than sandstone reservoirs, with more variability at all scales. Porosity and pore shapes can vary abruptly either because of deposition or later diagenesis and cementation. Calcite and dolomite, the two primary constituents of carbonate rocks, are more soluble than quartz. Large-scale dissolution creates stylolites that can dramatically decrease vertical permeability.

The pore surfaces in carbonate rocks have a greater tendency to be mixed- or oil-wetting than the typically water-wetting sandstone grains. With oil adhering to the grains, rather than sitting in the centre of the pores, the flow characteristics are different, and some of the standard logging correlations do not work.

Mechanically, carbonate rocks are stiffer than sandstones. This increases the tendency to fracture, which affects flow properties. The fractures tend to align, creating a preferential flow path. Sonic tools have been developed to show which fractures are open and conducting.

To find more reserves in carbonates or extend the life of fields, it is essential to determine the depositional environment and the texture of the rocks. Although carbonate classification − such as mudstone, wackestone, packstone, grainstone, boundstone and crystalline classes as proposed by R. J. Dunham in 1962 − is a first and necessary step, petroleum geoscientists have found it difficult to develop a system that provides a universal predictive tool for carbonate reservoir management.

Recent advances in wireline logging that provide high vertical resolution, on the order of 0.2 in., help resolve the small and abrupt changes in porosity in carbonate rocks. With high-resolution imaging, petrophysicists can identify thin-bed sequences and secondary porosity, and can characterize the rock types and facies (Dennis et al., 2003).

NMR logs now provide measurements of effective and total porosity, capillary-bound and producible porosity, permeability and pore-size distributions. Combining NMR with borehole imaging logs can aid in determining pore sizes from micro- to megapores, including fractures that intersect the borehole. NMR also has the potential to indicate wettability.

Reservoir simulators use cells large enough to require averaging of small-scale heterogeneities. This upscaling can be nontrivial, particularly for multiphase-flow properties. Heterogeneities at larger scales can be incorporated into models derived from log and seismic inputs.

Stimulating production in carbonate oil and gas fields is different from that in sandstone reservoirs, because of the highly reactive chemistry of the constituents. Acidizing to create wormholes, acid fracturing and matrix acidization provide means to improve flow from wells. Hydraulic fracturing in carbonates can be difficult if the formation is naturally fractured. The fracture is unlikely to grow in the predicted fashion.

Coiled tubing can be used to control the location of treatment, or an operator can preflush with ball sealers that preferentially plug high-permeability zones. Alternatively, chemical diverters, such as viscoelastic surfactants and foam, can be used. Significant improvements in surfactant technology have been made over the past few years.

7.   Technology for the Future

Experience has shown that additional reserves can often be found, accessed and recovered by appropriate application of existing technologies. As demand for energy continues to grow, however, new technologies will be necessary to increase exploration success and increase recovery.

In carbonate reservoirs, seismic imaging can be difficult and time-lapse seismic surveys even more so, for example due to overlying anhydrites. Improved data sampling and more data channels will increase the ability to find small accumulations in both carbonate and sandstone formations, extending basin life. Scientists are working to improve the use of seismic attributes to obtain better estimators for porosity, water and oil saturation, and pressure. Integration of these attributes with well-log data will provide more accurate estimations far from the wellbore. Use of multiazimuthal seismic acquisition has potential for determining fracture direction and orientation in carbonate reservoirs. This involves laying seismic lines in several directions over the same area. The importance of understanding preferential flow directions makes this a key technology to watch.

Recent results from a seven-year research project have shown the viability of an intelligent drillstring. This system incorporates high-speed data lines in each joint of drillpipe, along with an inductive coupling at each tool joint. Two-way communications have data rates approaching 2 megabits per second. In addition to communications with the bottomhole assembly, this system can connect hundreds of individual measurement nodes (Reeves et al., 2005). Application of this technology could provide a tremendous increase in the quantity of information that can be transmitted from a bottomhole assembly to surface.

Time-lapse seismic studies are beginning to expand from the North Sea to other parts of the world. In these new areas, baseline surveys should be obtained as soon as possible to facilitate determination of subsequent changes. One practice not yet implemented is repeat-logging surveys conducted during well workovers to tie changes in well data to changes revealed by time-lapse seismic data. Reservoir models capable of directly importing time-lapse seismic results are not yet available, but reservoir-management decisions could be made more rapidly with such a tool.

Faster, scaleable computers have created an opportunity for significant performance gains in reservoir modelling, which is an important tool for optimizing field recovery. The evolution of real-time reservoir management fosters the need for faster reservoir modelling to ultimately improve recovery rates. Reservoir engineers will be able more accurately to simulate flow in detailed geological models and around wells, model complex recovery processes and optimize coupled models of the subsurface and facilities. In addition, engineers and geoscientists will be better able to explore and quantify the impact of uncertainties in reservoir parameters by investigating many more dynamic scenarios than is currently possible.

The list of future technologies mentioned here barely scratches the surface of ideas under development. For example, PDO used a downhole array of permanent electrodes to monitor the approach of a flood front, but this technology has not yet been widely implemented (Van Kleef, et al., 2001). Developers are working to make logging and MWD tools that acquire more information, more quickly, and with higher resolution and deeper readings. For mature fields, the drive is to have the same tools as today, but with smaller diameter. The drill-bit seismic survey, an old idea that uses the noise of a bit as a wave source along with permanently instrumented seismic sensors in fields, is again being developed to provide a real-time update in determining distance and direction to a target.

For many years, the industry has increased productivity through the introduction of new technology, yet the future is likely to require greater gains in productivity than have been achieved in the past.

Ultimately, the ability to rapidly optimize management of a field requires implementation of an intelligent, digital oil field. Operation support centres, a current reality, are likely to develop into hubs for an intelligent oil field. Complex wells are being steered actively with important data streamed to surface; intelligent completion technology is proven; and remote monitoring and control of production are developing. The benefit of the intelligent oil field lies in the ability of these field technologies to exchange information and knowledge, and for that information to be communicated to the business side of the oil and gas industry.

We are now entering an era in which process change must take place to ensure the sustainability of carbon-based energy at an affordable price. The building blocks are already available, but the industry must focus on the business case for their deployment and be bold enough to move forward with implementation.

References

Al-Dossary, A.S., Salamy, S.P., Mubarak, H.K., Al-Aqeel, S.A. and Boyle, M.S. (2005), 'First Installation of Hydraulic Flow Control System (Smart Completion) in Saudi Aramco,' paper SPE 93183, presented at Asia Pacific Oil & Gas Conference and Exhibition, Jakarta, April 5–7, 2005.

Amin, A., Riding, M., Shepler, R., Smedstad, E. and Ratulowski, J. (2005), 'Subsea Development from Pore to Process,' *Oilfield Review* 17, no. 1 (Spring), 4–17.

Aronsen, H.A., Osdal, B., Dahl, T., Eiken, O., Goto, R., Khazanehdari, J., Pickering, S. and Smith, P. (2004), 'Time Will Tell: New Insights from Time-Lapse Seismic Data,' *Oilfield Review* 16, no. 2 (Summer), 6–15.

Atkinson, I., Berard, M., Hanssen, B-V. and Segeral, G. (1999), 'New Generation Multiphase Flowmeters from Schlumberger and Framo Engineering AS,' paper presented at the International North Sea Flow Measurement Workshop, Oslo, Norway, October 25–28, 1999.

Baldauff, J., Runge, T., Cadenhead, J., Faur, M., Marcus, R., Mas, C., North, R. and Oddie, G. (2004), 'Profiling and Quantifying Complex Multiphase Flow,' *Oilfield Review* 16, no. 3 (Autumn), 4–13.

Barkved, O.I., Kommedal, J.H., Kristiansen, T.G., Buer, K., Kjelstadli, R.M.,

Haller, N., Ackers, M., Sund, G. and Bakke, R. (2005), 'Integrating Continuous 4D Seismic Data into Subsurface Workflows,' paper C001, presented at the combined 67th EAGE Conference & Exhibition and SPE 14th Europec Biennial Conference, Madrid, Spain, June 14–16, 2005.

Barkved, O., Buer, K., Halleland, K.B., Kjelstadli, R., Kleppan, T. and Kristiansen, T. (2003), '4D Seismic Response of Primary Production and Waste Injection at the Valhall Field,' paper A-22, presented at the 65th EAGE Conference and Exhibition, Stavanger, June 2–5, 2003.

BP Statistical Review of World Energy (June 2005).

Copercini, P., Soliman, F., Gamal, M.E., Longstreet, W., Rodd, J., Sarssam, M., McCourt, I., Persad, B. and Williams, M.(2004), 'Powering Up to Drill Down,' *Oilfield Review* 16, no. 4 (Winter), 4–9.

Daveridge, S., Hiebert, S., Al Harthy, M., Griffiths, R., Edwards, J., AlHadrami, M., Taddei, G., Bazara, M. and Greiss, R. (2005), 'An Innovative Business Model to Leverage Innovative Well-Placement Technology,' paper OTC-17591-PP, presented at the Offshore Technology Conference, Houston, May 2–5, 2005.

Dennis, B., Akbar, M. and Petricola, M. (2003), 'Intrinsic Potential', *Middle East and Asia Reservoir Review*, no.4, 2–21.

Eiken O, (2003), 'Improvements in 4D Seismic Acquisition,' *World Oil* 224, no. 9 (September), 23–27, available at *www.worldoil.com/magazine/MAGAZINE_DETAIL.asp?A* (accessed August 17, 2005).

Hiebert, S., Mazrui, M., Al-Harthy, M., Harris, K., Shuster, M., Hadhrami, M.A. and Blonk, B. (2005), 'The Malaan Area of Northern Oman: PDO Opens up an Unconventional Play,' paper SPE 93793, presented at the SPE Middle East Oil and Gas Show and Conference, Bahrain, March 12–15, 2005.

International Energy Agency (IEA) (2005), 'Resources to Reserves – Oil & Gas Technologies for the Energy Markets of the Future', 70.

Jha, M., Tran, T., Hagtvedt, B., Al-Haimer, M. and Al-Harbi, M. (2005), 'Development of Low-Permeability Carbonate Reservoir by Use of Horizontal Wells in Mature South Umm Gudair Field in the Neutral Zone – A New Approach,' paper SPE 94249, presented at the SPE Production and Operations Symposium, Oklahoma City, Oklahoma, USA, April 17–19, 2005.

Le Turdu, C., Bandyopadhyay, I., Ruelland, P. and Grivot, P. (2004), 'New Approach to Log Simulation in a Horizontal Drain – Tambora Geosteering Project, Balikpapan, Indonesia,' paper SPE 88448, presented at the SPE Asia Pacific Oil and Gas Conference and Exhibition, Perth, Australia, October 18–20, 2004.

Manin, A., Jacobson, J.R. and Cordera, A. (2005), 'A New Generation of Wireline Formation Tester,' *Transactions of the SPWLA 46th Annual Logging Symposium*, New Orleans, June 26–29, 2005, paper M.

Nughaimish, F.N., Faraj, O.A., Al-Afaleg, N. and Al-Otaibi ,U. (2004), 'First Lateral-Flow-Controlled Maximum Reservoir Contact (MRC) Well in Saudi Arabia: Drilling & Completion: Challenges & Achievements: Case Study,' paper IADC/SPE 87959, presented at the IADC/SPE Asia Pacific Drilling

Technology Conference and Exhibition, Kuala Lumpur, September 13–15, 2004.

Pop, J., Laastad, H., Eriksen, K-O., O'Keefe, M., Follini, J-M. and Dahle, T. (2005), 'Operational Aspects of Formation Pressure Measurements While Drilling,' paper SPE/IADC 92494, presented at the SPE/IADC Drilling Conference, Amsterdam, February 23–25, 2005.

Rached, G., El-Emam, A. and Anderson, B. (2005), 'KOC Pioneers Single-Sensor Seismic,' *Hart's E&P* 78, no. 7, 68–69.

Reeves, M.E., Payne, M.L., Ismayilov, A.G. and Jellison, M.J. (2005), 'Field Trials Look at Viability, Potential of Intelligent Drill String,' *The American Oil & Gas Reporter* 48, no. 4 (April), 60–69.

Reinertsen, Ø (2003), 'Development of the Tampen Area, Value Creation in a Mature Region,' presented at Statoil Capital Markets Day, June 12, 2003, http://www.statoil.com/fin/nr303094.nsf/Attachments/gullfaks.pdf/$FILE/gullfaks.pdf (accessed September 28, 2005).

Saleri, N.G., Salamy, S.P., Mubarak, H.K., Sadler, R.K., Dossary, A.S. and Muraikhi, A.J. (2004), 'Shaybah-220: A Maximum Reservoir Contact (MRC) Well and Its Implications for Developing Tight-Facies Reservoirs,' *SPE Reservoir Evaluation & Engineering* 7, no. 4 (August): 316–320; also published as paper SPE 81487, presented at the SPE 13th Middle East Oil Show & Conference, Bahrain, April 5–8, 2003.

Shepard, S.F., Reiley, R.H. and Warren, T.M. (2001), 'Casing Drilling: An Emerging Technology,' paper IADC/SPE 67731, presented at the SPE/IADC Drilling Conference, Amsterdam, February 27–March 1, 2001.

Tchambaz, M. and Belhadj, B. (2003) 'Reservoir Monitoring with the CHFR in Water Injection Reservoirs, Libya Examples,' paper SPE 84598, presented at the SPE Annual Technical Conference and Exhibition, Denver, October 5–8, 2003. (CHFR is a mark of Schlumberger.)

Van Kleef, R., Hakvoort, R., Bushan, V., Al-Khodhori, S., Boom, W., de Bruin, C., Kabour, K., Chouzenoux, C., Delhomme, JP., Manin, Y., Pohl, D., Rioufol, E., Charara, M. and Harb, R. (2001), 'Water Flood Monitoring in an Oman Carbonate Reservoir Using a Downhole Permanent Electrode Array,' paper SPE 68078, presented at the SPE Middle East Oil Show, Bahrain, March 17–20, 2001.

Walker, I., McAllister, K. and Brown, G. (2004), 'Light Fantastic,' *Middle East & Asia Reservoir Review* no. 5, 4–21.

Zeybek, M., Ramakrishnan, T.S., Al-Otaibi, S.S., Salamy, S.P. and Kuchuk, F.J. (2004), 'Estimating Multiphase-Flow Properties from Dual-Packer Formation-Tester Interval Tests and Openhole Array Resistivity Measurements,' *SPE Reservoir Evaluation & Engineering* 7, no. 1 (February), 40–46; originally published as paper SPE 71568, presented at the SPE Annual Technical Conference and Exhibition, New Orleans, September 30–October 3, 2001.

CHAPTER 7

SOME ASPECTS OF THE CLIMATE CHANGE ISSUE

Benito Müller

1. Some Scientific Issues

1.1 The Phenomenon[1]

Global climatic changes are nothing new. The last 500 millennia have seen regular cycles in the Earth's climate, alternating between ice-ages and inter-glacial periods (Figure 1). Indeed, everything else being equal, evidence suggests that we are at the peak of one of these main interglacial periods, which accounts for the worry in the late 1970s about the onset of another ice-age (see, e.g., Hoyle, 1981). Yet these worries were not particularly acute. After all, the main cycle – with a temperature variation of 12°C – has a cooling period of over 80 thousand years. '*Après nous le déluge*' becomes less problematic at these time-scales, both as statement and as attitude.

This situation, however, has changed dramatically, as witnessed in the recent *Third Assessment Report* of the Intergovernmental Panel on Climate Change (IPCC) (see, for example, Depledge, 2002). The global average surface temperature – having increased by about 0.6°C over the twentieth century – is projected to increase between 1.4 to 5.8°C over this century, at a rate 'very likely [≥90%] to be without precedent during at least the last 10,000 years'. The threat of an impending ice-age has given way to concerns about much more immediate climatic changes in the 'opposite direction'. The reason is that in the course of the last century, mankind has unintentionally become a force to be reckoned with in influencing the Earth's climatic system. It graduated – or blundered – from 'climate-taker' to 'climate-maker'.

---

1   This section is based on Benito Müller, 'The Global Climate Change Regime: Taking Stock and Looking Ahead' in Olav Schram Stokke and Øystein B. Thommessen (eds.), *Yearbook of International Co-operation on Environment and Development* 2002/2003. London: Earthscan, 2002: pp. 27–38.

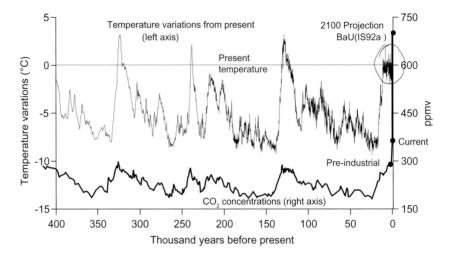

Sources. Pre-historic Temperature and CO_2 Concentrations: Petit et al. (1999); CO_2
Concentrations: Pre-industrial (= 280ppm), Current (1998 = 365ppm),
2100 Projections (= 540 - 970ppm, IS92a = 710ppm): IPCC TAR1

Figure 1: CO_2-Concentrations and Temperature Variations (from Present)

1.1.1 Fundamental Distinctions

The most general distinction between the causes of the current climatic
changes is thus between 'natural' on the one hand, and 'anthropogenic'
('human-induced', 'man-made'), on the other. A paradigm of natural
climate variations are the ice-age cycles of geological time scales, some
of which prove to be closely correlated with anomalies in the terrestrial
orbit (see, for example, Imbrie and Palmer Imbrie, 1997). Yet there
are other natural causes which can lead to changes in regional and
global climates.

Take the phenomenon of 'volcanic winters'. The sulphur dioxide
emissions of the volcanic eruption on the Aegean island of Thera
(Santorini) in 1628 BC (Manning, 1999), for example, have been used
to explain the average global cooling of 1.5°C over the following one
hundred years,[2] which, in turn, has been suggested as one of the key
factors in the downfall of the Minoan civilization during the first half

---

2   1647BC: +0.65°C, 1559BC: –0.9°C, relative to present. J.R. Petit, et al., (1999).
Data Source: 'Historical Isotopic Temperature Record from the Vostok Ice
Core' *<http://cdiac.esd.ornl.gov/ftp/ trends/temp/vostok/vostok.1999.temp.dat>*.

of the sixteenth century BC.[3] Other natural climate change events have been identified as having had equal, if not worse social impacts – the 3 to 5°C cooling following the Toba (Indonesia) eruption of about 73 thousand years ago apparently almost spelled the end of humankind (Rampino and Ambrose, 2000:71).

Anthropogenic causes, in turn, are largely based in human energy-use and agricultural practices relating to the emission of greenhouse gases. Rice cultivated under flooded conditions generates methane emissions into the atmosphere due to the decomposition of organic matter. Deforestation reduces the absorption of carbon dioxide (CO_2) by growing vegetation. However, the biggest anthropogenic cause of climate change by a long way is not these agricultural practices, but the use of fossil carbon – coal, oil and gas – as combustion fuels in all economic sectors: transport, domestic heating, industrial production, electricity generation, and so on.

There will obviously be differences in the relative shares of CO_2 emissions for these sectors within a country, but arguably the most significant differences are not within but between countries. In 1998, for example, the CO_2 emissions per head of population ranged from 20,000kg for the United States at one end of the spectrum, to least developed countries such as Sierra Leone with 110kg, at the other (Marland, Boden and Andres, 2002). Given the importance of energy in economic growth and the historic worldwide reliance on fossil energy sources, it will not be surprising to find (Figure 2) that over the last century, industrialized countries (the 'North' being OECD and the economies in transition of the former Soviet Union and Eastern Europe) have collectively emitted five times the emissions of the developing world (the 'South'),[4] a fact which gives some idea about the regional distribution of causal responsibilities for (potentially inevitable) anthropogenic climate change impacts.[5]

The reason for drawing the distinction between anthropogenic and

---

3   '...the eruption on Thera could have lowered annual average temperatures by 1 to 2 degrees across Europe, Asia and North America. ...the summer temperatures would have dropped more – suggesting years of cold, wet summers and ruined harvests', J. Cecil (2001). For more details on the eruption see Chapter 5 of Floyd W. McCoy and Grant Heiken (2000).
4   Source: World Resources Institute (WRI) 'Contributions to global warming map' <*http://www.wri.org/ climate/contributions_map.html*>.
5   However, one has to be cautious in interpreting such figures. If, for example, one is like me of the opinion that these responsibilities need to be compared in terms of average yearly per capita emissions, the Northern responsibility increases to fifteen fold that of the South.

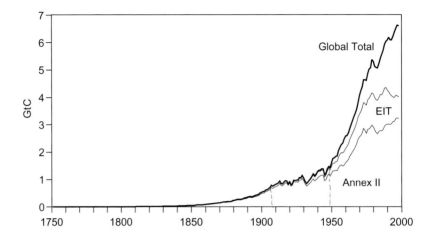

'Annex II' = 1990 OECD, 'EIT' = Economies in Transition (FSU and Eastern Europe)

Source: G. Marland, T.A. Boden, and R.J. Andres (2001), 'Global, Regional, and National Annual CO_2 Emissions from Fossil-Fuel Burning, Cement Production, and Gas Flaring: 1751–1998 (revised July 2001)', <*http://cdiac.ornl.gov/ftp/ndp030/region98.ems*>

Figure 2: CO_2 Emissions. Fossil-Fuel Burning, Cement Manufacture, and Gas Flaring. 1751–1998

natural causes lies in the possibility of attacking a root cause of the problem: while it is well within our ability to reduce greenhouse gas emissions, it is unlikely that our 'geo-engineering' skills will ever be able to control volcanic activity, let alone the terrestrial orbit around the Sun. However, people must not only be singled out as *causes* but also as *recipients* of climate change impacts. The fact is climate change is only a problem because of adverse impacts on life-systems. And this is true regardless of whether the impacts are anthropogenic or not.

As it happens, climate change impacts are divided not only with respect to their cause ('natural' versus 'anthropogenic'), but also relative to who or what they affect, namely 'social' or 'human impacts' on human systems ('Society'), on the one hand, and 'ecological ones' on natural eco-systems ('Nature') on the other. One and the same cause can obviously give rise to a variety of impacts, both on different social systems – social groups, countries or regions –, and different natural eco-systems, such as tropical rain forests or coral reefs. Giving rise to both types of impacts is common to many pollution problems. What distinguishes climate change is the nature and potential seriousness of its human impacts. They transform the issue away from a purely environmental into an environment- and development-related problem. Moreover, its anthropogenic components additionally introduce issues of interpersonal justice between those who have been causing the impacts and those who suffer them.

1.2   *Present Knowledge and Predictions: Decision-making under Uncertainty*

The debate about the scientific validity of findings – such as the one in the Third Assessment Report of the Intergovernmental Panel on Climate Change (IPPC, *www.ipcc.ch*), that

> In the light of new evidence and taking into account the remaining un-certainties, most of the observed warming over the last 50 years is likely [66–90%] to have been due to the increase in greenhouse gas concen-trations. Furthermore, it is very likely [90–99%] that the 20th century warming has contributed significantly to the observed sea level rise, through thermal expansion of sea water and widespread loss of land ice (IPCC, 2004, p.10).

– has been vociferous, although over the years the numerical strength of the 'opposing camps' has become more and more uneven. The great majority of the scientific community today tends to side with the IPCC and its findings, and reject what has become known as the 'climate sceptic' position, still upheld by a handful of individuals and institutions such as the Washington D.C. based CATO and American Enterprise Institutes (see, e.g. Moore, 2005). Yet the climate sceptics' dwindling number has by no means diminished the strength of their belief in the correctness of their views.

Indeed, two of the best publicized climate sceptics – Patrick J. Michaels and S. Fred Singer[6] – have recently contributed to a piece (Michaels, Singer and Douglass, 2004) which provides advice 'to all who worry about global warming, … chill out. The science is settled. The 'sceptics' – the strange name applied to those whose work shows the planet isn't coming to an end – have won.' The piece attacks the IPCC for having claimed erroneously that '1) we have reliable temperature records showing how much the planet has warmed in the last century; and 2) computer projections of future climate, while not

---

6   Michaels is a senior fellow in environmental studies at the Cato Institute, research professor of environmental sciences at the University of Virginia which is also where Singer is emeritus professor of environmental sciences. Michaels, for one, has been a contributing author to the IPCC.

perfect, simulate the observed behavior of the past so well that they serve as a reliable guide for the future.'

To be more precise, their contention is that 'as a consequence of greenhouse forcing, all state-of-the-art general circulation models predict a positive temperature trend that is greater for the troposphere than the surface. [...] However, the temperature trends from several independent observational data sets show decreasing as well as mostly negative values. This disparity indicates that the three models examined here fail to account for the effects of greenhouse forcings' (Douglass, Pearson and Singer, 2004).

The disparity referred to has been known for quite some time, indeed in a different article (in the same volume), Michaels and Singer and some other colleagues claim that their 'study thus makes unlikely some of the explanations advanced to account for the disparity' (Douglass et al., 2004) on the grounds that 'the disparity does not occur uniformly across the globe, but is primarily confined to tropical regions which are primarily oceanic' (ibid.)

The gist of the argument which is meant to 'settle the science' is thus that, because certain 'independent observational data sets' disagree with a prediction of climate models (based on greenhouse gas forcing), the science based on such models is wrong. What is clear is that the situation where model predictions are significantly out of tune with observations is untenable, and something has to give. However, given the highly complex nature of some types of 'observations' it is not as self-evident as Michaels and Singer seem to think that what has to go is the model.

As it happens, satellite observations of atmospheric variables are highly technical and inferred, which is why their validity deserves equally critical analysis as that of the models in question, particularly if these models manage to replicate very well a series of averages based on observations which are undisputed – even by the climate sceptics – namely surface temperature measurements.

And as it happens, this had been done, for example, in a *Nature* article by Fu et al (2004)[7] even before the two sceptical articles appeared in *Geophysical Research Letters*. Fu and Johanson themselves re-iterate their findings in a recent GRL article (Fu and Johanson, 2005), finding that – *pace* Michaels and Singer – 'tropospheric temperature trends in the tropics are greater than the surface warming and increase with height', and that the satellite data that failed to show this increase has a 'trend bias [which] can be largely attributed to the periods when the satellites

---

7   Note the reference to 'satellite-*inferred* temperature trends'.

had large local equator crossing time drifts that cause large changes in calibration target temperatures and large diurnal drifts', which actually may well explain the sceptics' findings that the disparity in question occurs mainly in the tropical (equatorial) regions.

Following Karl Popper's falsification methodology (Popper, 1935), Michaels and Singer tried to discredit the models that have been used to establish the existence of a greenhouse gas related 'anthropogenic fingerprint' in the undisputed observed raise in global average surface temperatures, and failed. Yet, this failure by the sceptics does, of course, not mean that the anthropogenic hypothesis is proven. Indeed, to those who espouse Popper's philosophy, such a proof is simply impossible, which is why one has to take with a pinch of salt the recent (19 February 2005) front page headline in the London-based *The Independent*:

'*The final proof: global warming is a man-made disaster*'. However, it is understandable why the article by Steve Connor (Science Editor of *The Independent* in Washington DC) about a recent *Science* paper (Barnett, et al., 2005) by Tim Barnett and colleagues at the Climate Research Division of UC San Diego's Scripps Institution of Oceanography should have been given such a spectacular heading: the study confirmed the man-made contribution to climate change on the basis of oceanic and not atmospheric data, thus increasing considerably the likelihood of the anthropogenic hypothesis.

Indeed, what the authors did was to replicate the observed changes in the oceans over the last 40 years with a model including anthropogenic greenhouse gas emissions and then test a number of hypotheses, such as whether the changes could be accounted for by the natural variability of the ocean system ('nature alone'), or by changes in solar or volcanic activity, and in each case the answer was 'no'. Of course, this still does not 'prove' the hypothesis, not just because a proof in the strict sense is in principle impossible, but also because there are unknown 'quantities' that have to be taken into account.

The key – as Dave Stainforth puts it in his recent article on the subject of uncertainty – is thus to deal appropriately with the 'known unknowns'. As concerns climate science, and more particularly climate modelling, these known unknowns, according to Stainforth, are threefold:

- 'natural variability. The climate system is chaotic, which means that small changes in one location at one point in time can lead to large differences at other locations at some future point in time. This is the familiar 'butterfly effect' whereby a butterfly flapping its wings in Indonesia is said to be able to affect whether a hurricane might hit Florida at some point in the future.

- Changing boundary conditions. The climate is affected by many factors which are considered to be separate from, or outside, the climate system. These include natural factors such as volcanic eruptions and solar output, and anthropogenic factors such as the emission of greenhouse gases.
- Scientific understanding of how the climate behaves and how it responds to changing boundary conditions such as a rapid increase in atmospheric concentrations of greenhouse gases' (Stainforth, 2005).

The study of such 'known unknowns' is the theory of probability, which is why it will be no surprise that much of the recent energy of the modelling community has gone into developing probabilistic climate models, such as the 'grand ensemble' models of the Oxford University based *ClimatePrediction.net* which by using thousands and thousands of (slightly) different model-runs are trying to generate a picture of the likelihood of climate change events.

In the run-up to the recent Gleneagles G8 summit, the Academies of Science of the G8 countries, as well as Brazil, China and India, published a declaration (Royal Society et al., 2005) which accordingly acknowledged under the heading '*Climate change is real*' that

> There will always be uncertainty in understanding a system as complex as the world's climate. However there is now strong evidence that significant [anthropogenic[8]] global warming is occurring. The evidence comes from direct measurements of rising surface air temperatures and subsurface ocean temperatures and from phenomena such as increases in average global sea levels, retreating glaciers, and changes to many physical and biological systems. It is likely that most of the warming in recent decades can be attributed to human activities. This warming has already led to changes in the Earth's climate.

In conclusion, the Science Academies' declaration – referring to the G8 nations being 'responsible for much of the past greenhouse gas emissions' and recalling the UNFCCC commitment by them 'to showing leadership in addressing climate change and assisting developing nations to meet the challenges of adaptation and mitigation' – calls on world

---

8   The declaration explicitly states (footnote 1) that it uses the UNFCCC definition of 'climate change', which is 'a change of climate which is attributed directly or indirectly to human activity that alters the composition of the global atmosphere and which is in addition to natural climate variability observed over comparable time periods'.

leaders to carry out a number of actions, first and foremost among them to 'acknowledge that the threat of climate change is clear and increasing'.

As it happens, the G8 summit at Gleneagles saw the adoption of a 'Gleneagles Plan of Action', as well as a Declaration by the key developing country participants, both of which will be discussed in the next section. However, in their final Communiqué, unusually signed by each of the G8 leaders, they did indeed declare 'climate change is a serious and long-term challenge that has the potential to affect every part of the globe' and that human activities 'contribute in large part to increases in greenhouse gases associated with the warming of the Earth's surface. And equally important, they all reaffirmed their commitment to the UN Framework Convention on Climate Change and its ultimate objective of stabilizing atmospheric concentrations of greenhouse gases at levels that avoid dangerous human interference with the climate system.

2.   Recent Political Issues

=> short of Climate discussion and Policies ; 1992

The international effort to combat the adverse effects of global climate change is guided by the United Nations Framework Convention on Climate Change (UNFCCC) – one of the conventions adopted at the Rio World Summit in 1992 – and its Kyoto Protocol, adopted at the 5th Session of the Conference of Parties (COP) to the UNFCCC in Kyoto in 1997. At the time of writing almost all countries recognized by the United Nations –189 out of 193, or 98 percent – have ratified the UNFCCC, and 152 had ratified the Kyoto Protocol to that Convention. The vast majority both of the industrialized Parties with assigned targets (namely 92 percent), and of developing countries without (74 percent) have ratified the Protocol, which entered into force on 16 February 2005.

2.1   *Key Developing Country Actors*

The Broad Coalition. Faced with a number of handicaps in their ability to participate as equal partners in international negotiations – insufficient resources to participate in sufficient numbers, inadequate analytic capacity and so on (see Müller, 2003) – developing countries have generally resorted to the strategy of coalition building under a multilateral umbrella, of 'finding strength in numbers'. In the United Nations context, the main coalition that had emerged for the purpose of addressing the common development interests of non-industrialized countries is the broad coalition called 'Group of Seventy-seven and China' (G77+China), whose membership since its creation in 1964 has risen from 77 to 132 UN members.

If there is communality of interests – if all the numbers 'pull in the same direction' – a broad coalition of this size can be remarkably successful, as witnessed in the climate change negotiations concerning 'common but differentiated responsibilities', and 'right to sustainable development'.[9] The significant growth of the G77+China membership since its formation is testimony to the attraction of this broad developing country coalition, but it also increases its frailty: an increase in numbers may be an increase in strength, but only if the numbers pull in the same direction – i.e. if a strong coincidence of interests is retained.

Starting the first Ministerial Meeting of the Group in Algiers (1967), a permanent institutional structure gradually developed which led to the creation of G77 Chapters in Rome (FAO), Vienna (UNIDO), Paris (UNESCO), Nairobi (UNEP) and the Group of 24 in Washington, D.C. (IMF and World Bank). The Group's work is coordinated by an influential chairman who acts as its spokesman. The chairmanship rotates on a regional basis (between Africa, Asia, and Latin America and the Caribbean) and is held for one year in all the Chapters. At the time of writing, Jamaica holds the Chair of the Group.

Narrow Coalitions. In the climate change context, the G77+China has been put under particular strain due to different interests within the coalition. The members of the Alliance of Small Island States (AOSIS), for example, 'are particularly vulnerable to climate change because a rise in sea level could destroy or render uninhabitable all or part of their territory, [while] the members of the Organization of Petroleum Exporting Countries (OPEC), stand to lose substantial revenue from measures to avert climate change' (Chasek and Rajami, 2003, p.254) which a number of them will find difficult to cope with given their low per capita income and dependence on these revenues.

There are other Groupings of developing countries – part of, or overlapping with the G77+China – that have played a distinctive role in the UNFCCC negotiations, not least the Group of Least Developed

---

9   'For example, the G-77 and China are united in arguing that environmental rules should not hinder their ability to develop. [...] Moreover, during the climate change negotiations the G-77 and China have maintained that the historical responsibility for climate change lies with industrial countries and that these countries should bear the main responsibility for correcting the problem.'[Chasek and Rajamani, 2003, p. 255]

Countries (LDCs). To begin with we take a look at the position of the three large regional leaders of the G77+China: Brazil, China, and India.

2.1.1   Large Developing Countries: Brazil, China, India

At COP8 (2002) in New Delhi, India's position was forcefully summarized in Prime Minister Vajpayee's High Level Segment opening address:

> India's contribution – indeed, the contribution of all the developing countries – to greenhouse gas concentrations in the atmosphere is very little, compared to that of the industrialized countries. This will be the case for several decades to come. Tragically, however, developing countries will bear a disproportionate burden of the adverse impacts of climate change. Hence, it follows that there is a need to pay adequate attention to the concerns of developing countries on vulnerability and adaptation issues in the Convention process. […] There have been suggestions recently that a process should commence to enhance commitments of developing countries on mitigating climate change beyond that included in the Convention. This suggestion is misplaced for several reasons.
>
> - First, our per capita Green House Gas emissions are only a fraction of the world average, and an order of magnitude below that of many developed countries. This situation will not change for several decades to come. We do not believe that the ethos of democracy can support any norm other than equal per capita rights to global environmental resources.
>
> - Second, our per capita incomes are again a small fraction of those in industrialized countries. Developing countries do not have adequate resources to meet their basic human needs. Climate change mitigation will bring additional strain to the already fragile economies of the developing countries, and will affect our efforts to achieve higher GDP growth rates to eradicate poverty speedily.
>
> - Third, the GHG intensity of our economies at purchasing power parity is low and, in any case, not higher than that of industrialized countries. Thus, the assertion that developing countries generate GHG emissions, which are unnecessary for their economies, is not based on facts (Vajpayee, 2002).

And India retained a somewhat combatant mood at the recent UN-FCCC Seminar for Government Experts (SOGE, Bonn May 2005), which was the fruit of protracted negotiations at COP10 in Buenos Aires concerning the way in which the post-2012 issue should at present be raised, if at all, in the UNFCCC context. In his presentation, the Indian expert, Surya P. Sethi, concluded that:

- Annex I [industrialised country] commitments not met – emissions still rising, transfers of finance/technology minimal.

- Low per-capita GHG emissions in India are due to sustainable lifestyles & not poverty alone

- India is doing enough in mitigation of GHGs. Technological and Financial barriers to achieving identified energy initiatives must be removed (Sethi, 2005).

By contrast, China – while also stressing its developing country status and low per capita emissions and highlighting its existing greenhouse gas mitigation measures – recognized the need for additional measures and urged 'the international community to engage in practical technological cooperation in the future so as to combat climate change effectively and promote global sustainable development.'[10]

Two months later, on 6–8 July, the Heads of China and India – joined by those of Brazil, Mexico, and South Africa – used the occasion of their participation in the G8 Gleneagles Summit Introduction to issue a *Joint Declaration*. While somewhat overshadowed by the G8's own *Gleneagles Plan of Action*, this Joint Statement is of considerable importance because it does indicate the areas on which there is a consensus among the key representatives of the developing world.

The second paragraph of the preamble, for example, contains a very strong general endorsement of multilateralism and, indeed the UN system:

> the Gleneagles Summit is an opportunity to give stronger impetus to [the process of UN reforms aimed at providing a greater voice to developing countries in UN decision-making], and to send a positive message on international cooperation. This should be achieved through the promotion of multilateralism, the enhancement of North-South cooperation, as well as through a renewed commitment to sustainable development and the harnessing of the benefits of globalization for all.

This sentiment was again reflected in the Joint Declaration articles on climate change. Article 12, for example, states that 'the United Nations Framework Convention on Climate Change (UNFCCC) and its Kyoto Protocol establish a regime that adequately addresses the economic, social and environmental aspects of sustainable development.'

With a reference to the Principle of Common but Differentiated Responsibilities of the Framework Convention, Articles 13 and 14

---

10 Abstract for the Seminar of Government Experts May 16–17, 2005 Bonn, Germany, Submitted by the People's Republic of China, *http://unfccc.int/meetings/seminar/items/3410.php*

urge industrialized countries to 'take the lead in international action to combat climate change by fully implementing their obligations of reducing emissions and of providing additional financing and the transfer of cleaner, low emission and cost-effective technologies to developing countries' and highlight the fact that the Convention and the Kyoto Protocol 'do not provide for any quantitative targets for emission reductions for developing countries'.

Yet, significantly, Article 14 also highlights the fact that the developing countries do have commitments under these treaties, namely 'to implement appropriate policies and measures to address climate change, taking into account their specific circumstances and with the support of developed countries'. Highlighting furthermore the fact that 'the Convention establishes economic and social development and poverty eradication as the first and overriding priorities of developing countries' Article 16 consequently concludes the G8 Summit should recognize that 'there is an urgent need for the development and financing of policies, measures and mechanisms to adapt to the inevitable adverse effects of climate change that are being borne mainly by the poor'.

In conclusion (Art. 18) the Joint Declaration urges the G8 leaders and the international community 'to devise innovative mechanisms for the transfer of technology and to provide new and additional financial resources to developing countries under the UNFCCC and its Kyoto Protocol'. For this purpose, the Declaration proposes a new paradigm for international cooperation that must ensure:

• Accessibility and affordability of climate friendly technologies to developing countries (requiring 'a concerted effort to address questions related to intellectual property rights')
• Additional financial resources (over and above current ODA) to enable developing countries to access such technologies
• Encouragement of North-South collaborative research on such technologies.

2.1.2 Oil-producing Countries

a) Organization of the Petroleum Exporting Countries (OPEC). The official OPEC position – judging from the statements made by OPEC's Secretary General to the Sessions of the Conference of Parties (COP) to the UN Framework Convention on Climate Change[11] – has been remarkably stable over the last five years at least. The two key issues raised by OPEC in the negotiations have been the rejection of (new)

---

11 *http://www.opec.org/home/Environmental%20Issues/Statements/cop10.htm*

developing country commitments, and what has become known as the 'impact of response measures', both of which have led to some controversy in the negotiations.

Developing Country Mitigation Commitments. In his statement to COP8 in New Delhi (November 2002), the then OPEC Secretary General Alvaro Silva-Calderon reminded the Conference that 'we need to keep our focus firmly on the principle of "common but differentiated responsibilities". Industrialized countries, whose activities over decades – and even centuries – have been responsible for the lion's share of adverse impacts on the environment, should recognize and honour their obligation to provide the lion's share of the response measures.' The same point had already been made, in a somewhat more poignant form, at COP6 (The Hague, November 2000) by Rilwanu Lukman when he declared that 'developing countries should not be roped into making commitments to emissions-reduction targets'. And it was re-iterated by Silva-Calderón (COP9, Milan, December 2003), and Maizar Rahman (COP10, Buenos Aires, December 2004), who both pointed out that 'new commitments [for developing countries] would affect the ability of many sovereign states to achieve sustained economic growth, develop their social infrastructures and eradicate poverty.'

Impacts of Response Measures. The positions of all Parties are to a large degree shaped by their wider, or more narrow economic self-interests. There is wide-reaching consensus – both among their supporters and detractors – that OPEC and oil producers in general, and Saudi Arabia, in particular, have been extremely successful in their climate change negotiations. One of their main achievements has been the introduction of the issue of impacts of response measures in several places in the language of both the UNFCCC and the Kyoto Protocol (see Box 1).

The term 'impacts of response measures' has become shorthand for 'adverse social and economic impacts of measures taken to reduce greenhouse gas emissions (particularly in industrialized countries) on the developing countries whose economies are highly dependent on the production and export of fossil fuel'. OPEC's position concerning these impacts has been shaped by concerns about their projected size and a sense of inequity.

Following a modelling study undertaken at the OPEC Secretariat by the director of its Research Division, Shokri Ghanem, the author himself delivered the official OPEC statement to COP5 (1999, Bonn) in which he said that 'the Kyoto Protocol, if fully implemented, would lead to a dramatic loss of revenue for oil-exporting developing countries, including OPEC's own Members. The financial impact on our countries has been estimated at tens of billions of US dollars per year, according

Box 1: Language pertaining to Impacts of Response Measures

FCCC

PREAMBLE

Affirming that responses to climate change should be coordinated with social and economic development in an integrated manner with a view to avoiding adverse impacts on the latter.

ARTICLE 4: COMMITMENTS

Art 4.1. All Parties ... shall:

(g) Promote and cooperate in ... research, systematic observation and development of data archives ... intended to further the understanding and to reduce or eliminate the remaining uncertainties regarding the causes, effects, magnitude and timing of climate change and the economic and social consequences of various response strategies;

(h) Promote and cooperate in the full, open and prompt exchange of relevant ... information related to the climate system and climate change, and to the economic and social consequences of various response strategies;

Art 4.8. In the implementation of the commitments in this Article, the Parties shall give full consideration to what actions are necessary under the Convention ...to meet the specific needs and concerns of developing country Parties arising from the adverse effects of climate change and/or the impact of the implementation of response measures, especially on:

(h) Countries whose economies are highly dependent on income generated from the production, processing and export, and/or on consumption of fossil fuels and associated energy-intensive products;

Kyoto Protocol

Art. 3.14. Each Party included in Annex I shall strive to implement the commitments mentioned in paragraph 1 above in such a way as to minimize adverse social, environmental and economic impacts on developing country Parties, particularly those identified in Article 4, paragraphs 8 and 9, of the Convention. In line with relevant decisions of the Conference of the Parties on the implementation of those paragraphs, the Conference of the Parties serving as the meeting of the Parties to this Protocol shall, at its first session, consider what actions are necessary to minimize the adverse effects of climate change and/or the impacts of response measures on Parties referred to in those paragraphs. Among the issues to be considered shall be the establishment of funding, insurance and transfer of technology.

to OPEC's calculations.' This was re-iterated in the following year by Lukman, when he told COP6 that 'independent studies estimate the loss at tens of billions of US dollars per year for OPEC's Members'.

The issue of inequity was raised at a very early stage. Ghanem himself stressed (COP5) that 'a sense of equity across all nations must prevail. In short, there must be no net winners and no net losers from these negotiations, as they run their course,' a sentiment reiterated most recently (COP9) by Silva-Calderón when declaring that 'We insist once again that oil-producing developing countries do not end-up as net losers from the climate change negotiations. We are still not satisfied that our legitimate concerns about the adverse impact of response measures on our hydrocarbon-dependent economies have been properly addressed.'

Apart from this egalitarian argument, there has also been a moral argument based on the 'principle of common but differentiated responsibilities' as put forward by Lukman who reminded COP6 that 'the established industrial nations bear the principal responsibility for the purported phenomenon of global warming, and not the developing countries. The onus, therefore, is upon the rich nations to minimize and finance the negative impact of their response measures on the poor countries of the south,' a sentiment re-iterated at the subsequent COP Sessions.

Other Issues. While there has been this remarkable stability in the official OPEC position on climate change issues, certain changes albeit sometimes in nuances can be detected in the official OPEC statements over the past five years. Thus while Lukman (COP6) made reference to both 'minimizing' and 'financing' the negative impacts of response measures, later statements only contain references to the former. Alí Rodríguez Araque's COP7 statement, however, did contain the acknowledgement that 'it was encouraging, therefore, to see that the Bonn Agreement included the establishment of a Special Climate Change Fund, to assist with the diversification of economies in countries which may suffer from the adverse effects of mitigation measures'.

An issue that was raised from time to time in Climate Change negotiations is the perceived unfair, if not illegitimate, fossil fuel consumption taxation in industrialized countries. As stated at COP7: 'And, while oil is taxed so heavily, other fuels are taxed at far lower levels and are sometimes even subsidised. The time is ripe to reconsider the entire philosophy of energy taxation, by restructuring fiscal systems to address broader concerns than the financial needs of governments, and to ensure consistency with international trade rules.' The view is that the need for emission mitigation provides an additional argument against these prevailing taxes and subsidies, which are seen to favour coal over the less polluting petroleum products. Another issue has recently made its appearance in the official OPEC position: carbon sequestration. At COP8, Silva-Calderón remarked that

> ...while there is the understandable call to develop renewables, the fact remains that the technology is still in its infancy. Therefore, while the renewable energy industry is being developed, all other available resources, which are friendly to the environment, must also be accessed, enhanced and utilised to tackle the dire problems of mankind and ensure sustainable development. Petroleum will feature prominently in this. Advances in technology continue to make oil and gas cleaner fuels. The successful development of carbon dioxide sequestration technology will ensure that fossil fuels, including oil, continue to serve the needs of mankind for the foreseeable future.[12]

And this sentiment was reflected in the assertion, at both of the subsequent Sessions, that 'proven reserves of oil and gas are sufficient to meet rising world demand for decades to come, while advances in technology help them meet the toughest environment regulations and make a substantial contribution towards sustainable development'.

And finally, at the same last two Sessions, OPEC reaffirmed its commitment to its policy of promoting clean fossil fuel technology and market stability 'in the interests of rich and poor nations alike, with secure supply, reasonable prices and fair returns for investors' (Silva-Calderon, COP9) with the proviso that 'to be effective in this, however, requires steady, predictable demand, built upon a clear, definitive vision of the evolution of the global environment in the years to come' (Silva-Calderon, COP9 and Rahman COP10).

b) Organization of Arab Petroleum Exporting Countries (OAPEC). After the ratification of the Kyoto Protocol by the Russian Duma on 22 October 2004, the OAPEC Editorial in the November edition of their Monthly Bulletin was dedicated to the 'UNFCCC and the Kyoto Protocol' (OAPEC, 2004). In it, the OAPEC General Secretariat tells the reader – in consonance with the OPEC position – that 'the Russian Federation's accession to the Protocol and the Protocol's implementation will have repercussions for all fossil fuel exporting countries, but especially OAPEC member countries'.

Yet, the editorial is quite circumspect as concerns the level of these repercussions:

---

12 Given the importance of this to fossil fuel producers, the issue of carbon sequestration is treated in a separate next Chapter by Professor Lackner.

however, it is difficult to predict with a high degree of accuracy a drop in oil demand due to the implementation of the protocol. Forecasts show that demand in developing countries will rise as they strive to implement development programs that enhance their peoples' living conditions. At the same time, demand growth rates are predicted to fall compared to the situation if the Protocol is not implemented. The Climox model shows that OPEC revenues will grow 65% between 1995 and 2010 in the base scenario, while the growth rate drops to 49% if the Kyoto Protocol is implemented [see Bartsch and Müller, 2000, 211].

Interestingly, the editorial does not stop at this point, but highlights some non-contentious options in which these expected adverse impacts of response measures could be minimized. For one, the editorial highlights that 'studies show that OPEC countries' revenue loss will be less if emissions trading is employed, although an OPEC study predicts that revenues will drop by a half if this mechanism is implemented fully'.

Moreover, the editorial also emphasizes that the loss of OPEC countries

will be reduced considerably if the clean development mechanism is implemented, which OAPEC member countries are allowed to employ in several areas related to oil projects, such as curbing flared gas, cutting emissions and pollution from various branches of the oil industry, using clean technology, producing clean fuel, and conserving energy and rationalizing consumption in energy consuming industries.

The editorial by the General Secretariat concludes by stressing,

the importance of member countries joining the Protocol so as to participate fully in the meetings of the Conference of Parties, which operates as an assembly for the parties to the Protocol. At its first meeting the Conference of Parties is expected to discuss several draft resolutions relating to the implementation of certain articles in the Convention and the Protocol that protect OAPEC members' vital interests. They should also take advantage of the clean development mechanism for the mutual benefit of petroleum exporting countries and consuming industrial countries.

Before the Russian Ratification, with its consequent entry-into-force of the Kyoto Protocol, the only OAPEC Member which had adopted the Protocol was Tunisia (22 January 2003), although Egypt had signed it in 1999. Since that time, all OAPEC Members eligible to adopt it – except Bahrain and Libya – have indeed acceded to it in rapid succession: Qatar (11 January 2005), United Arab Emirates (26 January), Saudi Arabia (31 January), Algeria (16 February), Kuwait (11 March).

2.2   The Lead Industrialized Protagonists

Among the 38 industrialized Parties – listed in Annex I of the UN-FCCC – who did put their signature to the emission targets of the Kyoto Protocol (specified in its Annex B, see Table 1), there is a by now deep-seated divide between the 35 (spearheaded by the EU) who have ratified the treaty, and the two (led by the USA) who have repudiated it, with only Monaco still making up its mind.

Table 1:   Kyoto Protocol. Status of Ratification and Annex B Targets*

Pro	%		%	Contra	%
Austria	92	Latvia	92		
Belgium	92	Liechtenstein	92	Australia	108
Bulgaria	92	Lithuania	92	USA	93
Canada	94	Luxembourg	92		
Croatia	95	Netherlands	92	**Undecided**	
Czech Republic	92	New Zealand	100		
Denmark	92	Norway	101	Monaco	92
Estonia	92	Poland	94		
EU	92	Portugal	92		
Finland	92	Romania	92		
France	92	Russia	100		
Germany	92	Slovakia	92		
Greece	92	Slovenia	92		
Hungary	94	Spain	92		
Iceland	110	Sweden	92		
Ireland	92	Switzerland	92		
Italy	92	Ukraine	100		
Japan	94	UK	92		

* = % relative to 1990 levels

2.2.1   The United States I: The Asia-Pacific Partnership on Clean Development and Climate

The most significant step with regard to the international climate change debate taken by the Kyoto repudiators – i.e. the USA and Australia – is no doubt the *Asia-Pacific Partnership on Clean Development and Climate*. Robert Zoellick, US Deputy Secretary of State, who formally announced the pact at the sidelines of the Association of South-East Asian Nations meeting in Vientiane, Laos, was at pains to emphasize that the agreement was not in direct competition to the Kyoto protocol ('We view this as a complement, not an alternative'[13])

According to a White House Fact Sheet, the Partnership 'will focus on voluntary practical measures taken by these six countries in the Asia-Pacific region to create new investment opportunities, build local capacity, and remove barriers to the introduction of clean, more efficient technologies [and] help each country meet nationally designed strategies for improving energy security, reducing pollution, and addressing the long-term challenge of climate change.'[14]

The aims of the Partnership were further elaborated in a *Vision Statement* according to which it 'will collaborate to promote and create an enabling environment for the development, diffusion, deployment and transfer of existing and emerging cost-effective, cleaner technologies and practices'. Areas for near-term collaboration mentioned included: energy efficiency, clean coal, integrated gasification combined cycle, liquefied natural gas, carbon capture and storage, combined heat and power, methane capture and use, civilian nuclear power, geothermal, rural/village energy systems, advanced transportation, building and home construction and operation, bio-energy, agriculture and forestry, and hydropower, wind power, solar power, and other renewables. Medium- to long-term collaborations were envisaged on hydrogen, nano-technologies, advanced biotechnologies, and next generation nuclear fission, and fusion energy.

According to *The Guardian* 'the existence of the pact, and the fact it was designed as an alternative to Kyoto, were disclosed by Australia's environment minister, Senator Ian Campbell'[15] the day before Mr Zoellick's official announcement. CNN added that 'Canberra and Washington had negotiated the new agreement for the past 12 months among the countries accounting for 40 percent of the world's

---

13 'Six nations agree new climate pact' CNN, 28 July 2005 *http://edition.cnn. com/2005/WORLD/asiapcf/ 07/28/sixnations.climate.ap*

14 White House Fact Sheet: President Bush and the Asia-Pacific Partnership on Clean Development, Office of the Press Secretary, 27 July 2005, *http://www. whitehouse.gov/news/releases/2005/07/print/20050727-11.html*

15 'US in plan to bypass Kyoto protocol' *The Guardian*, Paul Brown and Jamie Wilson in Washington, 28 July 2005 *http://www.guardian.co.uk/international/ story/0,,1537565,00.html*. 'He said: "It is quite clear that the Kyoto protocol won't get the world to where it wants to go. We have got to find something that works better. We need to develop technologies which can be developed in Australia and exported around the world – but it also shows that what we're doing now, under the Kyoto protocol, is entirely ineffective. Anyone who tells you that the Kyoto protocol, or signing the Kyoto protocol is the answer, doesn't understand the question." Kyoto would fail because "it engages very few countries, most of the countries in it will not reach their targets, and it ignores the big looming problem – that's the rapidly developing countries".'

greenhouse gas emissions. The pact was finalized during secret talks in Honolulu on June 20–21, a diplomat said, speaking on condition of anonymity.'[16]

While the Vision Statement stresses in its ultimate paragraph that 'the partnership will be consistent with and contribution to our efforts under the UNFCCC and will complement, but not replace, the Kyoto Protocol' – a fact emphasized by Zoellick in his official announcement ('We are not detracting from Kyoto in any way at all. We are complementing it.'[17]) – comparisons with Kyoto were inevitable, especially after Australian Prime Minister John Howard said: 'The fairness and effectiveness of this proposal will be superior to the Kyoto protocol.'[18]

However, neither of Zoellick's assurances – even after having been echoed by Alexander Downer, Australia's foreign minister[19] – convinced the European media that the Partnership was not intended as an attack against the Kyoto Protocol, indeed, more generally against the involvement of the United Nations in the effort to deal with global climate change, as can be gauged from the subsequent press headlines:

'US unveils alternative plan to Kyoto treaty' *Financial Times*
'Asia deal on table to counter Kyoto' *Financial Times*
'US in plan to bypass Kyoto protocol' *The Guardian*
'Le pacte climatique Asie-Pacifique « supérieur » à Kyoto (Canberra)'
    Le Monde[20]
'Bush startet Alternative zu «Kyoto»' *Neue Zürcher Zeitung*[21]
'Clima, accordo a sei parallelo a trattato di Kyoto' *Corriere della Sera*.[22]

Initial official reaction from outside the Partnership was less sceptical about the relation of the Partnership and the UN climate change regime. Thus the UK government 'welcome[d] any action taken by

16 'Six nations agree new climate pact' CNN, 28 July 2005 *http://edition.cnn. com/2005/WORLD/asiapcf/ 07/28/sixnations.climate.ap*

17 'US unveils alternative plan to Kyoto treaty', *Financial Times*, July 27 2005 20:11 *http://news.ft.com/ cms/s/c4ed87d8-fed1-11d9-94b4-00000e2511c8.html*

18 Alister Doyle, 'U.S.-led climate plan won't supplant Kyoto –experts', Reuters, 28 July 2005, *http://go.reuters.co.uk/newsArticle.jhtml?type=scienceNews&storyID=920 2317&section=news&src=rss/uk/scienceNews*

19 'U.S., Australia deny climate deal threat to Kyoto' Reuters Alert Net 28 July 2005 *http://www.alertnet. org/thenews/newsdesk/BKK265196.htm*

20 28 July 2005 *http://www.lemonde.fr/web/depeches/0,14-0,39-25409998@7-50,0. html*

21 28 July 2005, *http://www.nzz.ch/2005/07/28/al/newzzEBOI2VIA-12.html*

22 28 July 2005, *http://www.corriere.it/Primo_Piano/Esteri/2005/07_Luglio/28/kyoto. shtml*

governments to reduce greenhouse gases [...] The announcement from Australia and others certainly does not replace the Kyoto process. Kyoto represents a historic first step in world cooperation but needs to be built on post 2012 – that process continues in Montreal later this year. We made excellent progress on climate change at Gleneagles.'[23]

Barbara Helferrich, the European Commission's environment spokeswoman, in turn, welcomed the initiative but cautioned that it 'has to be seen in a global context. [...] If it is simply technology and clean coal, it is no substitute for agreements like the Kyoto Protocol and we do not expect it to have a real impact on climate change. There will have to be binding global agreements, but on what scale and what basis is yet to be decided.'[24]

2.2.2 The United States II: Domestic Sub-national Initiatives

However, there are a number of domestic activities happening in the USA that must be kept in mind, if only because most of the key environmental and other actions by the Federal government in Washington DC – with or without international aspects – are primarily due to precisely such domestic activities: given the current geo-political constellation, the only way in which Washington can be pressured into doing something is by its domestic constituencies. The activities currently underway are actually too numerous to discuss in this context, as even the listing of some of the activities and actions initiated at state level in the present year alone in Box 2 will demonstrate. These, and the numerous other current initiatives, not just at state, but at community level and in the corporate sector will make it very likely for the federal government in Washington sooner or later to be faced with sufficient pressure to harmonize these activities ('create a level playing field') to introduce binding domestic targets at the federal level.

2.2.3 The European Union

Probably the most important recent 'domestic' step in the EU was the launch of the European Emission Trading Scheme (EU ETS). In June 2003, the European Council and Parliament adopted (the Directive for) the EU ETS as the primary instrument for controlling industrial sector CO_2 emissions in Europe.

---

23 'US in plan to bypass Kyoto protocol' *The Guardian*, Paul Brown and Jamie Wilson in Washington, 28 July 2005 *http://www.guardian.co.uk/international/story/0,,1537565,00.html*.

24 'EU pushes binding climate deal' by Richard Black 28 July 2005 BBC, *http://news.bbc.co.uk/1/hi/sci/tech/4724877.stm*

Box 2: Sub-national Climate Change Activities initiated between January and July 2005

The U.S. **Conference of Mayors** – representing 1,183 cities from all 50 states – on 13 June votes unanimously to support the Climate Protection Agreement which mirrors the Kyoto Protocol's goal of reducing GHG emissions 7% below 1990 levels by 2012. Before the Mayors' Conference convened in June, 164 mayors from around the country had signed onto the agreement.

Arizona: (February) An executive order creates a state Climate Change Advisory Group, charged with developing recommendations to reduce Arizona's greenhouse gas emissions, culminating in the submission of a Climate Change Action Plan by June 2006. The Governor signs another executive order requiring new state-funded buildings to derive at least 10% of their energy from renewable sources, either directly or through the purchase of renewable energy credits.

California: (June) The Governor Schwarzenegger signs an executive order directing state officials to develop plans to reduce the state greenhouse gas emissions by 11% below current levels over the next five years (=2000 level), 25% by 2020 (=1990 level), and 80% by 2050.

Illinois: (July) The state Commerce Commission passes a resolution calling for both Renewable Energy and Energy Efficiency Portfolio Standards. The state utilities have agreed to acquire 2% of their electricity from renewable sources by the end of 2006 and reach 8% by 2013. They will also create new programs to reduce the increase in electricity demand 10% by 2008 with the ultimate goal of reducing the state's electricity demand growth by 25% in 2015.

Iowa: (April) The Governor signs an executive order instructing state agencies to increase their operational energy efficiency and renewable energy use. The order mandates a 15% improvement in energy efficiency at state facilities by 2010, and the procurement of hybrid or alternative-fuel vehicles for non-law enforcement state vehicles.

New Mexico: (March) The legislature passes three bills to promote energy efficiency and renewable energy investments. (June) The Governor signs an executive order setting greenhouse gas targets

---

Box 2: continued

for the state: 2000 emissions levels by 2012, 10% below 2000 levels by 2020, and a 75% reduction below 2000 emission levels by 2050. New Mexico is the first major coal, oil and gas producing state to set targets for cutting greenhouse gas emissions.

New York: (July) Governor Pataki signs into law the Appliance and Equipment Energy Efficiency Standards Act of 2005 which sets energy efficiency standards for appliances such as commercial washing machines; commercial refrigerators, freezers, and other commercial and household items. It is estimated that the standards will save consumers up to 2,096 GWh per year and up to $284 m savings, while reducing carbon dioxide emissions by 870 kilo tons

North Dakota: (April) The Governor signs into law a legislative package encouraging wind power, ethanol, and bio-diesel and allowing in-state generated Renewable Energy Credits to be sold to out-of-state buyers, as well as lowering the barriers to siting wind power and investing in new transmission.

Rhode Island: (July) The Governor signs the Energy and Consumer Savings Act, under which it joins Washington, Maryland, Connecticut, Arizona, New Jersey, and California in setting efficiency standards for household and commercial appliances.

Washington State: (April) The Governor signs a bill mandating that new public buildings meet the US Green Building Council's Leadership in Environmental Design (LEED) Silver standards. (May) The governor signs three bills, two with the aim of increasing supply and demand of renewable energy by ways of tax breaks to producers of solar equipment and 'feed-in' credits for solar and wind energy, and the third one adopting California's vehicle greenhouse gas emission standards for the state, making it the tenth state to do so. (May) Washington State joins Maryland, Connecticut, Arizona, New Jersey and California in adopting efficiency standards for 12 types of appliances.

Source: For more on these and other US domestic initiatives to combat climate change see 'State and Local News' at the Pew Center website: *http://www.pewclimate.org/what_s_being_done/in_the_states/news.cfm*

When it came into operation on 1 January 2005, it was immediately the largest ever emission trading scheme, covering all the largest 'point source' CO_2 emissions across the EU25 – namely power stations, cement manufacturing, iron and steel, pulp and paper, oil refining, glass and ceramics, and all other industrial facilities larger than 20MW thermal capacity – accounting for about 46 percent of European CO_2 emissions.

The EU ETS Directive specifies two phases: A Phase 1 2005–07 ('precursor period') in which member states retain the option of a conditional 'opt out' for the named sectors and facilities, and a Phase 2 2008–12 ('Kyoto period) when they are all mandatorily covered by the scheme, and governments have the option to 'opt-in' additional sectors and facilities. In Phase 1 (Phase 2) governments can auction up to 5 percent (10 percent) of permitted allowances to their domestic sectors.

The follow-up EU's Linking Directive, adopted in May 2004, allows companies also to use emission credits generated under Kyoto's project mechanisms towards compliance under the EU ETS, capping the volume of credits to be imported into the EU at 6 percent of total emissions (with an envisaged review when the cap is reached).

The Directive specifies a penalty to be levied in case of non-compliance, rising from €40/tCO_2 in Phase 1 to €100/tCO_2 in Phase 2. In addition, there is a requirement to make good the allowance shortfall by purchasing credits in the market, adding to the compliance incentives.

On 9 February 2005, the European Commission adopted a Communication ('Winning the Battle Against Global Climate Change'[25]) that set out its future policies of climate change. It also put forward a set of proposals designed to structure the EU's post-2012 international climate change negotiations.

- Broader international participation in reducing emissions. The EU should continue to lead multilateral efforts to address climate change, but identify incentives for other major emitting countries, including developing countries, to come on board. During 2005, it should explore options for a future regime based on common but differentiated responsibilities.

25 Commission of the European Communities, 'Winning the Battle Against Global Climate Change', Communication from the Commission to the Council, the European Parliament, the European Economic and Social Committee and the Committee of the Regions, COM(2005) 35 final, *http://www.europa.eu.int/comm/environment/climat/future_action.htm*

- Inclusion of more sectors, notably aviation, maritime transport and forestry since deforestation in some regions significantly contributes to rising greenhouse gas concentrations in the atmosphere.
- A push for innovation in the EU to ensure the development and uptake of new climate-friendly technologies and the right decisions on long-term investments into the energy, transport and building infrastructure.
- The continued use of flexible market-based instruments for reducing emissions in the EU and globally, such as the EU emissions trading scheme.
- Adaptation policies in the EU and globally, which require more efforts to identify vulnerabilities and to implement measures to increase resilience.[26]

The Communication, together with a Commission Staff Working Paper, gives a detailed account of these proposals. Based on the ultimate objective of the UN Framework Convention on Climate Change – namely

> to achieve, in accordance with the relevant provisions of the Convention, stabilization of greenhouse gas concentrations in the atmosphere at a level that would prevent dangerous anthropogenic interference with the climate system. Such a level should be achieved within a time-frame sufficient to allow ecosystems to adapt naturally to climate change, to ensure that food production is not threatened and to enable economic development to proceed in a sustainable manner.[27]

– the European Council adopted in 1996 the policy goal of limiting average global temperature increases to no more than 2°C of pre-industrial levels (see the EU 6th Environmental Action Programme[28]). Pending further scientific information, the EU is basing its decisions on the assumption that reaching the 2°C target would translate into a long-term greenhouse gas concentration level of 550ppm CO_2-equivalent, and that such a concentration level would translate into a global emission reduction of 15–20 percent by the year 2050 compared to 1990 emission levels (or by 50–60 percent compared to a 'business as usual' scenario).

Following sections on 'The Climate Challenge' and 'Benefits and Costs of Limiting Climate Change', the Communication identifies three challenges: The Participation Challenge, The Innovation Challenge,

---

26 'Climate change: Commission outlines core elements for post-2012 strategy' Commission press release, 9 February 2005, *http://www.europa.eu.int/comm/environment/climat/future_action.htm*

27 Article 2, UNFCCC.

28 *http://europa.eu.int/comm/environment/newprg/index.htm.*

and The Adaptation Challenge. With regard to The Participation Challenge, the Communication points out that the EU, alone, cannot solve the problem:

> even if the EU were to cut its emissions by 50 % by 2050, atmospheric concentrations would not be significantly affected, unless other major emitters also made substantial emission cuts. Therefore, effective action to tackle climate change requires widespread international participation on the basis of common but differentiated responsibilities and respective capabilities.

Highlighting that emission reduction measures – such as significant improvements in energy efficiency and the introduction of low carbon energy sources – do not necessarily pose a threat to economic growth, indeed may even contribute to sustaining rapid economic growth, the Communication contends that the recently adopted EU Action Plan on Climate Change and Development could be instrumental in supporting developing countries addressing these issues.

Referring to the well-known US argument that the absence of emission reduction targets for the large developing country emitters in the Kyoto Protocol gives them an unfair competitive advantage, the Communication recommends that

> the EU should support efforts to resolve this impasse. Indeed a relatively small group – EU, USA, Canada, Russia, Japan, China and India – accounts for about 75 % of world greenhouse gas emissions. It might be worthwhile to try to accelerate progress at the global level by discussing reductions among this smaller group of major emitters in a forum similar to the G8, in parallel with vigorous efforts to reach agreement in the UN context.

And in the Conclusions, 'the Commission recommends that the EU explore options for a post-2012 strategy with key partners. [...] In bilateral contacts with interested countries, including the large emitters, actions should be identified that they are ready to take within specified time horizons and conditions'.

Given that more than half of the countries mentioned in the Communication's 'relatively small group' are already part of the earlier-mentioned *Asia-Pacific Partnership on Clean Development and Climate* it would seem rational for the EU to join this Partnership. Yet it is not certain whether all the Partners would welcome such an enlargement – and not just because the EU presence in the Pacific is rather small (albeit very attractive, including Tahiti), for it is not altogether clear whether everyone in this Partnership would be happy about the EU's suggestion that

> the outcomes of bilateral discussions could then be fed into the UNFCCC negotiations, through commitments to act or to meet targets. The objective

is to establish a multilateral climate change regime post-2012 with mean-ingful participation of all developed countries and the participation of developing countries which will limit the global temperature increase to 2°C, and which is considered as a fair sharing of effort by all key players.

As concerns developing countries, probably the most important evolu-tion in the EU position has been that they are not going to ask for developing country emission reduction commitments in the post 2012 negotiations, as declared by the EU Presidency in the recent roundtable discussion in Ottawa, Canada.

2.2.4   *Russia*

Finally, a few words on the Russian situation. The Russian Federation had been centre stage from COP9 in Milan (December 2003) when one of President Putin's economic advisors started to question whether Russia would actually ratify the Kyoto Protocol, until 22 October 2004, when it did (which was sufficient to have it enter into force 90 days after). Since then, things have calmed down considerably. The most important questions being asked about Russia with regard to climate change is whether it will be ready in time (2008) to participate in the Kyoto flexibility mechanisms or not.

The key issues are the need for institutional capacity and the require-ment of an adequate national greenhouse gas inventory. According to PointCarbon,[29] a leading analyst of carbon market issues, it is expected that the rules for *Joint Implementation* (JI) – the project-based Kyoto mechanism applicable in Russia – will be adopted by the Russian government by the end of 2005, while the inventory is being built up with the help of the EU's technical assistance programme *TACIS* and the World Bank.

3.   A Key Economic Issue: Competitiveness

Irrespective of whether the future climate change regime involves an international emission 'cap and trade' regime, a key issue in the post-2012 debate will be that of (unfair) competitive disadvantages due to differentiated emission mitigation requirements. Indeed, competitiveness will be an issue for the EU even before 2012 in the debate on the 2008–12 National Allocation Plans for the EU ETS.

A clear definition of 'competitiveness' and how it can be measured

---

29 Point Carbon: '02.03.05 Russia to be Ready for JI by 2007' *http://www.point-carbon.com/article.php?articleID=6943&categoryID=470*

is essential for a reasoned discussion on the issue. Arguably, the most appropriate interpretation is in terms of 'profitability', where changes in competitiveness are gauged from changes in (relative) profitability. The question then becomes whether, and to what extent, profits are affected by differentiated mitigation targets for certain industrial sectors.

Carbon/energy-intensive sectors such as steel, aluminium, chemicals, cement, refineries and utilities are usually considered prime candidates for having profitability affected by mitigation targets. However, not all carbon/energy-intensive sectors are equally vulnerable. For example, if a local utility is not directly in competition with non-abating regions, there can be no loss in competitiveness. Even where there is competition, several factors are involved in determining the impact on profitability – such as the way in which the abatement is imposed (for instance, through carbon taxes or by 'grandfathering' emission allocations).

Many other low-intensity sectors (often generating by far the larger proportion of GDP, particularly in industrialized countries), meanwhile, have low energy costs compared to their turnover and may even gain in profitability and competitiveness by adopting further energy efficiency measures (Azar 2005).

Even if there is a significant risk of some carbon/energy-intensive industries relocating production to regions with weaker or no carbon constraints, stricter carbon policies are likely to promote the domestic development of carbon efficient technologies, which in turn will prove economically beneficial in the longer term. It stands to reason that the development of more energy efficient technologies – particularly in globalized markets for consumer goods such as automobiles and electric appliances – will set new standards in other parts of the world (Grubb, Hope and Fouquet, 2002; Müller, 2003).

'Early movers' in new technological innovations focused on energy efficiency could even gain a competitive advantage in new markets as other countries and regions follow with tightened carbon controls – also known as the 'Porter hypothesis' (Porter & van der Linde, 1995; Azar, 2005). The experience of the Danish wind power sector is an example of this kind of an advantage.

Therefore, one should not expect universal truths in the question on the impact of climate change mitigation measures on industrial sectors, but only answers to questions concerning specific economic contexts. Such answers will usually involve some modelling exercises. The UK Carbon Trust, just to name one – albeit important – example, relies heavily on modelling exercises (see Oxera, 2004) in their recent study concerning the impact of the European Emissions

Trading Scheme (EU-ETS) on industrial competitiveness. The results are worth considering in some detail in the present context (Carbon Trust, 2004).

3.1   The Carbon Trust Study (CTS)

The aim of the EU ETS – like any other such scheme – is to minimize mitigation costs, in this case to industry in the twenty-five member European Union (EU25). The CTS raises the point that if energy-intensive sectors in Europe face a significant reduction in profits as a result of the ETS, they may be tempted to move operations abroad, to places where CO_2 emissions are not controlled; or consumers may preferentially buy more goods from regions where emissions are not controlled. This would not achieve anything in terms of global emissions – and if the facilities concerned were less efficient than current European operations, it would even result in increased global emissions. The competitiveness implications of the EU ETS are thus a central concern, for environmental as well as economic reasons (Carbon Trust 2004, p.5).

The study focuses on a number of determinants of a sector's inherent potential exposure to the ETS:

- *Energy intensity*. Energy-intensive industrial sectors – regardless of whether covered by the ETS or not – will see their input costs rise if they fail to reduce their CO_2 emissions and/or energy consumption (indeed, according to the CTS the impact of the scheme on electricity prices represents a greater cost risk to many sectors than the direct impact of the scheme).
- *Ability to pass cost increases through to price* determined, *inter alia*, by the nature of the competition – 'In general, markets with more players are more competitive and costs affect sector pricing more directly'[p.6] – and the level of competition from outside the EU ETS.
- *Price-responsiveness of demand*. Sectors in which demand is not very sensitive to price will not suffer a significant loss in volume of sales when prices are increased, particularly if possible substitute products are exposed to similar cost uplifts as a consequence of the EU ETS.
- *Opportunity to abate carbon*. With increasing CO_2 prices, investment in abatement becomes more and more attractive a means to both limit exposure to the ETS, and potentially to benefit from the (energy) cost savings associated with abatement activity.

3.2 The CTS Sectors and Scenarios

The modelling work for the CTS involved an in-depth analysis of five industrial sectors in the UK (all except the last covered by the ETS):

1. *Electricity*, representing a large proportion of EU emissions and widely considered to be the key sector with unique characteristics, generally not exposed to international competition and seen by many as a possible winner from the ETS.
2. *Cement manufacture*, a highly energy-intensive sector with some degree of international/country-to-country competition.
3. *Paper* (newsprint) – part of the pulp and paper industry, a highly international sub-sector with material energy costs.
4. *Steel manufacture*, a highly energy-intensive sector with strong but differentiated international competition;
5. *Aluminium* (smelting) – a sector not part of the EU ETS but unusually dependent upon electricity, and a fully global commodity market.[p.7]

These sectors are treated as 'carbon price-takers' in the context of the EU-ETS, and the modelling exercise is centred around three price and allocation scenarios, of which the 'Kyoto scenario' is the most interesting in the present context.

For this scenario, during the second ('Kyoto') Phase of the EU-ETS (2008−12), the CTS assumes an allocation to companies similar to the national Kyoto targets, some use of Russian surplus allowances, and certain protective measures concerning 'business-as-usual' foreign credits under the Linking Directive (which links the ETS to other flexibility mechanisms).

The chosen carbon permit price was €10/tCO$_2$, and allocations were based on the principles of the UK National Allocation Plan (NAP) strengthened and extended to 2008−12. The NAP is led by cutbacks in the electricity sector, sufficient to achieve a national 20 percent reduction. Other sector allocations are not part of NAP. In the model, they reflect other UK Climate Change Agreement targets.

Conforming to the proposed definition of competitiveness earlier in this paper, the CTS sees the impact of the EU ETS on competitiveness as closely related to its impact on operating profit. As a measure of this, OXERA's Cournot model calculates the impact on the sector's total earnings before interest, tax, depreciation and amortization (EBITDA).

The impact itself is determined by five related factors, all measured in relation to current operating profits:

1. *Gross Carbon Costs.* At the outset, the model considers the potential 'gross carbon costs', that is to say the gross impact on the sector's production costs given the (assumed) carbon price, before taking into account allowance allocations or (product) price responses. In other words, it shows the potential cost that would arise from a pure carbon tax (on all emissions) at the same price levels. In the Kyoto scenario, this increases marginal costs of the UK electricity sector by 23 percent, which taken in isolation would be sufficient to offset the sector's operating profit.

2. *Net Value at Stake.* The next step in calculating the competitiveness impact of the ETS is to net-off the value of the 'grandfathered' free allowances. For the UK electricity sector the Kyoto scenario net value at stake is still in the region of one-third of the operating profit.

3. *Product Price Adjustments.* While being a 'carbon price taker', the sector may – depending on the exposure of the sector to lesser carbon constrained competition – actually be a product 'price-maker' and able to pass a significant part of the marginal cost increases on to the consumer by increasing the product price levels. The important fact here is that the marginal unit produced will incur a carbon cost, but those covered by the 'grandfathered' allowances will clearly not. However, an increase in price would apply to *every* unit produced, even those that are covered by these free allowances, which is why it is possible that carbon constraints can actually *increase* the sectors' profits, and thus its competitiveness. The UK utilities are indeed projected to be able to pass through 90 percent of the above-mentioned marginal cost increases in the Kyoto scenario, leading to a price increase for the consumer of 15 percent.

4. *Demand Adjustments.* Any price adjustments will have an effect on demand, which will vary depending on the 'elasticity' of the demand in question. Given the relatively 'inelastic' nature of electricity demand – people do not tend to vary their electricity use much in response to price variations – the Kyoto scenario impact on UK electricity demand remains at a modest 6 percent reduction.

5. *Abatement Adjustments.* In many cases, it will prove to be more profitable to abate the sectoral emissions than to buy in the permits not covered by the grandfathered allowances, not only because of reducing the permit purchasing costs, but often also energy costs. For the electricity sector, the main abatement option is fuel switching from coal to gas. The estimated impact of this for the UK sector under the Kyoto scenario yields a final position concerning the EU ETS impact of an *increase in profitability* (EBITDA) *of 63 percent.*

3.3 The Sectoral CTS Results

The electricity sector – used above to illustrate the modelling method – is, of course, typically atypical in the context of concerns about international competitiveness due to differentiated carbon targets, because there is, in most cases, no competitive international market in electricity. The results for the other sectors considered in the CTS – in particular the cement and steel sectors – may thus be more relevant to the international competitiveness issue. As it happens, none of the sectors studied in the UK has any change in the number of firms operating, and they are all, except aluminium smelting, projected to make carbon windfall profits, some of them, particularly cement and steel, significant (Cement +25%, Steel +17%), as listed in Table 2.

Table 2: Carbon Trust/Oxera Study: Headline results for Kyoto Period (2008–2012) (%)

Sector	marginal cost increase	price increase	marginal cost increase passed on to customers	change in quantity demanded	change in operating profit (EBITDA*)	'Net Value at stake'
Aluminium smelting	5	3	66	–6	–31	51
Cement (base line)	55	14	83	–4	25	1.9
Cement (competition**)	55	11	66	–8	13	1.9
Cold-rolled steel	7	3	67	–5	17	4

* Earnings Before Interest, Tax, Depreciation and Amortisation
** The base-line case assumes that the UK cement market is largely domestic, which is why a 'competition' scenario was also modelled, largely on the situation in Spain, assuming 30 percent non-EU imports.

Moreover, given the current carbon price of around €25/tCO$_2$ is more than twice the value assumed in the Kyoto scenario, it stands to reason that these windfall profits could be considerably larger than even the projected figures of the CTS, *if* the firms manage to pass the costs through to the consumer.

4. Summary

4.1. Science

Given the stakes of the perceived potential economic costs of tackling man-made climate change, it is not surprising that it has been among the publicly most heatedly debated scientific topics in recent times, rivalled possibly only by the debate on nuclear safety. However, a heated public debate does not necessarily imply a lack of scientific consensus. And while there are inevitable uncertainties – as in any other empirical debate – the vast majority of the relevant scientific community has come to the consensus taken up by the G8 heads at the Gleneagles Summit that 'climate change exists, that it is a serious and long-term problem, and to a large extent a man-made one'.

Indeed, as concerns climate change politics, the question of scientific accuracy has in a large part of the world become a non-issue. If anything, the focus has shifted to what sort of impacts can be attributed to man-made climate change, as witnessed in the recent debate about the connection with tropical storms in the aftermath of the devastations caused by Hurricane Katrina in the US Gulf coast. Most policy makers in the European Union and the other Parties to the Kyoto Protocol have adopted the scientific consensus that there is anthropogenic climate change and are now concerned mostly with how to tackle the problem.

4.2   Politics

International climate change politics at present is characterized by a polarization that goes far beyond the issue of climate change, namely the question of whether the world should be governed multilaterally through the United Nations or through bilateral/regional agreements. The position of the United States under the present administration has tended towards the bilateral/regional position. This has been witnessed not only in the tough stance taken up by the recently appointed US Ambassador to the UN but also in the more narrow context of climate change, in the establishment of the regional *Asia-Pacific Partnership on Clean Development and Climate*. Although this pact was officially not meant to replace the UN Kyoto Protocol, most commentators, particularly in Europe, saw it as an attempt by the USA and Australia to do precisely that.

The European Union – in the climate change context at least – has become the main champion of the UN approach and the Kyoto Protocol. In this, it has the support not only of China and India, but of most of the Kyoto Parties, industrialized or developing. For the EU, the Kyoto regime – at least with regard to its first commitment period (2008–2012) – is a fact and will not be undone. Indeed, the EU is committed to continuing its Kyoto-style Emission Trading Scheme ETS beyond this period.

The international political debate on climate change is indeed focusing more and more on the issue of what is to happen once the first commitment period of the Kyoto Protocol expires at the end of 2012. The lead protagonists have begun to put their cards on the table. The European Union is determined not only to keep the United Nations system at the heart of the international climate change process, but also to retain the broad architecture of the Kyoto Protocol, with binding absolute emissions caps and flexibility mechanisms at its core. The United States will, for the near to medium term at least, continue to reject this 'Kyoto architecture' and will instead focus its attention on voluntary bilateral regional agreements, such as the Asia-Pacific Partnership on Clean Development and Climate.

The question which of the two models will prevail essentially depends on economic factors. While the voluntary US model may at first sight seem to be more attractive to the business sector, one should not forget that with the introduction of the EU Emissions Trading Scheme, a considerable amount of assets have been created that would be lost in the absence of such a scheme, a fact which does give substantial support to the EU position. The main potential obstacle to the continuation of this model is whether it can be achieved without raising too many objections with regards to unfair effects on industrial competitiveness.

4.3   Economics

The issue of such impacts on competitiveness cannot be discussed in abstract, which is why the recent study about the effects of the EU ETS by the UK Carbon Trust is of considerable value. Contrary to the widespread opinion that the Kyoto Protocol would spell competitive doom for EU industrial sectors due to the absence of developing country emission reduction targets, this study shows that:

- It is possible that some low carbon/energy sectors gain in competitiveness under such an asymmetric regime ('reversed Dutch disease' phenomenon).
- Practically all of the prima facie most vulnerable carbon/energy-intensive industries in Europe will *not* face a worsening in their competitiveness. On the contrary, they will reap a (windfall) increase of it, if measured in terms of their profitability.
- However if, as is not unlikely, in the longer term permit price will be in the tens of Euros per ton of CO_2 there is the potential that the costs may not be passed through to the consumer which may indeed

lead to competitiveness problems for energy-intensive industries from companies that do not face similar carbon penalties.

It is therefore clear that any demands for protective measures by EU industry sectors (such as electricity, steel, cement and so on) will need to be very carefully and specifically analysed for in the near to medium term it looks as if only a few of them will be facing competitive disadvantages.

References

Adelman, M. (1993), *The Economics of Petroleum Supply*, MIT Press: Cambridge.

Alley, R.B. (2000), *The Two-Mile Time Machine: Ice Cores, Abrupt Climate Change and Our Future*, Princeton, NJ: Princeton University Press.

Azar, C. (2005) 'Post Kyoto Climate Policy Targets: Costs and Competitiveness Implications', *Climate Policy* (Special Issue on post-2012 Policy) Volume 5 Number 3, forthcoming

Barnett, T.W., D.W. Pierce, K.M. AchutaRao, P.J. Glecklei, B.D. Santer, J.M. Gregory, W.M. Washington (2005), 'Penetration of Human-Induced Warming into the World's Oceans', *Science*, 309(5732):284−7.

Barnetta, J., S. Dessaib and M. Webber (2004), 'Will OPEC lose from the Kyoto Protocol?' *Energy Policy* 32:2077–88.

Bartsch, U. and B. Müller (2000a), *Fossil Fuels in a Changing Climate: impacts of the Kyoto Protocol and developing country participation*, Oxford: Oxford University Press.

Bartsch, U. and B. Müller (2000b), Impacts of the Kyoto Protocol on fossil fuels', in Bernstein, L. and J. Pan (eds), *Sectoral and Economic Costs and Benefits of GHG Mitigation*, Bilthoven: Intergovernmental Panel on Climate Change.

Bernstein, P., D. Montgomery, T. Rutherford and G. Yang (1999), 'Effects of restrictions on international permit trading: The MSMRT model', *The Energy Journal* (special issue edited by Weyant, J.): 221–56.

The Carbon Trust (2004), *The European Emissions Trading Scheme: Implications for Industrial Competitiveness*, London: The Carbon Trust, June.

Cecil, J. (2001), 'Ancient Apocalypse: The Fall of the Minoan Civilisation', <*http://www.bbc.co.uk/history/ancient/apocalypse_minoan1.shtml* >.

Chasek, P. and L. Rajamani (2003), 'Steps toward Enhanced Parity: Negotiating Capacity and Strategies of Developing Countries', in Inge Kaul et al. (eds), Providing Global Public Goods: Managing Globalization, Oxford: OUP, for UNDP.

Depledge, J. (2002), *The Third Assessment Report of the IPCC*, Royal Institute of International Affairs Briefing Paper, London: Royal Institute of International Affairs.

Douglass, D.H., B.D. Pearson and S.F. Singer (2004), 'Altitude dependence of atmospheric temperature trends: Climate models versus observation' *Geophysical Research Letters*, Vol. 31, July [Received 29 March 2004].

Douglass, D.H., B.D. Pearson, S.F. Singer, P.C. Knappenberger, P.J. Michaels (2004), 'Disparity of tropospheric and surface temperature trends: New evidence' *Geophysical Research Letters*, Vol. 31, July [Received 9 April 2004]

Fu, Q., C.M. Johanson, S.G. Warren and D.J. Seidel (2004), 'Contribution of stratospheric cooling to satellite-inferred tropospheric temperature trends', *Nature*, Vol. 429, 6 May.

Fu, Q. and C.M. Johanson (2005), 'Satellite-derived vertical dependence of tropical tropospheric temperature trends' *Geophysical Research Letters*, Vol. 32, May.

Ghanem, S., R. Lounnas, G. Brennand, (1999), 'The impact of emissions trading on OPEC', *OPEC Review*, June, 23 (2):79–112.

Grubb M.J., C. Hope, and R. Fouquet (2002), 'Climatic implications of the Kyoto Protocol: The contribution of international spillover', *Climatic Change*, 54:11–28.

Hoyle, F. (1981), *Ice*, London: Hutchinson.

Imbrie, J. and K. Palmer Imbrie (1997), *Ice Ages, Solving the Mystery*, Boston: Harvard University Press.

IPCC (2004), *Summary for Policymakers: A Report of Working Group I of the Inter-governmental Panel on Climate Change*, www.ipcc.ch.

Manning, S.W. (1999), *A Test of Time: The Volcano of Thera and the Chronology and History of the Aegean and East Mediterranean in the Mid Second Millennium BC*, Oxford: Oxbow Books <www.rdg.ac.uk/~lasmanng/testoftime.html>.

Marland, G., T. A. Boden, and R. J. Andres (2002), 'Global, Regional, and National Fossil Fuel CO_2 Emissions', Carbon Dioxide Information Analysis Center, Oak Ridge National Laboratory, <http://cdiac.esd. ornl.gov/trends/emis/em_cont.htm >.

McCoy, F.W. and G. Heiken (2000), 'Volcanic Hazards and Disasters in Human Antiquity', *Special Paper 345*, Boulder, CO: Geological Society of America.

McKibbin, W., T. Ross, R. Shackleton and P. Wilcoxen (1999), 'Emissions trading, capital, flows and the Kyoto Protocol', *The Energy Journal* (special issue edited by Weyant, J.).

Michaels, P.J., S.F. Singer and D.H. Douglass (2004), 'Settling Global Warming Science', *Tech Central Station – Where Free Markets Meet Technology*, 8 August 2004 www.techcentralstation.com

Moore, T. Gale (2005), *Climate of Fear: Why We Shouldn't Worry About Global Warming*, (available at www.cato.org), or 'Climate Glacier Politics,' posted 16 May 2005 on www.aei.org, by Roger Bate a resident fellow at AEI.

Müller, B. (2002), 'The Global Climate Change Regime: Taking Stock and Looking Ahead', in Olav Schram Stokke and Øystein B. Thommessen (eds), *Yearbook of International Co-operation on Environment and Development 2002/2003*, London: Earthscan.

Müller, B. (2003), 'Framing Future Commitments: A Pilot Study on the Evolution of the UNFCCC Greenhouse Gas Mitigation Regime,' EV32, Oxford: OIES.

OAPEC, 'UNFCCC and the Kyoto Protocol', November 2004 *http://www. oapecorg.org/Editorial% 20November%202004.htm*

OXERA (2004), 'CO2 Emissions trading: How will it affect UK industry? Report prepared for The Carbon Trust, July 2004.

Pershing, J. (2000), 'Fossil fuel implications of climate change mitigation responses'. in Bernstein, L. and J. Pan (eds), *Sectoral and Economic Costs and Benefits of GHG Mitigation*, Bilthoven: Intergovernmental Panel on Climate Change.

Petit, J.R., J. Jouzel, D. Raynaud, N.I. Barkov, J.-M. Barnola, I. Basile, M. Bender, J. Chappellaz, M. Davis, G. Delayque, M. Delmotte, V.M. Kotlyakov, M. Legrand, V.Y. Lipenkov, C. Lorius, L. Pepin, C. Ritz, E. Saltzman, and M. Stievenard (1999), 'Climate and Atmospheric History of the Past 420,000 Years from the Vostok Ice Core, Antarctica', *Nature*, 3:429−36.

Polidano, C., F. Jotzo, E. Heyhoe, G. Jakeman, K. Woffenden and B. Fisher (2000), 'The Kyoto Protocol and developing countries: impacts and implications for mechanism design', ABARE Research Report. Australian Bureau of Agricultural and Research Economics, Canberra.

Popper, K.R. (1935), *Logik der Forschung*, Vienna: Springer; trans. *The Logic of Scientific Discovery*, London: Hutchinson, 1959.

Porter, M.E. and Claas van der Linde (1995), 'Towards a new conception of environment-competitiveness relationship', *Journal of Economic Perspectives*, Volume 9, Issue 4.

Rampino, M.R. and S. H. Ambrose (2000), 'Volcanic Winter in the Garden of Eden: The Toba Supereruption and the Late Pleistocene Human Population Crash', in McCoy and Heiken, 'Volcanic Hazards and Disasters in Human Antiquity', 71−82.

Royal Society *et al.* (2005), 'Joint science academies' statement: Global response to climate change', 7 June, *www.royalsoc.ac.uk/displaypagedoc.asp?id=13057*

Sethi, S.P. (2005), 'Indian Perspective', presented at the Seminar for Government Experts, Bonn, 16−17 May.

Stainforth, D. (2005), 'Modelling climate change: known unknowns', published by OpenDemocracy.net on 3 June (*http://www.opendemocracy.net*)

Vajpayi, Shri Atal Bihari (2002), Speech of Prime Minister Vajpayee at the High Level Segment of the Eighth Session of Conference of the Parties to the UN Framework Convention on Climate Change, New Delhi, 30 October *http://unfccc.int/cop8/latest/ind_pm3010.pdf*

CHAPTER 8

CARBON SEQUESTRATION

Klaus S. Lackner

1. Introduction

Energy is a cornerstone of the world's technological infrastructure. Yet, access to energy is taken so much for granted that little effort goes to developing new energy technology. Limitations in today's technology could, however, undermine the economic development of the twenty-first century. Even though there is plenty of energy in fossil, nuclear and solar resources, none of today's energy technologies are ready to meet the demand of a future world population of nine to ten billion people aspiring to a decent standard of living. Different technologies are unable to meet future world energy demand for different reasons. Many could not operate on the necessary scale; others would be too expensive or too damaging to the environment; and some would introduce long-term political and security risks. Unlike improving food production, water supply or medical services, overcoming constraints in the world energy supply does not just require better governance, it also requires better technology.

Fossil fuels, which today provide about 85 percent of all commercial energy,[1] will run into serious carbon constraints. While reasonable people can argue about the maximum tolerable carbon dioxide loading of the atmosphere, any reasonable limit is likely to be breached in the course of the twenty-first century unless decisive action is taken soon (Lackner, 2002). Stabilizing carbon dioxide concentration in the atmosphere at any agreed upon level requires not just reducing some emissions; it requires eliminating nearly all emissions (see Figure 1). An eventual agreement on the stabilization level will determine how much time is left for the transition; it will barely affect the long-term allowable emission rate.

---

1   *Annual Energy Outlook 2004: With Projections to 2025.*, U.E.I. Adminstration, Editor. 2004, Department of Energy, DOE-EIA: Washington DC.

Economic growth sufficient for the developing nations to catch up with today's developed nations makes the time available to turn around a lumbering energy infrastructure very short. Stabilizing at 450 ppm of carbon dioxide in the air is probably already out of reach unless the world were to endorse a worldwide 'Manhattan project'. 550 ppm (twice the pre-industrial level) would be a tough goal and even the possibility of stopping at 650 or 750 ppm is already made difficult by continued inaction. An optimal response even in that case would require immediate action.

Stabilization of atmospheric carbon dioxide would require a gradual phasing out of fossil fuels as a source of energy or, alternatively, the capture and storage of nearly all the carbon released in the consumption of fossil fuels. The latter is often referred to as carbon sequestration. According to the Oxford English Dictionary 'to sequester' means 'to set aside,' or 'to separate,' which applies well to the particular meaning the term has taken in the context of carbon management. Carbon sequestration encompasses a number of varied activities that all aim to keep carbon dioxide emissions from fossil fuel consumption out of the atmosphere or otherwise separated from the environment.

Since 'carbon sequestration' refers to a goal rather than a particular means of accomplishing that goal, the term is applied to a number of very different technologies and strategies. Carbon sequestration includes agricultural and silvicultural methods for collecting carbon; it includes means of introducing carbon dioxide to the ocean; technologies for storing carbon dioxide underground; or chemical conversion that transforms carbon dioxide gas into chemically inert carbonates. Carbon sequestration encompasses not only carbon or carbon dioxide storage options, but also technologies for capturing carbon – typically in the form of carbon dioxide – from industrial processes or directly from the environment. Finally, transport, handling, monitoring and management of carbon dioxide are part of the technology portfolio that is summarized under the term 'carbon sequestration.'

Economic growth is not running into energy resource constraints but into environmental constraints. Carbon sequestration could remove the most difficult constraint on the use of carbon based fossil fuels, and thus deserves more careful study. The next sub-section outlines how an enormous growth potential combined with vast fossil energy resources could collide with environmental concerns and precipitate carbon energy crises in the coming century. The subsequent sub-sections deal with the technology options that are available, and the policy strategies that could make such a transition a reality.

2.   Why Carbon Sequestration?

2.1   Protecting Economic Growth

Economic growth is highly desirable but it is inexorably linked to increased energy consumption. Energy and raw materials, like capital and labour are among the basic ingredients in producing goods and services. Even if energy consumption per unit of gross domestic product (GDP) is gradually reduced, GDP growth drives energy consumption. Efficiency improvements tend to reduce demands for labour and energy, but they do not eliminate them.

Planning for future energy demand must allow for growth in the developing nations. A peaceful world cannot be built on vast economic discrepancies without any possibility of eliminating them. Bringing the poor countries of the world to a standard of living taken for granted in the developed nations requires substantially more energy. At present, of the 6.5 billion people on Earth, one billion people live in rich countries,[2] while at least 1.6 billion people lack access to virtually all electric power, and have only minimal access to any commercial form of energy (Watkins, 2005). If the average GDP of the world would reach that of the USA today, world GDP would grow by about a factor of eight. General economic growth of the leading economies of 1.6 percent per year, with the rest of the world gradually closing the gap, would lead to even larger growth.[3]

Most business-as-usual scenarios, however, predict less growth in energy consumption, because they assume a reduction in energy intensity of GDP generation which could recover a factor of two to five over the course of the century. Therefore, typical business-as-usual scenarios lead to a three to fourfold increase in carbon dioxide emissions (Nakicenovic et al., 2001).

2.2   Maintaining Access to the Vast Fossil Carbon Resource

Fossil energy resources in aggregate are large enough to satisfy even the demand of ambitious economic growth for more than a century. While limits of oil and gas resources may become visible over the next fifty years, it is equally possible that additions to the resource base will keep pace with extraction. Between 1984 and 2004, the ratio of proven reserves to oil consumption has risen from 30 years to 40 years

---

2   Based on data by the US Energy Information Adminstration (EIA)
3   Lackner, K.S. and J. Sachs, in preparation. 2005.

(BP, 2005). However, the argument that fossil energy will not run out is based on tar sands, oil shales, coal and lignite, combined with the observation that fossil hydrocarbons are fungible.

Estimates of the available fossil carbon resource pool are in excess of 5,000 Gt (Rogner, 1997). If methane hydrates prove accessible, the total could be many times larger. For comparison, the total fossil carbon consumption since the beginning of the industrial revolution has been approximately 300 Gt of carbon.

Gasification, partial oxidation, steam reforming and water gas shift reactions allow the production of synthesis gas (a mixture of carbon monoxide and hydrogen) from virtually any fossil fuel. Fischer Tropsch reactions and similar synthesis processes in turn make it possible to convert synthesis gas into synthetic designer fuels (Dry, 2002).

As long as oil and gas are abundant, they provide the low-cost option for producing gaseous or liquid fuels. In the absence of these resources, synthetic fuels produced from lower grade hydrocarbons would enter the market. Based on past experience, coal derived liquids could be produced for $40 to $50 per barrel of oil equivalent, with more recent costs coming down (Steynberg, et al, 1999). Synthetic oil from tar sands is already much cheaper and can compete in today's market.

If output were to grow from thousands of barrels a day to millions of barrels per day, cost would likely come down. Experience with learning curves in other industries suggests that a reduction in processing costs by a factor of two is a conservative estimate. Resource limitations are unlikely to lead to drastically higher fuel costs over the course of the twenty-first century. The remaining obstacle is the management of carbon dioxide emissions. Thus, the development of carbon sequestration technologies is key to the long-term viability of all hydrocarbon resources ranging from high grade petroleum to low grade coal.

2.3 *Protecting against Environmental Risks*

Resource size does not limit access to fossil fuels, but environmental constraints do. Among them, concerns over carbon dioxide emissions are the most difficult to address. Carbon dioxide is the natural result of fossil fuel use and unlike conventional pollutants it cannot be eliminated by improving the cleanliness or efficiency of the energy extraction process.

Carbon dioxide is a greenhouse gas, it is physiologically active and it plays a critical role in the chemistry of the world's ocean. Doubling the carbon dioxide concentration in the air should be approached with caution, even if uncertainties remain in the detailed predictions.

Model calculations will always remain uncertain. Model independent conclusions may be less specific, but they too suggest that prudence demands action soon. The following can be said without recourse to detailed models:

- *The observed rise in atmospheric CO_2 is due to fossil fuel consumption.* Fossil fuel consumption has generated more CO_2 than necessary to explain the observed increase. The difference is mostly due to uptake by the surface ocean but the detailed balance between ocean and biosphere CO_2 is less certain than the fossil fuel input.
- *Carbon dioxide is a greenhouse gas.* Simple radiation balances known for at least a century predict a climate change more dramatic than observed. Radiative forcing from increased carbon dioxide combined with water vapour forcing consistent with maintaining a constant relative humidity results in a temperature increase in excess of observation. In model calculations, aerosols and thermal inertia can account for the difference, but this attribution is not entirely certain.
- *Warming is occurring.* Observations show a prominent warming trend over the last few decades. Warming is seen in air temperature recordings, ocean temperature, a worldwide retreat of glaciers, and the thinning of Arctic sea ice. The warming ocean provides large thermal inertia suggesting that there remains a substantial amount of unrealized warming, even without further increases in carbon dioxide concentrations.
- *Observed climate fluctuations demonstrate the vulnerability to change of human built infrastructure.* Whether natural or human induced, real fluctuations on various timescales show the economic impacts and human suffering caused by heat waves, droughts and extreme weather events. While attribution of individual events may be impossible, changes in probability of extreme events are likely and thus provide the basis for concern.
- *Carbon dioxide impacts transcend climate change.* For example, acidity and the carbonate to bicarbonate ratio in surface ocean water are largely determined by the atmospheric concentration of carbon dioxide. Changes in the carbonate chemistry of the surface ocean due to doubling atmospheric carbon dioxide have been shown to stunt coral growth (Langdon et al., 2000). It is likely although not proven that this slow down in growth will lead to the demise of coral reefs and associated biodiversity (Kleypas, Buddemeir and Gattuso, 2001).
- *Human contributions to mobile carbon pools are becoming substantial.* While the anthropogenic carbon flux is still small compared to photosynthesis

driven carbon fluxes between the air and biosphere, human activities are changing the size of the total mobile carbon pool on Earth. This pool of a few thousand Gigatons of carbon consists of carbon in the atmosphere, the biosphere, and the fraction of the ocean carbon that can be modified without significantly changing ocean pH. Human additions to this pool comprise roughly 10 percent of the total.

2.4   A Precautionary Approach

These observations form the basis for a precautionary approach that is not guided by a vague concern over moving into unknown territory, but by a concern informed by simple physical models that should capture the order of magnitude but still could be subject to a fortuitous cancellation of competing effects. Prudence would suggest that the world adopts strategies to prevent an uncontrolled build-up of carbon dioxide in the atmosphere, or a massive change in the carbon cycle and move toward stabilizing the amount of excess mobile carbon in the atmosphere, the ocean and the biomass.

The decision to move toward carbon management soon is made easier by the fact that net emissions of carbon will have to stop sooner or later in any event. What is uncertain is the total emission budget and thus the speed with which a transition to full carbon management will need to be made. Without halting net emissions, the amount of excess mobile carbon in the environment will continue to grow indefinitely and the atmospheric concentrations of carbon dioxide will continue to rise. Even at emissions much lower than those of today, carbon dioxide levels can grow well past all reasonable bounds (Figure 1). Uncontrolled growth in emissions could lead to atmospheric concentrations of carbon dioxide of 800 ppm or even 1000 ppm before the end of the century. With continued net emissions at a lower rate, carbon dioxide concentrations would climb more slowly but still would eventually reach levels far in excess of 1000 ppm. Therefore, stabilization of atmospheric concentrations of carbon dioxide at any level is tantamount to stopping net emissions to the air. Once the stabilization limit of the atmosphere has been reached, annual net emission allowances that would keep the CO_2 level in the atmosphere constant will drop, in a matter of decades, to below 30 percent of today's emissions (Houghton et al., 1995). Ten billion people sharing equally in such a budget would be allowed a per capita emission that is approximately 3 percent of the per capita emission in the USA today. For all practical purposes such an emission allowance can be considered as zero.

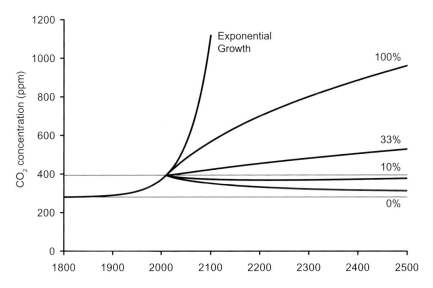

Figure 1a: Model Trajectories of Atmospheric CO_2 Concentrations for Simple
            Emission Scenarios

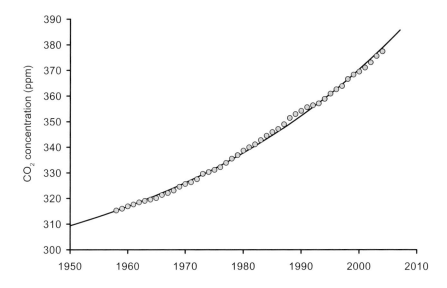

Figure 1b: Model Calibration against CO_2 Historic Concentration Trends

Box for Figure 1

A schematic model calculation showing that stabilization of carbon dioxide levels in the atmosphere is virtually equivalent to eliminating all carbon dioxide emissions. Part a) of the figure shows the partial pressure of CO_2 in the atmosphere based on a simple model calculation. At 2010, future CO_2 levels are branching off into five different and highly simplified emission scenarios. The rapid exponential growth scenario assumes that future growth in CO_2 emissions follows the same exponential growth which has been successfully fitted to the previous two centuries. The other three scenarios assume constant emissions from 2011 on. The different cases are labelled by the size of the emission relative to the rate of emission in 2010. We note that holding emissions constant at 7.3 GtC per year leads to a steady, nearly linear increase in CO_2 in the atmosphere. Dropping the rate of emission down to 33 percent of that, CO_2 concentrations keep rising, even though the per capita allowances for a world of 10 billion people would be down to about 3 percent of the actual per capita emission rate in the United States today. At 10 percent of the 2010 emissions, the initially high rate of ocean uptake will cause a slight drop of 25ppm in atmospheric CO_2 concentration, but in the long term CO_2 levels remain on the upswing, even though per capita emissions have been reduced to 1 percent of the US per capita emissions. Only at zero emission does the long-term downward trend due to ocean uptake prevail. The carbon dioxide concentration in the air is calculated as a linear superposition of impulse responses for every year's emission. The shape of the impulse response function is taken from Kheshgi and Archer (2004). It overestimates the actual removal of carbon dioxide from the atmosphere as the impulse response is that for 280 ppm. It is not adjusted to the slower response at higher CO_2 concentrations. In this simple model, past emissions are approximated by an exponentially growing rate of emissions with a characteristic time constant of 44 years and a rate of emission of 5.7 Gt/yr of carbon in the year 2000. These numbers are consistent with historic rates but have been fine tuned to reproduce the raise in CO_2 concentrations. The model implicitly assumes sources and sinks other than fossil fuels and oceans to first approximation cancel out. Part b) of the figure shows a comparison of the curve between 1958 and 2004 with data taken at Mauna Loa. In spite of its simplicity, the model remarkably well reproduces historic trends.

Under a stabilization scenario, fossil carbon can still be used to provide energy, but after the initial budget of carbon that need not be sequestered has been consumed, virtually all of the resulting carbon dioxide must be kept out of the mobile carbon pool or an equivalent amount of carbon must be removed elsewhere.

Because of the easy availability of fossil carbon, the problem of introducing excess carbon into the environment will not be self-limiting. Instead, worldwide economic growth has the potential of raising the current consumption by another order of magnitude. A future world population of 10 billion people, who consume energy like the United States does today, would require nine times as much energy as the world is consuming today. Additional growth in today's developed nations could dwarf this estimate.

Even a stabilization goal at relatively high values like 800 ppm would benefit from immediate action. This follows from a simple calculation assuming that aggressive action could be described as involving two phases; phase I will hold the rate of increase in carbon dioxide levels constant for the first 50 years, whereas phase II will reduce the rate of increase down to zero over the following 50 years (Pacala and Socolow, 2004). Starting immediately would lead to a built-in rise of about 100 ppm over the next 50 years, followed by another 50 ppm by 2100. This scenario suggests a stabilization level around 500 to 550 ppm. If carbon management is ignored for the next 50 years, emissions will roughly double, raising the increase between now and 2050 to approximately 150 ppm, followed by another 50 years at 200 ppm, and another 50 years adding an additional 100 ppm. In this case stabilization would occur near 850 ppm around the year 2150. Unconstrained growth by a factor of four would reach 850 ppm by 2100 and keep rising from there at more than 8 ppm per year.

Immediate action that is politically possible is not likely to be sufficient to stop CO_2 concentration at 500 or 600 ppm, but would have to be augmented with additional efforts as time goes on. By the same logic, a policy which procrastinates until 2050 would most likely have lost the chance to stabilize around 850 ppm.

There are very few options to reach this goal. One is to capture carbon dioxide and dispose of it safely and permanently. The alternative is increased reliance on energy resources that do not involve fossil carbon. Nuclear energy and solar energy provide two examples, which are unique in that they actually could potentially operate on the required scale. Carbon sequestration, which the remainder of this chapter discusses, is of particular importance as it would provide the means of retaining the dominant energy source of today in a future mix of energy options for a time scale which exceeds any reasonable planning horizon.

3.   Carbon Sequestration Strategies

Carbon sequestration encompasses a number of technologies and strategies that aim to reduce excess mobile carbon in the environment. Strategies fall into two broad categories, dilution and containment.

The strategy of dilution is to accelerate the distribution of carbon across the entire mobile carbon pool. Carbon will be bound in biomass, soil carbon, and most importantly dissolved inorganic carbon in the ocean. A dilution strategy minimizes the environmental impacts on each reservoir by reducing the size of the change in each subsystem. Adherents of this strategy argue that even without action, over time the carbon would redistribute itself in this fashion. Active intervention, by encouraging growth of trees, or by injecting carbon dioxide into the ocean is simply reducing the transient peak in the atmosphere and the surface ocean.

By contrast containment strategies aim to find receptacles for large volumes of concentrated CO_2 that can be isolated from the environment so that the mobile carbon pool is kept free from additional carbon loading. Such receptacles might be carbon dioxide lakes on the bottom of the ocean, geological formations, or surface mounds of solid carbonate minerals. Proponents of this approach must prove that the isolation of the carbon is virtually permanent and is neither subject to sudden catastrophic failure nor gradual leakage. Storage capacity and storage life time together determine the limits to this type of sequestration.

Dilution often, but not always, combines capture and storage into a single process. For example, increasing the standing biomass is a natural process which takes carbon dioxide from the air. Virtually all containment efforts require upstream capture and separation of carbon dioxide from other streams. Furthermore, there is usually a need to transport carbon dioxide from the point of generation to the point of disposal. Although such transport technology has not been implemented on the necessary scale, pipeline technology and shipping of cryogenic gases are well developed.

The strategy of dilution requires a good understanding of the environmental impact of increasing the mobile carbon pool. It is not enough to establish a tolerable limit of excess CO_2 in the atmosphere, but separate limits must be established for the surface ocean, the mid-depth ocean and the deep ocean. Similarly one needs to establish the maximum acceptable carbon loading in biological subsystems and soil. The lowest of all limits applies to the entire system. At a given loading, dilution increases the effective size of the reservoir. Extending the sink from the atmosphere and the surface ocean to the entire mobile carbon pool would increase the size of the sink by about a factor of three. Since the uptake capacity of the ocean saturates, this factor decreases with increased loading (Butler, 1991).

The components of the mobile carbon pool can be summarized as follows: carbon dioxide in the atmosphere, which increased from 560 GtC at pre-industrial levels to 760 GtC today; biomass, which holds about 600 GtC; soil carbon with 1600 GtC; and approximately 3400 GtC in the ocean (Schimel et al., 1995; Lackner, 2002). While the ocean holds 39,000 Gt of dissolved inorganic carbon, most of it must remain there to maintain charge neutrality of ocean water. The entire elasticity in the carbon pool at atmospheric doubling is probably less than 1,800 Gt of carbon, an amount that under business as usual could be injected into the air within the twenty-first century.

Reliance on the dilution principle would simply fill up all natural buffers without an ability to back out. It assumes that the capacity to tolerate excess carbon is comparable in all reservoirs, and it is still limited at about 3 to 5 times what could be accepted by the atmosphere. Reliance on the dilution principle will only work if the world abandons fossil fuel resources in favour of other forms of energy soon.

Containment strategies are only as good as the integrity of the containment. Carbon dioxide in its ambient state is a gas and in most instances one must maintain a physical barrier to prevent its spontaneous return to the atmosphere. For most storage modalities some leakage is unavoidable. However, on very long time scales, geological carbon cycles intervene by removing excess carbon from the mobile pool by forming carbonates. Leaky containment is limited in its capacity to no more than the dilution strategy. Only if containment lifetime is sufficient for geological processes to maintain stability (Lackner, 2002), can containment capacity exceed the dilution limit.

Geological carbon capture processes operate on various time scales from 5,000 to 200,000 years (Archer, Kheshgi and Maier-Reimer, 1997). For an imbalance in the range of hundreds of Gigatons, acceptable leakage rates may thus be well below 1 GtC/yr and could even be well below 0.1 GtC/yr.

The leakage rate L can be estimated as a function of the storage content C_s and the storage life time τ_0.

$$L = C_s/\tau_0$$

Given a maximum allowable leakage rate, the average storage time has to grow proportionally to the amount of stored carbon dioxide. If L is less than 1 GtC/yr, then the minimum storage time will exceed 100 years for 100 Gt of storage. Given the scale of the problem smaller storage would not even be worth considering. With allowable leakage rates at 0.1 GtC/yr and total storage capacity of 5,000 GtC, containment lifetimes need to be 50,000 years.

Even under very minimal expectations for the role of CO_2 storage, containment times should be measured in centuries. As actual storage volumes increase, leakage constraints become more stringent. They could reach several millennia by the end of the century, and eventually reach levels as high as 50,000 years. Lifetime constraints provide a clear criterion for selecting among different storage options. The overall capacity, the environmental impacts, and the safety of a particular storage system provide additional selection criteria.

4.   Carbon Sinks

4.1   Biological Fixation of Carbon

Biomass sequestration uses the growth of photosynthesizing organisms to capture carbon dioxide directly from the air and tie it up as organic carbon. The most obvious sink for this carbon is the living biomass itself. However, some of the carbon may move on into different storage like the soil.

4.1.1   Increasing the Standing Biomass
The total biomass reservoir contains about 600 GtC, most of it in the form of trees (Schimel et al., 1995). The rate of turnover in this reservoir is very large. More than 100 Gt of carbon move annually in and out of this reservoir. Only a small fraction is retained for multiple years, and an even smaller fraction is stored for a relatively long time. Wood in trees could be considered sequestered for decades and in some cases for centuries.

The expected change in the world's biomass carbon after equilibration with a higher carbon content in the atmosphere is small. Raising the standing biomass beyond this change in the equilibrium carbon content requires continuous maintenance of the biomass base (Hurtt et al., 2002). Increasing the standing biomass drastically will have ecological consequences and maintenance requirements grow rapidly as sequestration amounts increase. The increased fire hazards in the

United States Southwest are an example of the problems one might encounter.

The natural biomass cycle is too fast and the storage capacity in this reservoir is too small to deal with a large fraction of the total emissions expected over the next 100 years. Since most research focuses on uptake rates which can vary widely, the more obvious constraint on the size of the reservoir is less well known.

While inducing biomass production in the ocean through fertilization could prove useful in developing food sources through aquaculture (Jones and Otaegui, 1997), its impact on the carbon cycle is quite small. Ocean fertilization will be discussed under ocean sequestration options below.

4.1.2 Bio-fuels

Since biomass carbon is already part of the mobile carbon pool its use as a fuel or for other purposes does not affect the net carbon balance. While somewhat contrarian, this point of view has the advantage of simplifying carbon reporting issues. As long as biomass carbon remains in the mobile pool it does not change the overall carbon balance. Substitution of fossil fuels with bio-fuels either in power plants or in the transportation sector does not add to sequestration, but it displaces fossil carbon as would any other use of non-fossil energy.

On the other hand, removing biomass from the mobile carbon pool and storing it permanently is a form of sequestration. Since biomass is produced in a highly energetic state relative to carbon dioxide, it would, however, be advantageous to utilize this energy before one sequesters the carbon as carbon dioxide. Rather than storing logs in mines or on the seafloor, one should combust those logs in power plants, collect the resulting carbon dioxide and sequester it. The excess carbon offset that is being generated in a power plant, could then be applied against carbon emissions from cars or airplanes which are far more difficult to recapture (Keith, 2001).

Most bio-fuels are expensive, and biomass fuel production competes for land with agricultural food production (Kheshgi, Prince and Marland, 2000). Increased standards of living not only raise demand for fuel, but also demand for agricultural produces to support a more protein rich diet. This competition restricts the role for bio-fuels to the utilization of agricultural waste.

4.1.3 Soil Carbon

Soil carbon represents a larger and more stable carbon reservoir than living biomass. It is also accessed through biomass production. Soil carbon amounts to about 1400 Gt of carbon with typical lifetimes measured in months to decades. A small fraction, as for example carbon stored in peat-bogs, is very persistent. Even in most soils some fraction of the soil carbon is in very stable long-lived compounds. Development of soil carbon is a good way of storing additional carbon, but the question remains, how much can be stored without creating a drastic change in the overall ecological balance (Gower, 2003). Since fossil fuel emissions dominated greenhouse gas emissions in the past, reversing the depletion in soil carbon by itself cannot be sufficient to change the overall carbon balance.

Carbon sequestration in forests and forest soils raises difficult accounting and equity issues. If temperature or CO_2 changes cause some forests to increase their carbon stock, while others lose carbon, who may take credit and who is held liable for these changes?

The total changes available in soil carbon that stay within responsible environmental management, are likely to be very small compared to the excess carbon that need to be stored over the course of the twenty-first century.

4.2　Ocean Disposal of Carbon Dioxide

4.2.1　Dilution into the Ocean

The simplest approach to carbon sequestration in the ocean is to let ocean water equilibrate with the atmosphere. Over time this will happen by itself, but by injecting carbon dioxide from a ship to mid-depth, one short-circuits natural transport and accelerates equilibration between all the mobile reservoirs. This dilution strategy extends the natural CO_2 repository by the size of the ocean sink (Archer, Kheshgi and Maier-Reimer, 1997).

If the world had taken the emissions of the last two centuries of 300 Gt of carbon and distributed them according to the equilibrium conditions among ocean and atmosphere, roughly 240 Gt of carbon would have ended up in the ocean while 60 Gt would have remained in the air. Carbon dioxide in the air would have risen by about 30 ppm rather than the observed 90 ppm. Even without action oceans absorbed 120 Gt of carbon. Active sequestration with optimal mixing would have reduced the loading in the air and the surface ocean by approximately a factor of three. Doubling of the atmospheric carbon dioxide concentration would allow for a total consumption of about 1,800 Gt of carbon.

This estimate appears large, in part because it does not take into account the practical difficulties of evenly approaching equilibrium everywhere. It also assumes that driving the entire air-ocean system into a new equilibrium with a doubled concentration of carbon dioxide is acceptable. It does not consider ecological impacts, for example the sensitivity of deep ocean eco-system to pH changes. Changing course if environmental damage to the ocean becomes apparent is – with today's technology – virtually impossible. Once the air-ocean system has been brought into equilibrium, the new carbon level will persist for millennia.

4.2.2   *Priming the Carbon Pump by Fertilization*

Much of the ocean surface is nearly void of life for lack of nutrients. Fertilizing these ocean stretches would increase the oceanic uptake of carbon dioxide from air. Carbon is captured by photosynthesizing organisms near the ocean surface. As these organisms sink to greater depth, their remains are metabolized by bacteria and returned as dissolved carbon dioxide to the ocean water (Jones and Otaegui, 1997; Watson et al., 1994). This process consumes oxygen and thus limits the amount of fertilization that can be performed at any one site. Adding so much fertilizer that organic carbon will actually sink to the ocean floor in large quantities would be counterproductive, as it could easily precipitate drastic change in the ocean environment by creating anoxic conditions.

Carbon sequestration through ocean fertilization requires large modifications to the natural carbon cycle. The transfer of bicarbonate from the surface ocean to mid-depth by way of an organic intermediate would strengthen the already active biological carbon pump. Since the bulk of the captured carbon is released at relatively shallow depth, this leads to the circulation of carbon near the top of the ocean (Zeebe and Archer, 2005). The total possible worldwide uptake rate has been estimated at between 1 Gt/yr to 2 Gt/yr and the residence time of this carbon may not be very long.

Generally, ocean fertilization is designed to make substantive changes in the surface-ocean ecosystem. This may be of interest to ocean farming (Jones and Otaegui, 1997), but it has relatively little impact on the world's carbon balance.

As an aside, the formation of carbonates by biological systems leads to the emission rather than absorption of carbon dioxide. As long as the ocean water's alkalinity is fixed the removal of calcium carbonate lowers the total dissolved carbon dioxide that is stable in the water. A calcium ion dissolved in the ocean holds on to two bicarbonate ions. Once it precipitates out it removes one carbon atom in the form of a carbonate ion, leaving the second carbon in the second carbonate ion to be transformed to CO_2 and evade into the air.

4.2.3 Containment on the Ocean Floor

At great depth, carbon dioxide can form liquid pools on the ocean floor. At low pressure, liquid carbon dioxide is less dense than water, but it is far more compressible than water. For carbon dioxide to become denser than water requires near freezing temperatures and pressures typical of the deep ocean. Below 2,700 metres depth carbon dioxide will form pools on the ocean floor that are denser than water (IPPC, 2005).

At great depth, carbon dioxide not only sinks, but exposed to water it forms a solid carbon dioxide hydrate. Liquid carbon dioxide exposed to cold seawater at hydrostatic pressure in excess of 45 to 70 bars reacts with the seawater to form a stable form of ice that has cage structures in the crystal lattice occupied by carbon dioxide molecules. Approximately one CO_2 molecule is stored for every 6 H_2O molecules. Clathrates only form at elevated pressure and low temperatures. The melting temperature is a function of pressure, but under ocean conditions it is typically limited to temperatures below 8°C. Carbon dioxide is not the only gas that can form such clathrate structures, but similar gas hydrates are formed by methane and other gases. Clathrates only form in the presence of water that locally is saturated in CO_2. Clathrates without the presence of liquid CO_2 are not stable in ocean water.

Formation of clathrates generates heat. At 4.9 kJ/mol, the amount of heat released per mole of water is similar to the heat of formation of ordinary ice. This heat release needs to be accounted for when injecting large volumes of carbon dioxide into the sea, as heat transfer can limit the rate at which clathrates form. Similarly the decay of large sheets of clathrates will be limited by the cooling of the surrounding water that is associated with the melting of the clathrate. These heat transfer issues are important in considering the behaviour of ocean water in the presence of liquid carbon dioxide.

Deep ocean storage would result in lakes of carbon dioxide on the ocean floor. Below 4500 m the clathrate formed will float at the top of the lake as it is less dense than carbon dioxide at such great depth (Brewer et al., 1999). Between 2,700m and 4,500 m, the clathrate has a tendency to sink leading to a far more complex churning behaviour in the lake.

Storage on the ocean floor would have a virtually unlimited capacity for carbon dioxide, if carbon dioxide would not dissolve into the ocean within a few centuries creating an acidic plume with largely unknown ecological consequences. The same limitations that apply to dilution, also apply to storage on the ocean floor, with the additional caveat that the changes in water chemistry in the near zone are far more dramatic.

Storage in artificially contained systems on the bottom of the ocean is a different matter. It would be possible to form a lake on the bottom of the ocean and develop artificial containment, like plastic sheets or walls, or a covering of the surface of the lake under sediments or mud. Not much is known about such systems, but their potential for storing carbon dioxide could be very large. They would, however, require a sophisticated capability for remote or automated construction and earth moving under 3,000 m of water.

4.3. *Geological Storage*

Storage underground takes advantage of the fact that CO_2 is a fluid that can be injected into porous media and that reservoirs capable of holding vast amounts of CO_2 are indeed plentiful. The fate of the carbon dioxide in this pore space will depend on the type of formation, its temperature, its depth, the chemistry of the pore fluid and most importantly the geometry of the formation.

4.3.1 *Trapping Mechanisms*

Ideally carbon dioxide is introduced into a geological trap where impermeable layers above the location prevent the carbon dioxide from escaping. Such traps are either ancient natural gas traps or similar geological formations lacking natural gas storage. In most locations the carbon dioxide density will be less than that of pore water and therefore the carbon dioxide will rise toward the surface. On the way up it will get trapped in formations that are sealed to the top with impermeable layers. In some cases gas might find faults or cracks in the layer, or open such cracks by overpressure or chemical dissolution, or escape around the edges of such a layer until it has distributed itself so widely that a multitude of small traps successfully holds all fluid.

A second type of trapping that will occur in parallel is the dissolution of carbon dioxide into the pore water. Typically the dissolution of carbon dioxide into the brine will make it denser and cause it to sink. This helps to accelerate the interactions of carbon dioxide with pore water, as saturated pore water will naturally move away from rising carbon dioxide. Another process that may help in the dissolution process is the natural fingering instability of large bubbles of CO_2 that in the inhomogeneities of the rock tend to break up into smaller volumes with increased surface areas. In spite of these possibilities, the uptake of CO_2 in the brine is generally considered a slow process. It may, however, be of particular interest in formations of alkali rocks with high pH fluids in the pore space.

A third trapping process which can only happen after the CO_2 has been taken up by the brine is the chemical conversion of the carbonic acid into carbonate salt, by leaching additional alkalinity out of the surrounding minerals. Basalts and certain ultramafic rocks would be particularly well suited for such a transition.

Lastly, gravitational trapping can occur in certain locations below the deep ocean. Below ground carbon dioxide is nearly everywhere less dense than water, because the geothermal gradient prevents carbon dioxide from entering the pressure-temperature regime in which it is denser than the pore water brines. The exception is under the ocean floor below 3,000m where there is a window of a few hundred metres below the ocean floor, where CO_2 is cold enough to be denser than water. In such locations the CO_2 is trapped by gravity and cannot move spontaneously upward.

4.3.2 Capacity

The challenge of geological storage is the volumes involved, which are far larger than any past reservoir injections. Experience in reservoir engineering focuses on an operational regime in which the pore volume or pore pressure is reduced relative to those of the unperturbed system. Typically, rock is compressed rather than extended. Exceptions are technologies for increasing or stimulating flows through otherwise impermeable layers, like hydro-fracturing of rock.

Likely emissions in the United States over the next fifty years would produce a liquid volume sufficient to cover the entire country with a 5 cm layer of liquid the density of water. While it is commonplace that water and oil extraction has lowered the ground in places by several metres, raising the ground by several metres would not be acceptable practice.

Consequently, it is not clear what the ultimate capacity of geological storage will be. There is a substantial capacity that appears safe and permanent. Storage for about a thousand Gigatons of CO_2 is almost certainly available. The scale of prospective volumes, i.e., areas with proven saline reservoirs that are good candidates for carbon dioxide disposal is indeed very large, measured in many thousands of Gigatons of CO_2. It remains to be seen what fraction of these reservoirs is sufficiently stable and secure for long-term storage. This can only be determined by further research and by engaging the public into the discussion of what is safe.

4.3.3 Applications for Carbon Dioxide in Hydrocarbon Extraction

Geological storage can be combined with recovery of hydrocarbons.

Carbon dioxide is already used for tertiary recovery of oil in a handful of reservoirs. While most of this CO_2 comes from natural wells, in the future carbon dioxide may be derived from fossil fuel consumption. At present about 30 million tons of CO_2 are used per year in the USA for enhanced oil recovery (IPPC, 2005).

Carbon dioxide flooding is routinely performed in the West Texas oil fields. The carbon dioxide is brought in by pipeline from wells in Southern Colorado that are 800 km away. The economic value of the carbon dioxide flood is sufficient to support the pipeline operation and provide a revenue stream to the operators of the carbon dioxide wells in Colorado. Instead of using carbon dioxide from wells, it would also be possible to use carbon dioxide captured from the combustion of fossil fuels. An example of such an operation is already ongoing at Weyburn in Saskatchewan, Canada. The Weyburn field receives pipelined carbon dioxide from the Great Plains Synfuel Plant in Beulah, North Dakota. The carbon dioxide delivery creates an additional revenue stream for the Synfuel plant while providing the oil operations at Weyburn with a means of producing additional oil.

In the case of the West Texas field the price of carbon dioxide for many years has been around $15 per ton of carbon dioxide, which suggests that a substantial fraction of the cost of carbon dioxide capture and transport could be compensated for by the use of carbon dioxide for enhanced oil recovery.

While the economics of such operations varies from location to location, there is great potential for carbon dioxide sequestration that requires very little additional incentive to become economically feasible. Indeed the simple change in outlook that future supplies of CO_2 could increase might change the view on tertiary recovery opportunities for oil fields that are ready for CO_2 flooding.

Similarly, CO_2 could be used to maintain pressure in gas wells by underlaying them with CO_2 injection. Recovery of coal bed methane through carbon dioxide flooding is an additional option under active investigation. Coal adsorbs methane and in the case of deep unmineable coal it is still possible to extract the methane. Since carbon dioxide binds more strongly to coal than methane, it is possible to 'flush out' all methane by flooding the formation with CO_2. Unfortunately, CO_2 causes the coal to swell, lowering the permeability of the coal. The injection of CO_2 into deep coal seams does not deprive future generations of a potential resource. If a future mining operation, e.g. robotic mining with small units capable of removing thin seams, were to become an economic possibility, the fact that these seams have been depleted of their methane and are storing CO_2 is of minor consideration. Most of the carbon would be in the coal and the few percent of additional carbon bound in the carbon dioxide would only represent a small correction.

The energy sector provides a substantial reservoir for CO_2 disposal big enough to accept on the order of 100 to 300 Gt of CO_2. Even larger numbers which are quoted occasionally (Freund, 2001), appear doubtful as they would suggest that far more CO_2 can be stored than hydrocarbons have been extracted. Outside of energy extraction there are no large-scale applications of carbon dioxide that could lead to an appreciable reduction in carbon dioxide emissions. Most utilization is too short-lived, and too small in volume to make a difference.

4.3.4 Storage of Carbon Dioxide in Saline Reservoirs

Once the economically useful injection of carbon dioxide has been played out, there are large reservoirs of saline deep aquifers that could hold large volumes of carbon dioxide.

The first carbon sequestration project at Sleipner field takes advantage of such an aquifer. The difference to oil and gas reservoirs is that saline aquifers nearly always are open and it is more difficult to assure that there will be no path to the surface. On the other hand careful monitoring can assure that little leakage occurs. Below the ocean, a small leak into the ocean water would also be readily absorbed. On land, leakage hazards are more of a concern.

A natural gas leak in Hutchinson Kansas in January 2001 points to the hazard potential (Allison, 2001). A pipe entering a salt cavern used for temporary natural gas storage sprung a leak. Gas moved underground for tens of kilometres in a matter of days to blow out of century old, forgotten wells that had been used to bring salt brine to the surface. The high permeability pathways channelling the natural gas were not known prior to the incident. In a larger case, the turning over of Lake Nyos in 1986 (Kling et al., 1987) released the equivalent of a week's worth of CO_2 from a coal-fired Gigawatt power plant and caused the deaths of 1,700 people.

Such accidents point to the need for excellent reservoir characterization and long-term monitoring of the storage site. Starting with those which appear clearly safe, carbon management can point already to a very large reservoir base that will allow it at least decades of aggressive carbon dioxide disposal. Whether or not at the end of this time the safety and permanence has been established for the next generation of sites remains an open question. It is possible that the safe capacity will extend far into the future; it is also possible that by 2050 or 2070 new alternatives will be needed.

4.4   Mineral Sequestration

4.4.1   Basic Concepts

Eventually all carbon mobilized by human activities oxidizes to CO_2. The resulting carbonic acid weathers rocks to form carbonates from the alkali or earth alkali ions leached out of the mineral. A representative reaction is

$$Mg_2SiO_4 + CO_2 \rightarrow 2MgCO_3 + SiO_2$$

Even for CO_2 concentrations in ambient air the thermodynamic equilibrium favours carbonates over most silicates. However, the process is slow and proceeds naturally on geological time scales.

Formation of carbonates sequesters carbon permanently and irreversibly. The reaction is exothermic, in some implementations producing useful heat. The amounts of available alkalinity, i.e. calcium, magnesium, sodium or potassium bearing silicates completely dwarf even the most optimistic estimates of available reduced carbon. Mineral sequestration is not limited by mineral resources, but by the effort required in speeding the reaction up. Even specific minerals like olivine (in the form of forsterite) and its weathering product serpentine are abundant enough to provide sufficient storage for all of the carbon dioxide that could ever be produced. Even though serpentines and olivines are widely distributed, the largest deposit is in Oman. It by itself could absorb the entire world CO_2 production.

4.4.2   Underground and Ocean Carbonate Trapping

One approach to mineral sequestration is to send CO_2 underground and let it react with the rock matrix. In most cases carbonation is very slow, but there are certain minerals for which the reaction is reasonably fast. This approach has been embraced by Matter, Takahashi and Goldberg (2003), who suggest injecting CO_2 into basalt formations. Some twenty years later one could drill back to assure that the CO_2 has been bound up, maybe remove left over CO_2 and close off the reservoir. Sites for this approach could be in ancient flood basalts, or in volcanic formations like in Iceland. There volume, while potentially very large, is not well known.

The second option is to bring alkaline fluids up from underground and assure that they react with the CO_2, before they are reinjected. A variation of this option is to drive a circulation in an underground reservoir to accelerate mixing and dissolution after injection of carbon dioxide.[4] The energy demand of such a process would be miniscule relative to the energy generated in the production of an equivalent amount of CO_2.

Another option is to dissolve carbonate minerals in aqueous solutions and send the brine either underground or into the ocean. The formation of calcium bicarbonate with power plant CO_2 from limestone has been suggested by Rau and Caldeira to create a bicarbonate solution that can run off into the ocean.

Kheshgi (1995) has suggested introducing alkalinity, in the form of $CaCO_3$ or other bases, and directly into the ocean. With increased alkalinity the ocean would absorb additional CO_2 from the air, while at the same time reducing the pH impact of the increased partial pressure of CO_2 over the water. However, calcium carbonate is supersaturated in ocean water, which raises the possibility of carbonate precipitation which would defeat the dissolution of CO_2 in seawater and effectively release the CO_2 back to the air.

4.4.3 Industrial above Ground Sequestration

Finally one can consider industrial processes for mining olivine and serpentine and convert these materials into solid carbonates that are going to be disposed of in the mine from whence they came (Lackner et al., 1995).

Industrial carbonation processes are intrinsically more expensive than underground injection as one has to provide additional resources. The mining, crushing and grinding of raw material by itself does not add a prohibitive cost. Typical estimates for olivine and serpentine mines suggest that this part of the process plus mine reclamation and tailing processing will add less than $10 to the cost of sequestering a ton of CO_2.

The bulk of the cost in current process designs is, however, in the chemical reaction. Generally, minerals are slow to react and the entire process engineering focuses on the acceleration of the extraction of alkalinity or the acceleration of the carbonation process (IPPC, 2005).

Current process designs have been shown to work, but energy consumption and capital cost in chemically activating the magnesium silicate mineral are still too high. Estimates suggest cost for the entire process of $50 to $90 per ton of CO_2. These costs are costs per ton of CO_2 disposed and do not account for the additional CO_2 produced in the disposal process, which raises the cost per ton of CO_2 avoided by an additional 40 percent.

At present the best options use carbonic acid in the presence of

---

4   Keith, D.W., Private Communication. 2005.

sodium chloride and sodium carbonate bicarbonate to force the leaching of magnesium ions out of the silica matrix of olivines and serpentines. The reaction kinetics is improved by very fine grinding or by heat
treatment of the raw material.

Microscopically, the reaction is hindered by the formation of silica
layers on the outside of magnesium silicate grains. The precipitation
of magnesium carbonate in the surface silica layer also slows down
the kinetics.

The potential of mineral sequestration depends on the development of efficient and cost effective processes. Even a cursory review
of the existing literature suggests that the available option space has
not yet been explored carefully. Much of what has been done reflects
technology options that are simple and have not yet taken advantage
of modern process chemistry. This is not surprising because this field
of inquiry is brand new and even the most simple process designs have
not yet been fully explored.

4.5   Other Storage Options

There are several exotic storage options that are not mentioned above.
One of the earliest was the production of dry ice and storing the dry
ice in large blocks that have lifetimes of centuries. A major cost in this
design is the cost of producing dry ice.

Other options include the storage of large volumes of liquid CO_2
in salt domes. This raises serious safety issues concerning the stability
of a liquid filled cavity in plastic salt. The CO_2 filled cavity is buoy-
ant and thus moves upward to a point where the gas is released. The
potential for release of many cubic kilometres of CO_2 would be a
major hazard.

Ocean disposal already deals with the formation of clathrates on the
ocean floor. An interesting variation of this theme would be to form
clathrates at the bottom of thick ice sheets. Greenland glaciers or Antarctic ice-sheets are thick enough to create conditions near the bottom
where CO_2 is denser than ice and thus would not escape. Moreover
thermodynamics favours the formation of clathrates. For ice near freezing temperature the clathrate formation generates just enough heat to
replace the water it has consumed with fresh meltwater; the process
is thus self-sustaining as long as liquid carbon dioxide is present. The
clathrates are thermodynamically stable at temperatures a few degrees
above the melting point of ice. Hence the bottom of a thick ice sheet,
which may be reaching the melting point of ice because of its exposure
to geothermal heat, would be a stable zone for clathrates. Because

clathrates are denser than ice, they would gravitate to the bottom of the ice sheet were they would have life times of hundreds of thousands of years (Longhi, 2005; Sevier, 2005). Since there are lakes embedded in the Antarctic Ice as large as Lake Superior, there is plenty of space to store substantial amounts of carbon dioxide. A major obstacle to sub-ice storage is the extremely hostile environment that would hamper the delivery of carbon dioxide to such a remote site.

4.6   Storage Roadmap

Storage in well defined containment has many advantages over other alternatives. Therefore underground injection and mineral sequestration provide better alternatives than dilution of carbon dioxide into the ocean. Nevertheless, dilution in the ocean, biomass-soil, and atmosphere system could store on the order of 1000 to 2000 Gt of carbon. This is a risky approach but on a small scale it buys time.

By contrast, geological disposal does allow for large storage capacity and it eliminates rather than dilutes the impacts on the overall carbon cycle. At present, geological disposal is by far the cheapest of all options and thus it is the most likely to be the starting point for the implementation of carbon sequestration.

A special opportunity is afforded by combining carbon dioxide storage with reservoir management for stabilizing the pressure in oil and gas wells and for the development of tertiary recovery from oil from wells that otherwise would not remain economic. In that case one can offset the cost of carbon capture against the benefit of enhanced oil recovery.

In all underground storage there are issues to consider but they are nearly always site specific. Geological sequestration in saline aquifers and in enhanced oil and gas recovery could start immediately and provide a substantial sink for carbon. Whether the sites are large enough to cope with the demand for storage for all carbon from all sources is at present an open question.

Mineral sequestration provides another alternative, which can assure permanence and safety but will have to be reduced in cost by at least another factor of three before it can be considered as a reasonable backstop technology.

5.   Capture from Concentrated Sources

Before carbon dioxide can be stored or disposed of it needs to be separated from the product stream it is a part of. It then needs to be compressed and transported to the site of disposal.

Approximately half of all carbon dioxide emissions occur at large facilities, where capture can operate on large concentrated streams. A Gigawatt coal-fired power plant produces roughly 1000 tons of carbon dioxide per hour. Power plants are responsible for more than 30 percent of total emissions, but steel furnaces, cement kilns, refineries and ammonia plants also create large, rich streams of carbon dioxide.

In a few cases not much would need to be done to capture carbon dioxide. For example, many fertilizer plants and some refinery operations produce concentrated streams of pressurized carbon dioxide ready for disposal. Without an economic incentive for avoiding CO_2 emissions, these plants simply vent the carbon dioxide.

In most plants, however, carbon dioxide is just one component of an exhaust gas that is at ambient pressure. Examples include coal-fired and natural gas-fired power plants, steel mills and cement kilns.

In a few instances plants dispose of solid carbon, like petroleum coke, which is usually not disposed of in a form that will assure long-term sequestration. Rather than let the waste stream gradually oxidize, one should consider the possibility of sequestering this small sub-stream as well.

5.1   Technology Options

Typically, the first step in CO_2 capture is to separate CO_2 from the other exhaust gases prior to pressurization. Since a major part of the capture expense lies in the compression of the carbon dioxide stream, any process that ends up compressing the entire gas stream is wasteful. For a conventional coal-fired power plant compression for the entire flue gas stream would consume all electricity generated at the plant.

Separation of carbon from the fuel can occur at various stages in the process (IPPC, 2005): post-combustion separation of CO_2 from a flue gas; pre-combustion separation of carbon from the fuel via gasification followed by combustion of carbon free fuel gases like hydrogen; or power generation utilizing oxygen rather than air. The latter is accomplished either through an upstream air separation unit or through an energy conversion device, like a solid oxide fuel cell, with intrinsic air separation.

Post-combustion separation requires little change in the energy conversion device. Carbon dioxide capture becomes an end-of-pipe process operating on the off-gas stream. For dilute streams it is possible to scrub with a standard sorbent like an amine, which readily absorbs the bulk of the CO_2 present in the flue gas and which can be recovered in a steam heating process at affordable costs. Other variations of this concept involve the use of inorganic alkaline scrubbing solutions, or solid minerals that can form carbonate, ranging from calcium hydroxide, to lithium zirconate.

There is a large body of literature on these approaches (IPPC, 2005), and it is generally accepted that the process of capture will consume about 20 to 30 percent of the energy available in the power plant. This additional energy consumption is a dominant part of the cost (Herzog and Drake, 1996) causing additional production of CO_2 which amplifies the cost of disposal.

Another approach separates oxygen from air and mixes the oxygen with enough flue gas to maintain combustion conditions similar to those one would have with air (Andersson, Johnsson and Strömberg, 2003). Such oxyfuel combustion requires little change to the power plant design. Energy penalties are comparable to those in downstream flue gas scrubbing. If the plant is at or near the disposal site it could avoid most flue gas scrubbing and deliver all off-gases to an underground disposal facility. By disposing of all off-gas, the oxyfuel plant can eliminate all emissions to the air.

Current oxygen separation plants would consume approximately 30 percent of the power output of the oxyfuel power plant. Today the best option is cryogenic air separation, but future designs may take advantage of even larger economies of scale, and make use of the fact that the heat capacity of the liquid oxygen can be used to cool the incoming air.

Flue gas scrubbing and oxyfuel combustion are suitable for retrofitting existing power plants. However, compared to other alternatives their cost penalty is large. Typical estimates suggest a downrating of the existing plant by 30 to 40 percent in electric output, while fuel consumption remains unchanged. In addition there are capital and maintenance cost, yielding capture costs of between 4 and 8 cents per kWh.

Retrofitting current power plants can only be a stopgap measure. Past generations of coal-fired power plants were built under the assumption that fuel and fuel related costs are small. Since carbon sequestration greatly increases the total cost proportional to fuel use, old plant designs are far from optimal. Optimization would be more than fine tuning and likely result in vastly different designs with far lower capture costs than are anticipated today.

5.2   Integrated Gasifier Combined Cycle Plants (IGCC)

Integrated gasification and combined cycle plants provide great opportunities for integrating carbon dioxide capture into the plant design. IGCC plants consume coal, gasify it and run a turbine on the gasification products. By shifting the synthesis gas mixture to pure hydrogen, it is possible to run a turbine without carbon dioxide emissions and to capture the carbon dioxide at high temperature and high pressure, before it leaves the gasifier portion of the plant.

IGCC plants provide a first step toward power plant designs that combine high efficiency with carbon dioxide capture. IGCC plants are slightly more expensive then conventional power plants, but they create far less pollution. Since they gasify the fuel rather than combust it, the amount of pure oxygen that would be required per unit of carbon input is smaller than in oxyfuel combustion. The higher efficiency reduces the oxygen combustion per kWh even further.

IGCC plants already exist. The growth of this technology has recently obtained a boost as General Electric has announced that it will produce complete IGCC plants. The conversion of such plants to carbon neutral plants is reasonably straightforward. In effect all one needs to do is to shift the synthesis gas that is produced to pure hydrogen. The cost of carbon capture at the IGCC plant is less than two cents per kWh or $23 per ton of CO_2 (IPPC, 2005).

Thus, the gasification technology in an IGCC plant could equally well provide the basis for hydrogen plants that produce a carbon free fuel for consumers, while using fossil carbon energy in its production. The ability of gasification not only to drive power plants but also produce synthesis gas and hydrogen as output increases the flexibility of these plants for future combined systems of chemical and power generation (Williams, 2004).

5.3   Zero Emission Coal Plants

In the past, carbon dioxide has been considered an unavoidable gaseous emission from energy conversion processes using fossil fuels. Carbon sequestration changes this view. Once committed to carbon dioxide capture it becomes possible to avoid gaseous effluents and design a power plant with no emission to the atmosphere (Yegulalp, Lackner and Ziock, 2001). Such zero emission power plants need to keep nitrogen from the air away from the fuel. The residual off-gas is condensable water mixed with a small volume stream of impurities. Some of the impurities can be condensed out, the remainder leaves the plant with the carbon dioxide.

To avoid the import of nitrogen in a zero emission coal plant, oxidation of carbon is accomplished with pure oxygen, or with oxygen bound to chemical sorbents that are used for chemical looping, or the separation of air is intrinsic to the energy conversion process as in a solid oxide fuel cell.

Capping the flu stack eliminates scrubbing flue gases to air quality standards. Some gas scrubbing will still occur, but its purpose is to protect the next processing unit from inputs it cannot handle. The scrubbing effort will be determined from an economic balance between engineering sufficient tolerance to impurities into the processing unit vs. engineering more efficient scrubbers.

The abovementioned oxygen blown combustor could become the first zero emission plant. By assuring that the oxygen is pure one can eliminate the flue stack and avoid all emissions to the air. Rather than approximating an air-like combustion environment, one might begin to optimize the recirculation loop and create a more efficient hybrid system between gasification and oxygen combustions.

IGCC plants are less well suited as zero emission plants. Impurities that came through the gasification process as well as nitrous oxides produced during the combustion of hydrogen in air would be released to the atmosphere. A gasifier plant specializing in the production of ultra-clean hydrogen rather than electricity may qualify as a zero emission plant.

The ultimate zero emission plant would use fuel cells, most likely solid oxide fuel cells. Since in these cells the charge carrier is oxygen they can oxidize fuel mixtures other than hydrogen. They intrinsically separate oxygen from air in the process of producing electricity. While fuel cells are not yet ready for commercialization in power plants, long-term projections aim for sufficiently low costs. The cost requirements in automobiles are far more stringent than in power plant applications.

The first plant designed for zero emission was proposed by the Zero Emission Coal Alliance (ZECA) (Lackner and Ziock, 2001). The ZECA plant aimed for producing nearly pure hydrogen upstream of the fuel cell by using a calcium oxide based water gas shift reaction. The calcium oxide not only absorbs carbon dioxide, but this carbonation reaction also provides the heat necessary to support the water gas shift and steam reforming reactions that create the carbon dioxide in the first place. The solid oxide fuel cell generates enough waste heat to calcine the product calcium carbonate and recycle all input materials except for the carbon that has been converted in the process to carbon dioxide.

The ZECA process is one of a larger family of designs (Lackner and Yegulalp, 2005). All of them first convert solid coal into a gaseous fuel in a heavily endothermic reaction. The heat of reaction is provided either directly or indirectly by the waste heat from the solid oxide fuel cell. The conceptually simplest approach is based on the Boudouard reaction, which converts one mole of carbon and one mole of carbon dioxide into two moles of carbon monoxide fuel. The carbon monoxide is oxidized in the fuel cell to carbon dioxide, half of which is recycled to produce more carbon monoxide and half of it is sent off for disposal. This last step requires a technology for separating carbon dioxide from other gases. While much remains to be done to make these technologies a reality, they promise extremely high conversion efficiencies, up to 70 percent, while at the same time eliminating all emissions to the air and collecting the carbon dioxide in a concentrated pressurized stream.

5.4. *Natural Gas Based Power Plants*

Until recently, it appeared that coal-based power plants would have an advantage in a carbon constrained world over natural gas-fired plants, because of the much higher concentration of carbon dioxide in the exhaust of a coal-based power plant. Natural gas burning turbines prefer operating with excess air and the exhaust of the turbine will be at the lowest possible pressure making carbon dioxide capture more difficult. An alternative has been proposed by Alstom (Griffin et al., 2005). In this case the working fluid for the turbine is oxygen depleted air. The oxygen is removed upstream as it traverses a high temperature mixed oxide membrane to oxidize the fuel at the other side. The advantage of this approach is that the combustion products are created at high pressure and remain at high pressure, while the turbine expands out the preheated, pre-compressed and oxygen depleted air.

5.5. *Biomass Plants*

Biomass could substitute for low grade fossil fuels in all of these designs. With carbon capture still in place this would create excess carbon credits as biomass consumption is intrinsically carbon neutral. However carbon dioxide is entirely fungible, and if in the future it makes economic sense to deal with the carbon dioxide from a coal plant, it makes equal sense to deal with the carbon dioxide from a plant that consumes biomass. The carbon credits could be used elsewhere in the economy to offset carbon dioxide emissions from other activities which may be more difficult to control.

6.   Capture from Distributed Sources

Stabilization of the concentration of carbon dioxide in the air cannot be achieved unless all energy consuming sectors contribute to emission reductions. The transportation sector which accounts for nearly a third of the total can not be exempt. Improved mileage of vehicles is a good first step. Ultimately emission from this sector must be captured as well.

Emissions from fuel consumption could reach 25 billion tons of CO_2 per year (5 billion cars at 5 tons of CO_2 per year). Under such a scenario stabilization would not be possible unless these emissions are avoided or otherwise compensated for. One option to deal with carbon from the transportation sector is upstream decarbonization and carbon free energy carriers for the transportation sector. However, there are no good substitutes for liquid hydrocarbons for energy storage. Gasoline and diesel are unmatched in energy density, low weight tanks, and convenience of handling.

Decarbonizing the transportation sector would mean providing either electricity or hydrogen for transportation. Onboard storage of large amounts of electricity has proven difficult. Electric charging from the roadbed may become possible with the advent of hybrid cars that can drive several kilometres without access to recharge. Such a transition would require major technological advances and today is not under serious consideration.

A transition to hydrogen fuels would move the problem of carbon management to the hydrogen producing plant where it could be more easily accommodated. However, affordable hydrogen storage of sufficient capacity on board of a vehicle is still lacking. Internal combustion engines based on hydrogen are certainly feasible, but fuel storage on board of a vehicle is insufficient. The drive toward fuel cell cars is necessary because only ultra-efficient power systems can even hope for sufficient storage.

The transportation sector could keep using carbon based fuels but compensate for the emissions by collecting an equal amount of CO_2 from the air. This would have minimum impact on the existing infrastructure. Biomass used as power plant fuel as discussed above could play this role. Power plants that combust biomass while sequestering CO_2 would create excess carbon credits that could be applied to the transportation sector. While it may be more economic to substitute biomass for other low grade fuels, biomass could in principle also be converted to transportation fuels. In that case, there is no net emission of carbon dioxide from the use of this fuel.

Industrial plants for the purpose of extracting carbon dioxide directly from the air would overcome the capacity limits of biomass capture. While challenging, such an approach is not as difficult as it sometimes is thought to be (Lackner, Ziock and Grimes, 1999).

Most sorbents that can extract carbon dioxide from flue gas streams also are capable of collecting carbon dioxide from air. Compared to flue gas, the carbon dioxide content of the air is much lower, but the disadvantage appears smaller when one compares the CO_2 concentration after extraction. The extraction of carbon dioxide from air does not aim at removing all carbon dioxide, whereas the capture at the plant should aim for extracting most if not all CO_2 from the flue gas stream. If the power plant exhaust is stripped down to 1 percent of the initial CO_2 content, the final CO_2 concentration is only three times larger than that of CO_2 in ambient air. Required binding energies scale logarithmically in the remaining rather than the initial concentration in the exhaust stream. Thus the additional energy penalty compared to flue gas scrubbing is small and practically irrelevant. Most sorbents considered for use in power plants are able to pull carbon dioxide out of ambient air. Choices are more limited in air extraction not because of sorbent strength, but because of environmental concerns over accidental release of sorbent material into the environment.

Initial analysis shows that the major cost of carbon dioxide capture from the air is not in the air handling but in the recovery of the sorbent. The air handling system is small. Wind mills that can displace a certain amount of fossil carbon emissions are typically several hundred times larger than the CO_2 collection device that would recover an equal amount of CO_2. Since the concentration of CO_2 in the air is still large enough to contain the cost of contacting the air, the big cost in the process remains in the sorbent recovery, which is not very different from the sorbent recovery in a power plant.

For a number of possible air extraction sorbents the recovery involves the transference of the CO_2 from the original sorbent to $Ca(OH)_2$ which results in the production of $CaCO_3$. The energy penalty of the sorbent recovery is then set by the calcination of calcium carbonate into calcium oxide. Most other energy demands in the process will be covered by the energy recovered from the transformation of CaO to $Ca(OH)_2$. To avoid additional CO_2 emissions from the calcination processes, one should use pure oxygen in the calcination. Separating oxygen from air will require electricity, which at least in the foreseeable future will result in CO_2 emissions. These emissions are only a few percent of the CO_2 that has been captured, and thus are tolerable. Using today's cost of oxygen and coal, the combined costs of fuel and oxygen are well below $20 per ton of CO_2.

The total energy penalty is about 40 percent of the energy content of the fuel. This is large but comparable to the energy penalty one would incur in a transformation of coal energy to hydrogen.

7.   Institutional and Policy Issues

The biggest stakeholder in a debate on managing carbon is the general public, who eventually will have to pay for the additional expenses that arise from carbon capture and storage. How much carbon dioxide sequestration will be introduced depends critically on the public's view of the consequences of continued emissions of carbon dioxide. Without a consensus that increasing levels of carbon dioxide need to be curtailed, only small no-regret steps can be taken. Actions commensurate with the need of stabilizing carbon dioxide concentrations in the atmosphere will have to await consensus.

Fortunately, there are only very few sources of carbon that enter into the environment. These are all fossil fuels, which enter the anthropogenic carbon cycle the moment they are extracted from the ground. In addition there are smaller contributions from the calcination of limestone, mainly in cement kilns but also in steel furnaces and power plants. The biggest remaining source is fossil carbon dioxide produced for enhanced oil production or as by-product of natural gas production. The injection of carbon into the mobile carbon pool could therefore be managed at a small number of entry points.

The two groups, who for proper remuneration can actually accelerate the introduction of carbon capture and storage, are the producers and the equipment builders. Equipment builders can modify the technology, the producers can become the focal point for collecting the monies required to make CO_2 capture feasible. Paying for externalities upstream is far more practical than at the end of the tailpipe of every consumer activity. However, moving the point of collection upstream does not absolve consumers of their ultimate responsibility.

For example, such a strategy would make it possible to introduce carbon permits, cap and trade options upstream without interfering with a multitude of complex market transitions. Apart from a public consensus this would require only the cooperation of a very small group of producers of fossil fuel resources, while harnessing financial resources for the introduction of carbon sequestration from the entire economy. If a permit is introduced at the well or at the mine mouth, the economic impact will move through the entire economy.

Economic decision making would then internalize the cost of carbon emissions and re-align the balance between fossil carbon consumption, energy efficiency and alternative energy sources in order to optimize costs and benefits. If carbon sequestration were rewarded with carbon sequestration certificates that could be traded in lieu of permits, one would create a market dynamics which decouples carbon fuel consumption from carbon emissions. As permits are phased out in favour of certificates of sequestration, one can approach stabilization of atmospheric carbon dioxide without stopping the use of fossil fuels (Lackner, Wilson and Ziock, 2000). Permits in this view become a transitory phenomenon to allow the gradual transition toward a net zero carbon economy. Oil producers who provide carbon dioxide storage space would in effect become carbon neutral.

In the long term, the most efficient way of dealing with the carbon problem would be to raise the cost of introducing carbon into the mobile carbon pool. How much fossil carbon is still used in a zero emission economy is hard to predict, but the arguments laid out above suggest that fossil fuels could well maintain a competitive edge. Economic drivers will decide whether carbon free energy carriers are more advantageous than taking carbon dioxide back from the air. Markets will also balance the relative contributions from non-fossil energy sources, and energy efficiency improvements with additional carbon capture and storage.

As the constraints on carbon dioxide emissions from the transportation sector come to bear, the types of solutions that will be implemented will have great influence on the future of oil. In the long term oil would benefit from the ability of collecting carbon from dilute sources. By contrast, a transition to carbon free energy carriers, be it electricity, hydrogen, or even ammonia, would ultimately remove the premium that is currently paid for oil. Petroleum fetches a premium because of its chemical proximity to gasoline, kerosene and diesel. Hydrogen just like electricity is so different from any primary energy resource that any source of energy will do equally well in its production. On the other hand, the advantages of liquid hydrocarbon fuels are such an important consideration that they could tip the balance in favour of solutions that collect carbon dioxide from the atmosphere. In such a scenario, the advantages of oil would remain.

For market forces to rearrange all the parts of the infrastructure requires time. Introducing new property rights on such a large scale needs to be done gradually. As a result, the politically acceptable price for general carbon permits will be too low to make a difference for a long time to come. Directly influencing the technology that goes into energy conversion systems from power plants to automobiles may thus have a more immediate impact on the carbon balance.

Even though the number of energy consumers is very large, a large fraction of all energy consumption is influenced by a very small number of manufacturers of fuel consuming machinery that converts fossil fuels into other energy carriers or performing energy intensive tasks. Utility operators are a small group, but they are far more numerous than makers of utility equipment, who are also involved in the construction of other big energy consuming plants like steel plants, cement kilns and boilers used in myriad applications. The production of the world's automobile and truck fleet is also concentrated in the hands of a few manufacturers. The world's air plane production is even more concentrated and so is the production of heavy construction and earth moving equipment. Thus a large fraction of the production of machinery that today is responsible for the vast majority of all carbon dioxide emissions is performed by a very small group of companies.

By providing the appropriate incentives for this group, public policy can greatly influence carbon emissions on a short time scale.[5] Carbon reductions would be obtained by inducing a transition to higher efficiency (i.e., hybrid engines in automobiles, rapid introduction of integrated gasifier combined cycle or other high efficiency plants in the electricity sector), through the introduction and development of carbon capture and storage systems in government-industry collaborations, and through introduction of non-fossil energy alternatives.

Getting a head start on market forces is important because of the large inertia in the energy sector. With the world's rapid economic growth, an enormous amount of energy infrastructure is going to be built over the next few decades and this infrastructure will lock in emissions for the next fifty to seventy years. Advancing the development of less carbon intensive technologies or technologies that are more easily retrofitted to carbon management would greatly enhance the flexibility of the long-term response.

Multilateral international treaties relating to the carbon management problem are an important ingredient in assuring that progress is made in distributing the social, environmental and economic costs of fossil fuel use in an equitable manner. Countries have common but differentiated responsibilities and also have different capabilities to effect change. The countries that must achieve the largest reductions in emissions are

---

5   Sachs, J. and K.S. Lackner, A Robust Strategy for Sustainable Energy. 2005, in preparation.

the developed industrial countries, whereas the biggest potential for carbon dioxide reduction occurs in countries with most rapid growth. Rather than having to replace an existing infrastructure with a new one, in the developing nations exists the opportunity of introducing a new energy infrastructure that avoids emissions from the start. An equitable solution would require the industrial nations who have the largest emissions to pay for and take credit for reductions that are more easily accomplished in the economies that grow the fastest.

8.   A Roadmap for Carbon Sequestration

The above discussion suggests a potential roadmap for managing carbon emissions over the next century.

First there needs to be a large effort of educating policymakers and the public about the need to deal with the carbon problem. This has to be done in a sensible manner avoiding polarization and fearmongering, but clearly laying out the risks and costs of ignoring human impacts on the carbon cycle and comparing them with the economic and opportunity cost of managing carbon. Once it is accepted that stabilization has to be the goal and that it will require eliminating virtually all emissions to the atmosphere, the precautionary step of acting sooner rather than later will not raise the cost of the process dramatically.

No-regret options would make it possible to start right away. Increased energy efficiency, introduction of energy alternatives wherever they make sense, and the general reduction of the environmental footprint of energy consumption in the generation of SO_x, NO_x, heavy metals, and fine particulate, will all contribute to a better starting point for carbon management.

In addition, the sequestration of carbon dioxide in enhanced oil recovery provides a win-win situation where the economics of tertiary oil recovery is nearly sufficient to support the development of carbon dioxide capture opportunities in nearby fossil fuel power plants, chemical plants or fertilizer plants. These examples will produce additional technological know-how and assure the public that managing the carbon from fossil fuel consumption need not be prohibitively expensive.

Even while the support of the public is still hesitant, the small number of players who control the carbon intensity of the energy sector could position the world energy infrastructure in a better way over the next few decades. As environmental problems develop, public support for rebuilding the energy system will follow. Power generators and manufacturers of power consuming machinery may not be able to absorb large costs for new technologies while they cannot yet pass the cost on to the consumer, they can prepare for the transition and position themselves in a manner that reduces future liability, and minimizes the cost of the transition that is going to occur. Public policy needs to encourage such actions.

Positioning for long-term trends means making choices in day-to-day decisions which incorporate a long-term trend toward carbon management, even if it is of little importance today. For example, by looking at different options for mitigating pollution one should choose those which in the long term are compatible with carbon management.

Policy measures that do not seriously affect energy cost should start considering carbon management as an option. For example, in the USA and Europe renewable energy portfolio standards have been used to encourage wind and solar energy. Under such rules, utilities must have a minimum amount of alternative energy in their portfolio, which in effect supports higher prices for renewable energy. Without endorsing such a policy, one could agree that from a carbon management perspective one should also include zero carbon or reduced carbon emissions in such a portfolio standard.

It is in the best interest of coal, gas and oil producers to recognize that carbon emissions, when properly managed, could create new opportunities for their business and would increase the value of their resource base. This points to the need for developing carbon sequestration technologies as a means of keeping fossil fuels viable. While carbon capture and storage is already providing a path forward to the providers of fossil fuels for the electricity sector and large industrial fossil fuel consumers, technologies for the producers of transportation fuels are still further away. Introducing and directing R&D efforts to improve this situation should be of high priority.

Putting a price on carbon is another important step, already in progress in parts of the world. An initially low price will start to squeeze out obvious waste and increase efficiency. Furthermore, commercial activity will commence only if clear rules of accounting can be established, and only if disputes among parties over how much CO_2 has been deposited and by whom can be settled. Learning by doing will only commence once action has been taken.

The next step is to focus one's attention on the easy examples of carbon capture and carbon storage, combined with a replacement of some fossil energy sources with non-fossil energy sources. This could be done today.

In addition, research to improve the technology base needs to be promoted, so that technology is ready when the public will be ready.

There should be substantial emphasis on those energy resources and those energy conversion technologies that could operate on the tens of Terawatt scale which will be required to satisfy world energy demand. At present the three likely candidates are fossil fuels with carbon sequestration, nuclear energy and solar energy.

In the next twenty years, the world should transition away from large power plants that are coal based and unable to deal with their own carbon dioxide emissions. New plants should either be nuclear plants or alternatively they should be able to capture their carbon dioxide and prepare it for permanent disposal. In countries that do not yet have the consensus to deal with the problem today, it would be advantageous to move forward with a strategy where CO_2 readiness in the new infrastructure is emphasized. While the general public may still need time to be convinced, industry leaders should realize that ignoring the likely arrival of carbon constraint could strand vast amounts of capital if plants are not designed from the beginning to allow for the transition once it is demanded.

Technologies for carbon capture at concentrated sources could be introduced virtually immediately. Over the next fifty years novel, ultra-efficient designs could make the cost of capture quite small. Alternatives like nuclear power broaden the palette of options even more. What is left out in this balance is the transportation sector, unless air capture processes are developed.

After capture attention must focus on disposal of carbon. In the short term there is a large potential for geological storage. In the initial phase carbon dioxide disposal can be combined with reservoir engineering for maximum extraction of oil and gas resources. During this phase, part of the capture cost will be offset by carbon dioxide use in tertiary oil recovery. While there are a few fields worldwide which are using carbon dioxide for enhanced oil recovery today, the potential for this option will grow dramatically as the cost of the carbon dioxide gradually drops to zero and may even become negative. The total reservoir size of these options is substantial but it will be filled in a matter of decades. Far larger reservoirs will be needed to support continued carbon storage. The size of saline aquifer reservoirs that satisfy all safety and leakage constraints are currently unknown but they will certainly last for the first and possibly the second half of the twenty-first century.

A successful demonstration of mineral sequestration, particularly above ground mineral sequestration would have an impact on the policies surrounding carbon sequestration even if it could not compete with geological storage for the foreseeable future. A reduction in cost by about a factor of three would settle the question of whether or not carbon based energy sources need to be phased out. If mineral sequestration can cap the cost of carbon storage it will automatically eliminate the limitations on the size of the available storage, it will automatically remove the limitations on storage times, and it will provide an insurance against the future need of having to retrieve CO_2 that was stored underground and proved to be in a place that leaks too much to leave it there. Liability in that case could be limited to the cost of bringing the CO_2 back to the surface and storing it again in mineral sequestration. At present mineral sequestration is still too expensive to even perform this function.

Taking together all the various options for carbon capture and storage, one can discern the development of a new technology option for developing a sustainable world energy infrastructure. Carbon sequestration could be phased in immediately. The sole obstacle on this path is a lack of public consensus and a lack of the public will to move forward. Trends in Europe suggest that this might be changing soon. In the longer term much work on the appropriate policies and the appropriate technologies remains to be done. There are clearly options on the drawing board, but it remains to be seen whether they will be realized.

9.   Conclusion

Concerns over climate change and environmental impact of excess carbon in the environment are likely to curtail the allowable CO_2 emissions over the course of the twenty-first century to a total that is far lower than what would be desirable from the standpoint of economic growth or what would be possible based on the fossil carbon resource size.

Nuclear energy and carbon capture and storage are in principle capable of reducing emissions from the large concentrated carbon dioxide sources by more than an order of magnitude. Renewable energy could also contribute to this reduction. Elimination of emissions from large concentrated resources addresses one half of all emissions.

The remainder of distributed emissions requires a number of distinct strategies. These strategies will aim to either decarbonize the distributed energy sector, or alternatively compensate for emissions in this sector by removing carbon from the mobile pool elsewhere.

Decarbonization would imply rigorous electrification of the energy consumption sector or a transition to hydrogen. For stationary consumption that can be reached by wires and pipelines, both strategies are viable. What is left is the mobile energy consumption sector which is comprised of automobiles, trucks, ships, earth moving equipment and airplanes. Today this sector is the domain of petroleum because of the intrinsic advantage of liquid hydrocarbon fuels.

While in the short run, coal will be most affected by a transition to a carbon constrained world, in the long term the role of oil is more uncertain. On a century scale, the demands for carbon constraints cannot be satisfied while ignoring the transportation sector. Even greatly improved energy efficiency is likely to be overwhelmed by increased worldwide demand for additional transportation. Thus it will become necessary to curtail carbon dioxide emissions from vehicles or collect an equivalent amount of carbon from the mobile carbon pool. The latter approach would be fully compatible with the use of oil. I am involved in the development of air capture technologies because of the great advantage liquid hydrocarbon energy carriers have in the transportation sector.

Without carbon sequestration stopping the accumulation of excess carbon in the environment becomes extremely difficult and the probability of failure is large. The outline of a carbon sequestration strategy is visible today, but much remains to be done before carbon sequestration can operate on the global scale. In order to succeed it is important to start now.

References

Allison, M.L. (2001), 'Hutchinson, Kansas: A Geologic Detective Story', *Geotimes*:14–18.

Andersson, K., F. Johnsson, and L. Strömberg (2003), 'Large Scale CO2 Capture – Applying the Concept of O2/CO2 Combustion to Commercial Process Data', *VGB Powertech*, (10):1–5.

Archer, D., H. Kheshgi, and E. Maier-Reimer (1997), 'Multiple timescales for neutralization of fossil fuel CO_2', *Geophysical Research Letters*, **24**(4):405–08.

Brewer, P., et al. (1999), 'Direct experiments on the ocean disposal of fossil fuel CO2', *Science*, **284**(5416):943–5.

BP Statistical Review of World Energy (2005), London: BP plc.

Butler, J.N. (1991), *Carbon Dioxide Equilibria and Their Applications*, Chelsea, Michigan: Lewis Publisher.

Dry, M.E. (2002), 'The Fischer-Tropsch process: 1950–2000', *Catalysis Today*, **71**(3–4):227–41.

Freund, P. (2001), '*Progress in understanding the potential role of CO_2 storage. P.*' in 5th International Conference on Greenhouse Gas Control Technologies (GHGT-4), Cairns, Australia.

Gower, S.T. (2003), 'Patterns and Mechanisms of the Forest Carbon Cycle',

Annual Review of Environment and Resources, **28**(1):169-204.

Griffin, T., et al. (2005), 'Advanced Zero Emissions Gas Turbine Power Plant', *Journal of Engineering for Gas Turbines and Power*, **127**(1):81−85.

Herzog, H.J. and E.M. Drake (1996), 'Carbon dioxide recovery and disposal from large energy systems', *Annual Review of Energy and the Environment*, 21:145−66.

Houghton, J.T. et al. (1995), *Climate Change 1994, Radiative Forcing of Climate Change and an Evaluation of the IPCC IS92 Emission Scenario.*, ed. J.T. Houghton, et al., Cambridge: Cambridge University Press.

Hurtt, G.C., et al. (2002), 'Projecting the future of the U.S. carbon sink', *PNAS*, **99**(3):1389−94.

IPPC (2005), W.G.I.o.t., *IPCC Special Report on Carbon dioxide Capture and Storage (To appear soon)*.

Jones, I.S.F. and D. Otaegui (1997), 'Photosynthetic greenhouse gas mitigation by ocean nourishment', *Energy Conversion and Management*, **38**(Supplement 1): S367-S372.

Keith, D.W. (2001), 'Sinks, Energy Crops and Land Use: Coherent Climate Policy Demands an Integrated Analysis of Biomass', *Climatic Change*, **49**(1−2):1−10.

Kheshgi, H.S. (1995), 'Sequestering Atmospheric Carbon Dioxide by Increasing Ocean Alkalinity', *Energy*, **20**:915−22.

Kheshgi, H.S. and D.E. Archer (2004), 'A nonlinear convolution model for the evasion of CO2 injected into the deep ocean', *J. Geophys. Res.*, **109**: C02007.

Kheshgi, H.S., R.C. Prince, and G. Marland (2000), 'The Potential of Biomass Fuels in The Context of Global Climate Change: Focus on Transportation Fuels', *Ann. Rev. Energy and Environ.*, **25**:199−244.

Kleypas, J., R. Buddemeier, and J.-P. Gattuso (2001), 'The future of coral reefs in an age of global change', *International Journal of Earth Sciences*, **90**(2):426−37.

Kling, G.W., et al. (1987), 'The 1986 Lake Nyos Gas Disaster in Cameroon, West Africa' *Science*, **236**:169−75.

Lackner, K.S., H.-J. Ziock, and P. Grimes (1999), 'Carbon Dioxide Extraction from Air: Is it an Option?', in *Proceedings of the 24th International Conference on Coal Utilization & Fuel Systems*, Clearwater, Florida.

Lackner, K.S., R. Wilson, and H.-J. Ziock (2000), 'Free-Market Approach to Controlling Carbon Dioxide Emissions to the Atmosphere: A discussion of the scientific basis', in *Conference on Global Warming and Energy Policy*, Fort Lauderdale: Kluwer Academic/Plenum Publishers.

Lackner, K.S., et al. (1995), 'Carbon Dioxide Disposal in Carbonate Minerals', *Energy*, **20**:1153−1170.

Lackner, K.S. (2002a), 'Can Carbon Fuel the 21st Century?', *International Geology Review*, **44**(12):1122−1133.

Lackner, K.S. (2002b), 'Carbonate Chemistry for Sequestering Fossil Carbon', *Annu. Rev. Energy Environ.*, **27**(1):193−232.

Lackner, K.S. and H.-J. Ziock (2001), 'The US Zero Emission Coal Alliance', VGB Powertech, *International Journal for Electricity and Heat Generation*, **12/2000**:57–61.

Lackner, K.S. and T. Yegulalp (2005), 'Thermodynamic foundation of the zero emission concept', *Minerals and Metallurgical Processing*, **22**(3):161–7.

Langdon, C., et al. (2000), 'Effect of calcium carbonate saturation state on the calcification rate of an experimental coral reef', *Global Biochemical Cycles*, **14**(2):639–54.

Longhi, J. (2005), 'Phase equilibria in the system CO2-H2O I: New equilibrium relations at low temperatures', *Geochimica et Cosmochimica Acta*, **69**(3):529–39.

Matter, J.M., T. Takahashi, and D.S. Goldberg (2003), 'Carbon sequestration in aquifers associated with ultramaific to marc rocks', *Abstracts of Papers of the American Chemical Society*, **226**: U605-U605 141-GEOC Part 1.

Nakicenovic, N. et al. (2001), *Special Report on Emission Scenarios*, Intergovernmental Panel on Climate Change (IPCC).

Pacala, S. and R. Socolow (2004), 'Stabilization Wedges: Solving the Climate Problem for the Next 50 Years with Current Technologies', *Science*, **305**(5686):968–72.

Rogner, H.-H. (1997), 'An Assessment of World Hydrocarbon Resources', *Ann. Rev. Energy Environ.*, **22**:217–62.

Schimel, D., et al. (1995), *CO₂ and the Carbon Cycle*, in *IPCC Special Report on Climate Change 1994: Radiative Forcing of Climate Change and an Evaluation of the IPCC IS92 Emission Scenarios*, J.T. Houghton, et al., Editors. 1995, Cambridge University Press: Cambridge. pp. 35–71.

Sevier, D. (2005), 'Sequestration under Greenland or Antarctic Ice', in *Erice Meeting, 2005*.

Steynberg, A.P., et al. (1999), 'High temperature Fischer-Tropsch synthesis in commercial practice', *Applied Catalysis A: General*, **186**(1–2):41–54.

Watkins, K. (2005), *Human Development Report 2005*, New York: United Nations Development Programme, 388.

Watson, A.J., et al. (1994), 'Minimal effect of iron fertilization on sea-surface carbon dioxide concentrations', *Nature*, **371**:143–5.

Williams, R.H. (2004), *Toward Polygeneration of Fluid Fuels and Electricity via Gasification of Coal and Biomass*, Princeton Environmental Institute, Princeton University.

Yegulalp, T.M., K.S. Lackner, and H.-J. Ziock (2001), 'A review of emerging technologies for sustainable use of coal for power generation', *The International Journal of Surface Mining, Reclamation and Environment*, **15**(1):52–68.

Zeebe, R.E. and D. Archer (2005), 'Feasibility of ocean fertilization and its impact on future atmospheric CO₂ levels', *Geophys. Res. Lett.*, **32**:L09703, doi:10.1029/2005GL022449.

CHAPTER 9

THE FUTURE TECHNICAL DEVELOPMENT OF AUTOMOTIVE POWERTRAINS

Olivier Appert and Philippe Pinchon

1. Introduction

The transport sector is dominated by oil; cars, trucks, aircraft and ships are powered almost exclusively by petroleum products. If transport depends overwhelmingly on oil, conversely, oil's future destiny as a major energy source depends on technological developments in car and truck powertrains.

The main purpose of this chapter is to present key technological facts in order to assess the threat that the invention of a new fuel or engine may or may not pose to oil in the future.

The chapter is organized as follows. In Section 2, we discuss the basic facts about road transport and the future evolution of transport worldwide. In Section 3, we address the main issues raised by transport growth, particularly oil dependence and pollution, and the main challenges that the expected growth in demand for oil is likely to impose on sustainable development. Section 4 focuses on the impact of road transport on both local and global atmospheric pollutions and analyses some of the solutions used to deal with these problems. In Section 5, we discuss the major technologies that are already in production or are likely to emerge in the future. It is argued that although these technologies will bring about many improvements in terms of vehicle performance, fuel consumption and emissions, the high cost of production may limit their usefulness. Section 6 provides a tentative roadmap for the release of the innovations for motorcar powertrains. Section 7 is a summary of the main conclusions.

2. Road Transport

2.1 The Basic Facts

Oil constitutes the major source of the world's primary energy. In 2002, it represented 36% of the total primary supply (or about 3.6 Gtoe) compared to 45% in 1973 (IEA, *Key World Energy Statistics*, 2004 edition). Followed at a distance are coal and natural gas which in 2002 accounted for 23.5% and 21.29% respectively of world total energy supply. Nuclear, hydro, and 'combustible renewable and waste' account for the remaining 20% of global (commercial and non-commercial) primary energy supply.

The transport sector is clearly dominant in petroleum product consumption accounting for 50% of the total in 2002 compared to 37% in 1971 (see Figure 1). It is also expected that the bulk of the increase in oil demand in the next decades will come from the transport sector. OECD countries are the main drivers of this petroleum product consumption, absorbing around 75% of the 1.75 Gtoe consumed globally by this sector. The main consumers within OECD are the United States, the EU and Japan which account for 55% of the 1.75 Gtoe consumed. The share of other forms of energy used by the transport sector is a meagre 1.9%, with electricity accounting for 1%, biomass 0.5%, coal 0.3% and natural gas 0.2% (IFP Panorama 2005).

Road travel dominates the transport sector, representing 90% of all passenger journeys and 75% of all freight hauled. It accounts for some 81% of the energy consumed by the transport sector (see Figure 2)

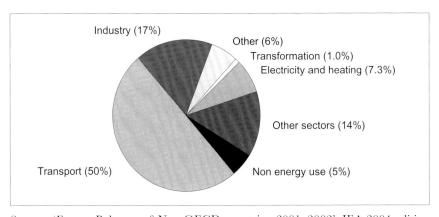

Source: 'Energy Balances of Non-OECD countries, 2001–2002', IEA 2004 edition

Figure 1: World Petroleum Product Consumption in 2002: 3.8 Gtoe

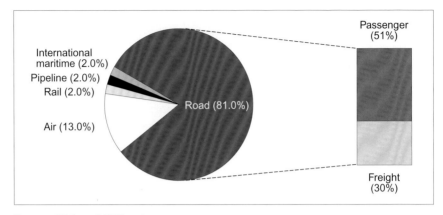

Source: IEA and IFP estimates

Figure 2: Energy Consumption Breakdown for the Transport Sector, 2001

and, despite recent advances in energy efficiency, is still the most energy-intensive mode of transport (per tonne of product hauled and per passenger carried per kilometre).

There has been an increase in per capita motor vehicle ownership over the years mainly because of GDP growth and substantial improvements in infrastructure and technology. In the last 25 years, the vehicle fleet has more than doubled in OECD countries which account for 80% of the world fleet. In 2002, there were nearly 600

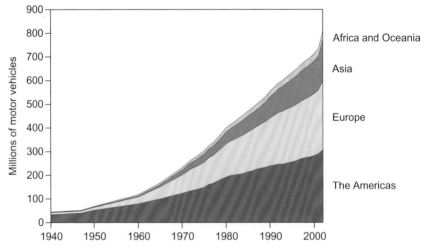

Source: IFP Panorama 2005

Figure 3: World Motor Vehicle Fleet, 1940 to 2002

million private automobiles and 209 million light trucks registered in the world (see Figure 3). In the three key OECD regions — the USA, Japan and the EU — road represented 96% of the 13,760 billion passenger-kms travelled in 2000. Air, rail, tramway/metro and waterway transport account for the rest. Only Japan, which has the requisite infrastructure, reports a larger proportion of public transport than other regions. Since the 1970s, the average household transport budget in the OECD countries has remained fairly constant at about 13% of total household income.

Freight transport is influenced, among other things, by global economic growth and world trade. In the last two decades, world trade (expressed in value) has expanded very quickly compared to GDP. The former increased by 170% but the latter by only 50%. There was also an increase in total distances travelled that was greater than the increase in total tonnage hauled. The road and air segments of freight transport measured in tonne-kms have posted an increase of 120% in the last 20 years, the strongest growth worldwide. Other transport modes (pipe, rail, international and domestic maritime) witnessed more moderate increases, ranging from 50% to 80%. Nevertheless, maritime transport continues to dominate international freight.

In the three key OECD areas (USA, Japan and EU), domestic haulage relies on road transport for 36% of the volume carried, expressed in tonne-kms. Europe and Japan show a preference for maritime and road transport solutions. The freight breakdown between these modes of transportation is more evenly distributed in the USA, where road transport is not economically viable for long hauls and where other modes, especially rail, are — and will continue to be — essential for domestic freight haulage. On the other hand, the road segment is still winning market share from the rail sector in Europe and Japan, where the geographical scale is smaller.

The growing reliance on road transport limits the impact of energy conservation policies in the areas of environmental protection and oil dependence. One key factor in this paradoxical trend is that other modes of transport that consume less energy are not sufficiently competitive and/or lack the necessary infrastructure. These alternatives are becoming less and less adaptable to current economic requirements for three main reasons:

- Residential areas are widely distributed and not always close to railway stations.
- The cost of motor fuel per vehicle-km has significantly decreased since the last oil counter-shock and this has boosted private car

ownership. In 2004, however, the soaring price of oil renewed concerns over the economic implications, and revived a debate about alternatives to gasoline and diesel fuel, in particular bio-fuels.

- Rising consumer demand for specialized, customized products, manufactured 'just in time' led to the production of a greater range of smaller cars. For instance, customers can now choose from several dozen versions of a given model (70 for the Renault *Mégane*, 92 for the Peugeot *307*). The delivery waiting time has been reduced to one month thanks to industrial 'just in time' management systems which are now widely adopted. Road transport is perceived by customers as the most competitive and much freer from the industrial and distribution constraints that affect alternatives such as slow waterway transport, or rail which is only suitable for large-scale flows and is subject to scheduling rigidities.

2.2   *Evolution of Transport Demand and Its Consequences*

Various studies such as those by the International Energy Agency or more recently by the World Business Council for Sustainable Development (WBCSD), conclude that the transport and demand for mobility for people and goods will increase continuously throughout the next 30 years. This fundamental trend is explained by a number of inter-related factors. The main drivers remain demographic and economic growth.

An annual average world economic growth of about 3% is expected during the coming decades. This growth will not be uniform throughout the world; many discrepancies will be explained largely by the differences in local economic conditions. Growth should be particularly high in developing countries, for example in East Asia, Eastern Europe or South America. China and India are expected to be the champions with an average annual economic growth rate in the range of 4.5 to 5 percent. At the same time the GDP of OECD countries should continue to grow although at a more moderate rate estimated at about 2% per year.

Economic growth induces an increase in transport demand through several influencing factors. First of all, it is worth noting that the desire for individual mobility is inherent to human nature, and that people want to satisfy this desire as soon as they can afford it. Therefore, personal mobility closely follows average individual incomes and is strongly correlated to a nation's wealth. But personal mobility is not intended to satisfy only a desire for leisure travel. It is also a very important factor in the quality of life. This often leads to having the family house at a great distance from the workplace. Greater personal mobility may provide opportunities for better jobs and career prospects. Economic activity requires more efficient, fast and easy transport facilities for both people and goods. Some production organizations like the famous 'just in time' one are already based on flexible, reliable, fast and low cost transport. Moreover, international trade which is a very important motor of the world economy relies on a transport network that is able to carry people and goods from any location to the most remote areas.

Consequently, economic development will bring a demand for significant growth of personal and goods transport in the coming decades. This is illustrated in Figure 4, which was published by the WBCSD in

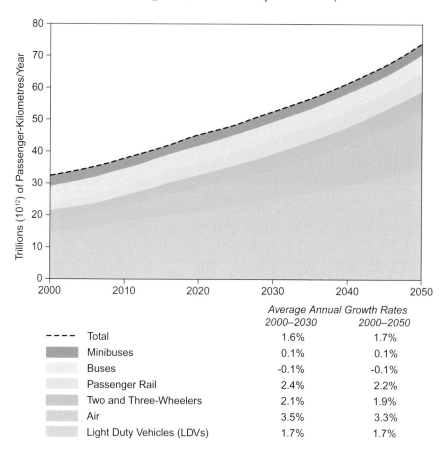

| | | Average Annual Growth Rates | |
		2000–2030	2000–2050
- - - -	Total	1.6%	1.7%
	Minibuses	0.1%	0.1%
	Buses	-0.1%	-0.1%
	Passenger Rail	2.4%	2.2%
	Two and Three-Wheelers	2.1%	1.9%
	Air	3.5%	3.3%
	Light Duty Vehicles (LDVs)	1.7%	1.7%

Source: Sustainable Mobility Project, 2004 (WBCSD)

Figure 4: Personal Transport Activity by Mode

their Sustainable Mobility Project (SMP) report in 2004. According to this study, an average annual growth rate of about 1.6 % is expected during the next three decades for personal transport activity. Most of this growth will take place in non-OECD countries which will represent more than 50% of personal transport activity worldwide in 2030 and this should double in 50 years. Growth in freight transport is even higher. The annual number of tonne-kilometres transported worldwide is expected to increase from 15×10^{12} to 45×10^{12} in 50 years – that is an increase by a factor of 3.

Note that in Figure 4, growth of personal transport is expected to be mainly in air travel and light duty vehicles. According to the WBCSD study, the passenger car will still represent by far the most important transport mode in 2030. This conclusion is confirmed by many other studies and can be explained by the convergence of a number of advantages for the users. Travelling by car is easy, not very expensive and rather convenient for local connections. Furthermore it is a flexible mode of transportation adapted to daily house-to-work journeys as well as to family holidays. Some residents in major countries, particularly in North America, have based their way of life on easy connections by car, in large cities with wide avenues and a distributed habitat. This way of life is likely to be followed by many developing countries.

The impact of growth is likely to be felt most in the production of vehicles. In OECD countries, there has been exponential growth in the last 30 years. In the medium term, developing countries (especially China and to a lesser extent India and Indonesia) are expected to follow this pattern. It is already happening in China where, in the last six years, vehicle production has increased nearly nine-fold to 4.4 million vehicles in 2003 – equivalent to more than 20% of European production. Since the vehicle ownership rate stands at only ten vehicles per 1000 inhabitants in China, this is just the beginning. According to the baseline *World Energy Outlook 2004* of the International Energy Agency scenario (i.e. no major changes in technology or in the behaviour of major players), the non-OECD motor vehicle fleet will triple by 2030 to about 550 million, but remain 25% lower than the OECD fleet. By 2030, the world fleet will double, reaching nearly 1.3 billion vehicles.

Showing a similar upward trend, transport energy demand should reach 3.2 Gtoe by 2030, with oil accounting for 95% of this total. The transport sector is expected to represent 59% of total oil demand by 2030 – compared to 50% today and 37% in 1971 – absorbing nearly two-thirds of the increase in oil demand by that date.

3. Growth Coherent with Sustainable Development

The growing importance of the road transport sector raises a number of questions. Will the world be able to ensure a sufficient energy supply at an acceptable economic cost to sustain its development during the decades to come? Will we be able in this context to reduce local atmospheric pollution? What measures can be adopted to help arrest the increase or even to reduce the greenhouse gas emissions related to transport?

Faced with these challenges, public policies are expected to play a major role in influencing private investment decisions as well as the end users' choices: decision makers will use financial incentives, taxes, regulations and norms in the best interest of society. The market will also be a very important driver: in this case, the end users and customers will base their decisions on the perception of a trade-off between the advantages and drawbacks of any particular means of transport.

Naturally, public policies will also give much consideration to the common interest of society, in addition to individual requests. Most public authorities throughout the world are determined to prevent uncontrolled growth of transport activity which would yield harmful side effects. This means that there is an urgent need to promote the use of low-consumption, low-emission vehicles; and it is vital to support R&D focused on reducing per-unit consumption in conventional vehicles and on alternative solutions. Certain alternative vehicles and technologies whose good environmental performance is recognized are already available. As for freight transport, which generates nearly 40% of CO_2 road transport emissions, there is less room for change, because it is so closely linked to economic growth. Of course, this sector will be able to benefit from any technology advances made relative to the private car. In the meantime, it is vital to stimulate growth for other, less energy-intensive transport modes by making them more competitive and more in line with industrial needs.

3.1 Air Quality

Control of the impact of road transport on air quality is carried out in particular through regulations related to the noise and atmospheric pollutants emitted by passenger cars and heavy duty vehicles. Over time these regulations have resulted in very tough emission limits. Thus it is anticipated that the antipollution measures that have been and will be implemented to cope with the past, present and future regulations will lead to an improvement of air quality despite the increase of traffic and the world vehicle fleet. However, this will be a gradual improvement, following the implementation of regulations and the replacement of old vehicles which emit high levels of pollutants. As a consequence, the process of cleaning the air in developing countries will take more time since the implementation of air quality standards will be delayed.

3.2   Land Use

The land used for transport infrastructures and occupied by the vehicles themselves is another factor to be considered. In particular, it becomes more and more difficult to build new transport infrastructures in developed countries because of the public reaction to expropriation and the opposition of non governmental pro-environmental organizations.

3.3   Oil Dependency

The energy supply is also an important concern. Energy used by the dominant modes of transport (in particular, road and air) is concentrated on fuels produced almost exclusively from oil (more than 98% in 2004). Since no alternative fuel exists that can substitute in transport on a large scale for oil-derived fuels, there will be no immediate change. As oil reserves are by nature limited, it is likely that a production plateau will occur sooner or later and clearly well before 2050. Indeed, a 'business as usual' projection based on transport demand would require by that date a world daily oil production which appears inconceivable.

3.4   Greenhouse Gases

In addition to increasing oil dependence, the transport sector raises major concerns about climate change. Two factors weigh heavily in the world CO_2 emissions balance: electricity production and transport, which account for 41% and 21% of the market respectively (Figure 5). According to the baseline WEO scenario, the transport sector will see its CO_2 balance rise by 78% by 2030.

Burning such a large amount of carbon would lead to an increase of greenhouse gases proportional to the fuel consumption and this would almost certainly be considered unacceptable by the world community. Figure 6 shows the evolution of greenhouse gases (GHG) emissions estimated by WBCSD on a well-to-wheel basis with a 'business as usual' scenario. Light duty vehicles, freight trucks and air would continue to

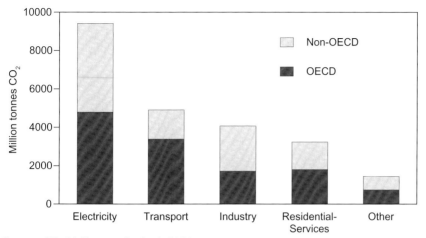

Source: World Energy Outlook 2004

Figure 5: World CO_2 Emissions Breakdown, 2002 (Total: 23.7 $GtCO_2$)

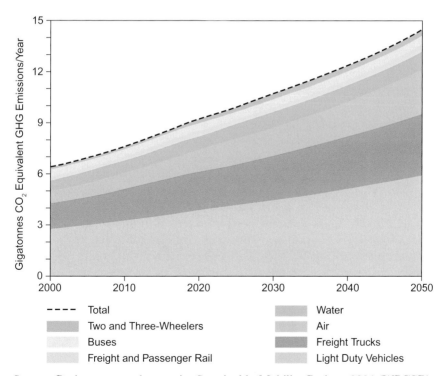

Source: Business-as-usual scenario. Sustainable Mobility Project, 2004 (WBCSD)

Figure 6: Transport-related Well-to-Wheels CO_2 Emissions by Mode

be the major contributors to GHG emissions which would more than double in 50 years.

It is clear from Figure 6 that the business-as-usual scenario leads to GHG emissions and worldwide oil consumption at levels which appear to be unsustainable in the long term. The objectives of the Kyoto agreement for example imply a reduction of GHG emissions below current levels. It is unacceptable that the development of world road transport in the coming decades will result in a doubling of oil-derived fuel consumption and GHG emissions. These considerations lead to the question: what are the most efficient ways to curb GHG emissions?

One possibility is to reduce either the total volume of transport activity or its modal mix. This is difficult as both are influenced by consumer demand and led by the economic growth and development that no one wants to see declining. Rather than reducing transport demand, this is more a question of optimizing the use of the various transport modes in order to improve the energy efficiency or the GHG balance when several interconnected modes of transport are employed. However, to bring about any radical change − for example to promote faster development of rail transport which depends less on oil supply − would be difficult because of the huge amount of investment and public funding it would require. In fact, the modal mix of transport activity is not expected to change drastically when compared to the present situation.

Another possibility is to promote the use of fuel with low or no carbon content. This has a double advantage since it decreases the dependency on oil and reduces the CO_2 emissions in the atmosphere. A well-known example is hydrogen which releases no CO_2 during its combustion. However, the mass industrial production of hydrogen has still to be developed in a way that avoids excessive emissions of CO_2 during the process, if it is produced from a fossil primary energy source. A lot of work is being carried out which could result in the introduction of fuel cells burning hydrogen in vehicles within the coming decades. However there are many hurdles and barriers to be overcome before a significant proportion of passenger cars are equipped with fuel cells. Natural gas also has many advantages, for example it reduces by 22% the CO_2 emitted for a similar amount of energy released. Furthermore, gas fields are relatively well distributed worldwide and distribution networks already exist. Natural gas can be used directly as a fuel in improved vehicles or transformed into liquid fuels through the Gas to Liquid process (GTL). In addition, synthetic fuels can potentially be produced from any fossil fuel, including heavy oil, tar sands or even coal. In that case, the techniques for capturing and storing CO_2 which are currently under development will have to be implemented in order to prevent any net increase of CO_2 emissions. Transport fuels can also be produced from biomass. Due to the absorption of atmospheric CO_2 during the plant growth, and depending on the production process and the raw materials used, the net decrease of CO_2 estimated on a Well to Wheel basis, can range between 40 and 90 percent. For example, ethanol is already widespread in Brazil and North America. Fatty Acid Methyl Ester fuels are also considered for the supply of synthetic diesel fuels, especially in Europe. Wooden materials can also be processed to produce synthetic fuel: a diesel fuel of excellent quality can be produced by the Biomass to Liquid process (BTL) but at a relatively high cost at present. These alternative fuels could represent a significant share of vehicle fuels in 30 or 50 years (both the IEA and WBCSD consider it plausible that 33% of the diesel and gasoline pool could be displaced by 2050).

However, one of the most important considerations is how to reduce the amount of energy required by the vehicle to perform a given amount of transport activity. This involves both the most effective use of the modes of conveyance and an improvement in the energy efficiency of individual vehicles. These two factors will be influenced by the cost of energy which will probably rise continuously during the coming decades in line with energy demand. Pricing of energy and public incentives should be the main drivers. In the next sections we will explain how technology can bring significant progress to the fuel efficiency of road vehicles. The downsizing of internal combustion engines, the hybridization of vehicles and perhaps later the implementation of fuel cell vehicles, will all help to bring about an improvement in the average fuel efficiency of road vehicles.

Finally, according to the simulations carried out by the WBCSD in the framework of the SMP project, and those carried out internally by the Institut Français du Pétrole, it seems that it will be possible to curb GHG emissions from road transport so as to reach a level of GHG in 2050 similar to that in 2000. This is obtained through a combined set of measures: the scrapping of old cars and their replacement by high efficiency vehicles (lower weight, downsized engine), the implementation of hybrid vehicles for passenger cars, a change in customers' preferences towards high efficiency vehicles, the introduction of biofuels with a high CO_2 reduction potential, and so on. In any case, and even if very strong measures are eventually decided on, actual trends mean that CO_2 emissions from transport activities are expected to peak around the year 2020.

In conclusion, curbing these emissions will require the implementation of new efficient technologies as well as some major changes in transport organization. Public incentives will be necessary to make consumers accept that higher prices are necessary to finance the technological solutions required for engines and fuels in order to control the energy supply and GHG emissions in a sustainable way.

4. Transport and Atmospheric Pollution

4.1 Reduction of Local Pollution Caused by Motor Vehicles

Nearly all motor vehicles are equipped with internal combustion engines fed by liquid fuels (gasoline or diesel) produced from oil. The combustion of these hydrocarbons generates a number of pollutants, emitted from the exhausts of engines when the process of combustion has ended. Some of these pollutants are regulated. This is true of carbon monoxide (CO), nitrogen oxides (NO and NO_2, noted NOx), unburned hydrocarbons (HC) and particles of soot. CO, HC and soot are the result of incomplete combustion while oxides of nitrogen arise from the reaction between the oxygen and the nitrogen of the air heated at high temperature. The transport sector contributes mainly to the emissions of NOx and CO (Figure 7) and remains an important emitter of particles and unburned hydrocarbons. The effect of these

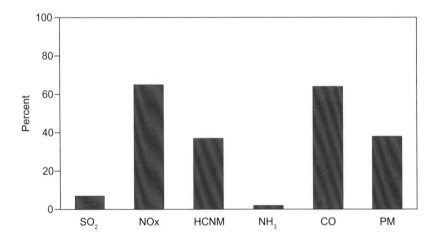

Source: ADEME , 2001

Figure 7: Road Transport Contribution to Global Pollutant Emissions in the EU

pollutants is felt first at a local level, in particular in urban areas. The first effect comes from the primary pollutants which are emitted directly from the exhaust and the pollution takes place close to the sources of the emissions, in particular in areas of dense and heavy traffic. This is especially critical in long narrow streets that are badly ventilated and lined with tall buildings preventing the dispersal of atmospheric pollutants. In some climatic conditions, characterized in particular by a long period of sunshine and little wind, primary pollutants can be at the origin of a secondary pollution that is provoked by a chemical transformation in the atmosphere. This is the case for ozone which is produced by a set of numerous complex chemical reactions, activated by the sunlight and involving NOx and volatile organic compounds (VOC), a part of which are unburned hydrocarbons. Ozone is the cause of most of the summer pollution peaks which generally occur in the neighbourhood of big cities.

For this reason, when drawing up vehicle emission regulations standard driving cycles characterized by typical urban and suburban traffic have been used for vehicle certification. These cycles correspond to engine low load running conditions. For example, the average speed of a vehicle on the whole European cycle is 32 kph and it begins with a start-up under cold conditions. The anti-pollution systems are thus conceived to be completely effective during difficult operating conditions and must be representative of the actual use of vehicles, even if the effective representativeness of the regulated driving cycle profile is regularly reassessed.

Since the 1970s, emission standards have been subjected to successive tightening, notably in Europe and in the USA. At the request of the Public Authorities emission limits have been reduced by a factor of 10:1 to 100:1, depending on the pollutant, within the last three decades (see Figure 8). It is worth noting that these ambitious targets were effectively reached and this is largely attributable to the remarkable progress achieved by technology.

Considerable attention has been given to research and development by car manufacturers and research laboratories. The greatest progress was in the following areas:

- First of all the after-treatment of pollutants by catalytic systems. For example, the so-called '3 way' catalyst allows for the simultaneous elimination of NOx, CO and HC (Figure 9) with more than 99% efficiency when the catalyst is warm. However, these systems require the combustion of an air-fuel stoichiometric mixture (fuel/air equivalence ratio equal to 1).

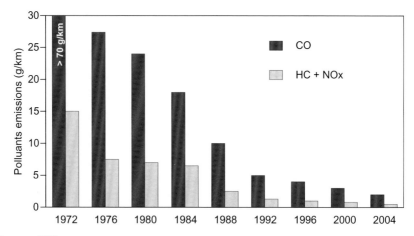

Source: IFP.

Figure 8: Evolution of European Emission Standards (Emission Limits for Passenger Cars

- The electronic control of most of the key parameters of the engine, in particular injection, ignition and air supply
- The injection of the fuel. The availability of high pressure fuel direct injection systems for example has enabled a huge improvement in the performance of the diesel engine.
- The combustion system. Benefiting from progress in numerical modelling, combustion chambers are continually being improved.
- The quality of fuels. The fight against motorcar pollution and the progressive tightening of emission standards have been accompanied by an evolution of fuel specifications. Parameters such as benzene content, aromatics content, sulphur content are now regulated.

It must be recognized that the policy based on the two axes of tight regulation and technological progress has been rather successful. It has resulted in a decrease in emissions of atmospheric pollutants related to motorcars during the past decade, and projections made within the framework of the European program Auto-Oil indicate that the global emissions of toxic pollutants will continue to fall in Europe (Figure 9). The rate of decline of these emissions is essentially controlled by the following factors:

- The application of stricter emission standards following technical progress. The interval of time between the previous and the new standards is about five years.

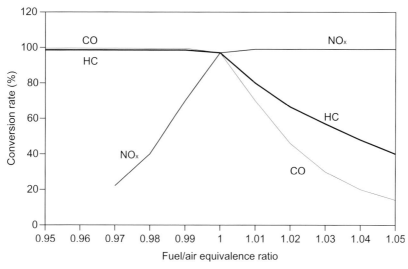

Source: IFP.

Figure 9: Working Principles of a 3-way Catalyst.

- The rate of renewal of the car park. This rate is generally slow, for example the average vehicle age before scrapping is more than 13 years in France.
- The simultaneous increase in the size of the car park and the intensity of road traffic.

4.2 Reduction of the CO_2 Emissions

It is clear from Figure 10 however that emissions of CO_2 did not follow the same path as those of the atmospheric pollutants. The CO_2 emissions are directly linked to a vehicle's fuel consumption and there are no realistic solutions today to eliminate them directly in the exhaust. The growth of CO_2 emissions over the past years illustrates several factors: the increase of the automobile park, the increase of road traffic and of the average weight of vehicles – the latter caused in particular by improved safety and comfort systems. The stabilization of emissions foreseen for the next ten years comes directly from measures taken by all car manufacturers to reduce the average CO_2 emissions of vehicles marketed in Europe.

CO_2 is one of the major greenhouse gases connected to the process of global climate warming. The European Commission has committed itself to reduce its emissions by 8% between 1990 and 2012 within the

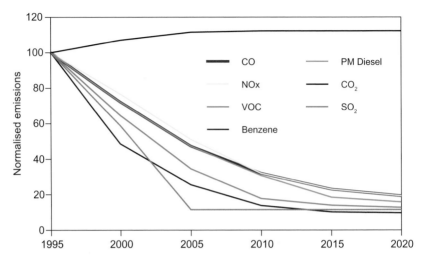

Source: European Commission, Auto Oil 2000

Figure 10:   Predicted Pollutants Emissions of Passenger Cars in Europe

framework of the Kyoto agreement. Road transport contributes 22% of CO_2 emissions in Europe and as a result the main associations of car manufacturers − among which the ACEA (Association of the European Car Manufacturers) − have committed themselves to reducing the average emissions of vehicles marketed in Europe to 140g/km in 2008. This corresponds to a 25% reduction compared to 1995 and a supplementary reduction to 120g/km is also envisaged by 2012. At the same time however, consumers will continue to demand more safety and comfort which will generally induce an increase in the weight of vehicles. Although the US federal administration has not yet ratified the Kyoto agreement, which is intended to reduce human emissions of greenhouse gases, there is some evidence that this could change in the future. For example, with a long history of environmental leadership, California has recently decided (in bill AB 1493 signed by the Governor) to implement emission standards aimed at controlling emissions of greenhouse gases by passenger cars in the future. Near-term standards, phased in from 2009 through 2011, would reduce CO_2-equivalent emissions of vehicles marketed in California by between 22 and 24%; mid-term standards, phased in from 2012 through 2014 would achieve a 30−32% reduction in comparison to the present situation. California will probably be followed by other US states.

The main ways to reduce CO_2 emissions on such a scale are (i) to develop fuels with a low carbon content − more precisely, fuels whose life-cycle leads to reduced CO_2 emissions – or (ii) to reduce the fuel consumption of vehicles. Indeed, for a given fuel, the CO_2 emissions are directly linked to fuel consumption. From this point of view, the combustion of natural gas produces emissions about 23% lower than those produced by an oil-derived fuel for the same energy release. Biofuels are also of interest because plant growth consumes some of the atmospheric CO_2 which leads to a favourable net balance. Among possible biofuel candidates are fatty acid methyl esters – which are produced from vegetable oils –, ethanol and methanol. However, even if biofuels are sure to play an important role in the future, it can hardly be hoped that they will replace fossil fuels at a rate of more than a few percent. For example, in a recent directive the European Commission envisages rates of replacement of the order of 6% in 2010 and 8% in 2020.

Any further reduction of fuel consumption will require additional technological progress. Figure 11 illustrates various ways to reduce consumption by cars during the standard European NMVEG driving cycle, characterized by mixed urban and suburban travel. We demonstrate the fuel consumption decrease (expressed in percent) generated by a 20% improvement in a particular characteristic of the vehicle.

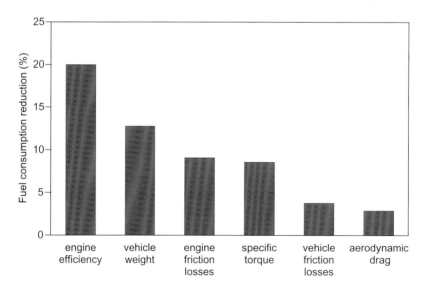

Source: IFP calculations

Figure 11: Main Ways to Reduce the Fuel Consumption of Passenger Cars: Effect of a 20% improvement on each of the main controlling parameters

Improving engine efficiency is found to be the best way to reduce fuel consumption as the latter gains the full benefit of any engine improvement. Weight reduction is also very effective but is limited by the trend towards greater comfort and safety as well as an increasing consumer demand for cars in a higher range, generally heavier. Optimization of the use of conventional materials and of new lighter materials (chassis or engine blocks made of aluminum, intake collector made of plastic) will help to reduce vehicle weight. Reduction of internal friction is also of interest because it acts directly on the efficiency of the engine. This method has been widely investigated during the last ten years and has already enabled engines of the latest generation to benefit from much lower levels of consumption than those of their predecessors. Reduction of the friction losses was made possible by refining the design of the engine, including all the moving parts and the plan of the lubrication circuit. The application of materials with a low friction coefficient and the use of lubricants with a low viscosity have also contributed to this result.

For some time now, many have considered that an increase of the engine specific torque (ratio of the maximum torque to the capacity) results in an increase in the pure engine performance. Nowadays, this is considered to be one of the most effective ways to reduce fuel consumption. This is because there is a reduction of engine capacity while maintaining performance (this is known as the 'downsizing' approach). This solution allows the internal losses of an engine to be reduced considerably which has a direct effect on efficiency. On the other hand, a reduction in the friction of the transmission line of the whole vehicle has a relatively low impact because the levels of mechanical efficiency today are excellent and thus there is a small remaining margin for progress.

The gains to be expected from a reduction of the aerodynamic drag of the vehicle are also relatively small because of the fairly low average speed of the standard driving cycle. On the contrary this parameter dominates for the high speeds observed on highways (130 km/h).

In conclusion, the most effective and direct way to achieve a significant reduction of motor vehicle fuel consumption is to improve the powertrain system which consists of the engine and its transmission. The two main methods of progress are the increase of engine efficiency and of the specific torque and power in order to reduce the engine capacity while maintaining performances at the same level.

The evolution of the transmission system is also a very effective way to reduce fuel consumption. The various systems proposed (continuously variable transmission, automatic gear box, hybrid systems, and so on) are all based on almost complete decoupling between the vehicle speed and the engine rotating speed. This allows greater optimization of the engine operating conditions according to the vehicle's use.

Figure 12 shows how average CO_2 emissions of European vehicles will evolve according to the ACEA commitment as well as those recorded since 1996, on the basis of cars actually marketed in Europe. This shows that until now, the reduction targets have been met. These results are mainly attributable to the large increase in diesel car penetration (which represented about 50% of the market share in 2004). This trend is particularly marked for the heaviest vehicles. Another explanation is to be found in the improved efficiency of gasoline and diesel engines. We note in particular, in the case of the diesel engine, a fast transition from the indirect fuel injection technology to the direct injection one, which corresponds to an average efficiency improvement of approximately 15 percent. The diesel engine has benefited from turbo-charging technology and the downsizing approach, and this trend of replacing gasoline engines by diesel ones is certainly going to continue. However, there are limits linked on one hand to the additional cost of the diesel engine technology, which is not suitable for low cost vehicles, and on the other hand to possible tensions over diesel oil supply if demand increases at a similar rate to that of the last few years. The challenge of reducing motorcar CO_2 emissions will thus continue, in particular for gasoline engines.

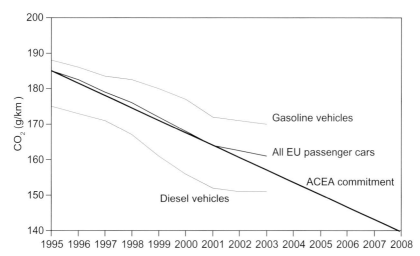

Source: European Commission, Report (COM 269) to the European Parliament, 2005

Figure 12:   Evolution of CO_2 Emissions of European Passenger Cars

5.   Future Engine Technologies for the Motorcar

5.1   Diesel Engines

Because of a different combustion principle, the direct injection diesel engine has fuel consumption and CO_2 emissions approximately 25% lower than those of the gasoline engine for the same performances. In addition this property possesses another potential for progress via the 'downsizing' approach – that is the reduction of capacity while maintaining performance. So, the main European car manufacturers have launched on the market diesel engine families of relatively small capacity (typically 1.2 l or 1.5 l) but with very high specific performances (specific torque of 150 Nm/1 and specific power of 50kW/l). By replacing engines of a higher capacity, a reduction of fuel consumption ranging from 5 to 10% is achieved. This was made possible by the development of two key technologies   high pressure fuel direct injection and variable geometry turbo-charging (VGT). The diesel fuel injection system with common rail[1] reaches maximum injection pressures of 1600 to 1800 bars while the challenging technology, based on unit injectors, exceeds 2000 bars. The high injection pressure ensures a fast introduction of the fuel at full load together with excellent atomization. The turbo-charging recovers the energy available at the exhaust on a turbine which is used to drive a compressor, and this increases the air intake pressure. The increase of the intake air flow rate allows more fuel to be burned and produces more energy from combustion. When the turbine is equipped with a variable geometry system, directional fins deflect the entering gas stream. Their orientation, piloted by the Electronic Control Unit (ECU) and varying with operating conditions, improves recovery of the exhaust gas energy on the whole range of engine speeds. Thus the engine benefits from a higher torque at low and high speed.

The real challenge for the future of the diesel engine is not linked to the fuel efficiency level, which is already excellent, and further development of the injection side and turbo-charging technology will lead to more progress. In fact, the difficulty for the diesel engine will be to cope with future tight emission standards within an acceptable cost.

---

1   With a common rail injection system, the fuel is first pressurized in a rail, common to all the engine cylinders, and then sampled and injected in the combustion chamber using an electronically controlled valve. High pressures and very precise injection timing are achieved independently of the engine speed.

The emission standards in force in Europe for 2000 (EURO III) and for 2005 (EURO IV) (Table 1) are indeed different for vehicles equipped with diesel and with gasoline engines. The diesel engine, because of its 'lean burn' combustion cannot benefit from the '3 way' catalyst to reduce the oxides of nitrogen. This explains why the emission standard limits for NOx are higher for the diesel than the gasoline engine. However, reducing NOx emissions by means of optimizing combustion will probably no longer be sufficient to reach the levels demanded by the standards expected to be in force beyond 2005. Also, reduction of the emissions of soot particles will require the implementation of very efficient exhaust gas after treatment devices.

Table 1:   Emissions Standards for European Motorcars

g/km NMVEG	EURO III year 2000	EURO IV year 2005
Gasoline		
CO	2.3	1.0
HC	0.2	0.1
NOx	0.15	0.08
Diesel		
CO	0.64	0.5
NOx	0.5	0.25
HC+NOx	0.56	0.3
PM	0.05	0.025

Source: European Commission

US Federal requirements concerning the Federal Tier 2 legislation which is being phased in from 2004 through 2009 does not distinguish between gasoline and diesel cars as far as NOx emissions are concerned. It requires that a manufacturer's vehicle fleet of light duty vehicles/light duty trucks on average meets a NOx limit of 0.07g/mile in model years 2007/2009. Moreover, the particulate emission standard has been set to a very low level (0.01 g/mile for bin 1 to 6)

Thus, many innovative technologies are in preparation for an effective and radical treatment of diesel engine emissions. An overview of the main technologies will now be given.

The injection systems will benefit from the application of piezo-electric command devices. The high precision of this command allows a distribution of the injected fuel quantity in multiple injections (5 to 7 different injections) during the thermodynamic cycle. By modulating the process of combustion, this technique enables a strong reduction in the emissions of CO, NOx and particles directly at the source. Thus, the small quantity of fuel which is injected first (the pilot injection) aims to reduce noise and NOx; the one injected just after the main injection aims to reduce soot emissions. By increasing the injection pressure (1800 to 2000 bars) it will be possible to reduce the injector holes diameter to the order of 100 μm. The resulting improvement of the fuel atomization also allows a reduction of soot emissions. In addition, fuel injection systems suppliers are also developing injectors with variable section holes. The small section holes are used at part load to ensure the quality of the atomization and the higher diameter size is used at full load to increase the fuel mass flow rate required for maximum power.

Further improvements to *the combustion chambers* will enable them to take advantage of future fuel injection equipment. The quality of the combustion and the emission of pollutants are indeed very sensitive to the geometrical details of the combustion chamber. Thus the shape of the piston must be drawn with high precision. The considerable progress that has been achieved over the last few years by the three-dimensional modelling of the combustion will be of benefit to research departments.

Significant attention is being given today to developing the *after-treatment systems* of exhaust gases with the aim of eliminating the NOx and particles of soot emitted by the diesel engine.

For a reduction in the emissions of NOx, two main technologies are considered: selective catalytic reduction (SCR) and NOx traps. The SCR is especially envisaged for application to heavy duty vehicles. It requires the installation of a supplementary tank filled with a specific reducing agent, generally urea. Urea is injected into the exhaust line, upstream to a catalyst which releases ammonia further to a hydrolysis reaction. The ammonia is produced in stoichiometric proportions to reduce NOx molecules − in such a way that their elimination is almost complete within the SCR catalyst and so that all the ammonia is consumed during the reaction. A NOx sensor is therefore necessary in order to ensure control of the urea injection. The efficiency of this system is rather good: more than 90% of the NOx is eliminated when exhaust gases are in the correct temperature window (200°C to 550°C). An important advantage of this device is that its efficiency is independent from the system of combustion; the main drawback lies in the obligation to install a relatively voluminous tank of urea.

The NOx trap works on the basis of an alternation between periods when the trap stores the NOx in the form of nitrates and periods during which the NOx is desorbed in order to be reduced in a specific catalyst of a 3-way type. Regeneration of the trap is done in very precise conditions of temperature and air-fuel mixture proportions. Exhaust gases must be the result of the combustion of a mixture in excess of fuel with regard to the stoichiometry so that the reduction of NOx can be realized. The electronic control plays a fundamental role here because it has to ensure that the succession of periods with poor and rich burn gases with regard to the stoichiometry does not lead to an increase in emissions of the other pollutants (HC, CO, soot) or to any engine torque jolt. This system requires good control of the exhaust temperature to ensure optimal storing of the NOx, between 250°C and 550°C. It also requires the use of fuel without sulphur (<10 ppm), otherwise the trap saturates very quickly with sulphates and becomes ineffective. In optimal conditions, elimination of the NOx is higher than 90% at the price of an average over fuel consumption of about 2 to 3 percent.

The second important technology, which is already found in many diesel vehicles of car manufacturers such as PSA Peugeot Citroen or Daimler Chrysler, is the particulates filter. The efficiency of filtration of these systems, generally based on ceramic monoliths (made of silicon carbide, cordierite, and so on) is higher than 95 percent. The main difficulty lies in regeneration of the filter after a period of soot loading. The stored carbon mass has to be burned out without pro-voking excessive thermal stress in the ceramic material. Spontaneous regeneration of the filter is achieved for a temperature higher than 600°C, which is rarely reached during the normal service of diesel vehicles. As a consequence, all the particulate filter technologies applied to the motorcar propose some devices to enhance filter regeneration. For example, most of them use control of the direct fuel injection to achieve the optimal conditions of temperature upstream to the filter and then to facilitate the regeneration. In particular, an increase of the exhaust gases temperature is obtained via a delay of the main injection or via a post injection during the expansion stroke.

The addition of a catalytic additive to the fuel allows the temperature of soot combustion to be lowered to approximately 450°C. The best known additives are based on cerium and iron oxides. The system of PSA Peugeot Citroën includes a supplementary tank of additive which is automatically added to the fuel during fuelling. The advantage of this approach lies in its reliability; the main drawback is the accumulation of metallic ashes in the filter which necessitates a maintenance operation for cleaning after 120,000 km. However, PSA announced recently that it was soon going to increase this distance to 200,000 km.

A filter with continuous regeneration (CRT: Continuously Regenerating

Trap) proposed by the company Johnson Matthey is based on the use of NO_2 as an agent for oxidizing soot and does not require any specific procedure for the regeneration. An oxidation catalyst is placed upstream to the filter and transforms NO into NO_2, which in turn is used to oxidize the soot as soon as it is stored in the filter. The inconvenience of this system is that it requires an engine exhaust gases temperature of the order of 300°C which is rarely reached in city driving conditions. Besides, it is necessary to maintain the rough emissions of NO at a sufficient level to be able to oxidize soot, which is a difficult constraint to overcome. Finally, the duration of the regeneration period is relatively important. A filter with continuous regeneration is thus envisaged very little for motorcar use. It is on the other hand a good candidate for application to heavy duty vehicles because of their conditions of use with highly loaded engines. This filter also requires the use of a fuel with a low sulphur content in order to avoid the generation of sulphate particles on the oxidation catalyst.

The temperature of regeneration can be lowered using a filter whose walls are coated with an oxidation catalyst. Once again the temperature increase at the necessary level for the regeneration is obtained by an action of the Electronic Control Unit on the fuel injection. This type of filter is also very sensitive to the fuel sulphur content which must be limited (< 50 ppm) to avoid the undesired emissions of sulphate particles.

The coupling of the NOx trap with the particulates filter will permit the elimination of both pollutants simultaneously. However, it raises the problem of the respective location of each device in the exhaust line. It must remain compatible with the temperature conditions necessary for both systems to ensure their efficient operation and sufficient durability. Moreover both devices, NOx traps and filters, present specific phases of periodic regeneration, operated according to different processes. The electronic control strategies thus have to guarantee an optimal setting of these phases, by avoiding fatal interactions and by looking for synergies. Generally it is preferable to position the NOx trap upstream of the particulates filter so as to maintain well adapted temperature conditions.

The 'ultimate' solution is the so-called '4-way' catalyst which eliminates four pollutants (NOx, HC, CO, particles) simultaneously. It involves a filter coated with a NOx trap type catalyst which achieves soot filtration and the trapping of NOx on the same substrate. The Toyota Motor company has marketed the DPNR (Diesel PM and NOx Reduction) system which works according to this principle. The problem with this approach is that active sites — which ensure the trapping and reduction of NOx — must be left accessible to the surface of the filter and action is needed to prevent too much soot from covering them. Particular attention is therefore given to the porosity of the ceramic material in order to limit this phenomenon. In addition, the conception of the ceramic substrate has to prevent an excessive increase of the pressure drop through the filter due to the presence of the catalytic coating.

New processes of combustion are also studied with the aim of bringing about a considerable reduction of emissions of NOx and soot at the source without increasing fuel consumption and CO_2 emissions. The basic principle of these new processes is to make the air-fuel mixture much more homogeneous in the combustion chamber. By lowering the temperature of combustion and avoiding zones too rich in fuel, the formation of NOx and soot is significantly reduced. While in a traditional diesel engine combustion is controlled by the rate of injection, here it is mainly the process of auto-ignition that must be controlled.

For example, the IFP process NADI (Narrow Angle Direct Injection) is based on a combustion system consisting of a specific injector whose injection cone angle is very narrow (60° against 150° for conventional systems) and an adapted combustion chamber in the piston. This system allows the fuel to be injected very early in the engine cycle without risk of the cylinder walls wetting, so avoiding dilution of the fuel in the lubricant. It also allows an increase of the time period dedicated to the homogenization of the air-fuel mixture before the auto-ignition occurrence. An adaptation of the compression ratio and the rate of recycling of burned gases is used to ensure a quasi-optimal timing of the combustion, by modulating the speed of the reactions of auto-ignition. The NADI process allows the emissions of NOx to be divided by a factor of 10 to 50 and the emissions of soot particles by a factor of 5 to 10 (Figure 13). The drawbacks of this type of approach are a tendency to an increase in noise emissions, CO and unburned hydrocarbons. These problems are solved by a multiple injection strategy fitted for noise control and by the use of an oxidation catalyst for the elimination of unburned hydrocarbons and CO. A coupling between these new combustion processes, reducing emissions at the source and an advanced after-treatment system, should thus allow the diesel engine to cope with future emission standards despite their increasing severity.

5.2 Gasoline Engines

The after-treatment system which equips gasoline engines is based on the 3-way catalyst that simultaneously eliminates CO, HC and NOx.

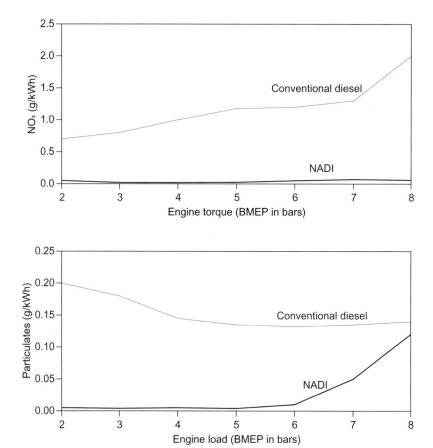

Figure 13: Particulates and NOx Emissions. Comparison between Conventional Diesel and HCCI Combustion (NADI Technology). 1500 rpm

Source: IFP

Since its introduction this device has benefited from many improvements both from the point of view of efficiency and durability, and it is particularly successful today. For example, the vehicles that were approved in California as passing the SULEV regulations (Super Ultra Low Emissions Vehicle, with a limit of NOx six times lower than EURO IV) are not very different from conventional vehicles. The after-treatment of exhaust gases is carried out using a 3-way catalyst which is generally associated with a device controlling the internal turbulent air flow and allowing stability of combustion to be improved during cold start-up. It is completed with an insulated exhaust manifold to ensure a faster warm-up of the catalyst, injection of air in the exhaust to accelerate the catalyst light off at low temperature and a hydrocarbons trap. The system can also be equipped with a light off catalyst located close to the exhaust valve and finally a very efficient fuel injection control system. It is thus possible to reach extremely low levels of emissions with the 3-way catalyst technology, especially when it is combined with an electrical preheating before the engine start-up, to improve even further its fast light off. This latter solution has already been produced on some large or powerful vehicles.

These various devices naturally have a cost but they are extremely efficient.

Really the most important challenge for the gasoline engine is the reduction of CO_2 emissions and consequently the improvement of its fuel efficiency.

Table 2: Technologies Considered for the Reduction of Fuel Consumption (GDI: gasoline direct injection, VVT: valves with variable timing, VVL: valves with variable lift, VCR: variable compression ratio).

(for the same vehicle weight and power)	*Fuel consumption improvement (non additive) by comparison to the standard motorcar combustion engine (gasoline, 3W catalyst)*
Diesel Engine	
• Direct injection , turbo-charging, downsizing	20 to 25%
Gasoline Engine	
Advanced combustion	
• Direct injection (GDI), stratified charge combustion	10 to 15%
• Auto-ignition Combustion (CAI)	10 to 15%
Variable Valve Actuation	
• VVT, intake VVL + exhaust VVT	7 to 10%
• electromechanical valves («camless»)	10 to 13%
Turbo-charging and reduced capacity	
• Indirect injection	10 to 15%
• GDI + VVT, intake VVL	18 to 20%
• GDI + VVT, intake VVL + exhaust VVT	25%
• GDI + VVT, intake VVL + exhaust VVT + VCR	30%
Non Conventional Engine	
• hybrid vehicle (gasoline or diesel)	20 to 50%
• fuel cell (low temperature with or without onboard fuel reformer)	20 to 50%

Source: IFP

Table 2 presents various possible technologies to lower the fuel consumption of gasoline engines in order to reduce the gap with diesel engines, estimated today at approximately 25 percent. Most of the proposed solutions lead to a reduction of the losses related to gas transfers and heat transfers to the walls, the effect of which is all the more negative at low loads.

The first approach is to use direct fuel injection in the combustion chamber rather than indirect injection. The Renault company is the first European car manufacturer to launch on the market a vehicle equipped with a direct fuel injection system engine. Although working with a stoichiometric air/fuel homogeneous mixture, this engine, using the principle of a strong re-circulation of exhaust gases, combined the advantages of reduced fuel consumption with an increase of the maximum torque. Another approach is a 'stratified charge' combustion. The control of the injection allows a rich cloud of fuel to be located in the neighbourhood of the spark plug and for the rest of the chamber to be filled with air. This type of combustion requires a precise optimization of the chamber geometry and the characteristics of the injection. The stratification of the combustion is realized via an injection occurring close to the time of ignition, in the neighbourhood of the Top Dead Centre of the piston. Figure 14 gives an example of the improvement

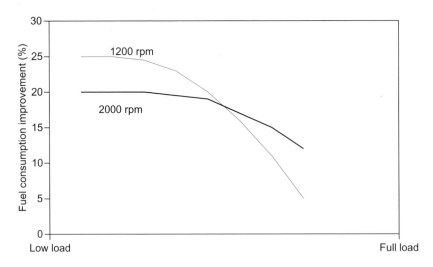

Source: IFP

Figure 14: Fuel Consumption Improvement with Direct Gasoline Injection and Stratified Charge as Compared to Conventional Technology (Stoichiometric Combustion)

in fuel consumption on an engine of 1.2 l capacity. The decrease in this case is about 20% at low load. On the standard driving cycle, expected improvements using the direct injection technology combined with the stratified charge combustion are of the order of 10 to 15 percent. The inconvenience of this process is that it is no longer compatible with the '3-way' catalysts because it works with a lean mixture: it is therefore necessary to equip the exhaust with a NOx trap which tends to degrade the fuel consumption and requires low sulphur fuels(< 10 ppm). In Europe, PSA Peugeot-Citroën and Volkswagen have already marketed vehicles equipped with this type of engine.

New combustion processes like CAI (Controlled Auto Ignition) aim for the same results as direct injection and stratified charge combustion but without the NOx trap technology. The air-fuel mixture is ignited by auto-ignition and not by the spark plug. This is provoked by keeping in the cylinder warm gases that were produced by the combustion in the previous cycle. Dilution of the mixture by a large amount of residual gases results in reduction of heat losses and the lowering of NOx emissions to such an extent that a system of after-treatment is no longer necessary. This process targets a fuel consumption reduction of about 10 to 15 percent. Current improvements aim to widen the CAI mode operating range of the engine and control the transitions between CAI and conventional combustion during the transient phases.

Variable distribution systems are also interesting solutions for the reduction of fuel consumption. These devices, controlled by an electronic control unit, can control the valve timing or even the intake and exhaust valves lift. It suggests an optimal tuning according to the operating conditions of the engine. The fuel consumption gains which can be obtained (between 7 and 13%) depend on the potential of the system. The simplest system, which is already widespread, is used only to vary the angular setting of the camshaft which commands the intake valves. If combined with a double lift system actuated by a hydraulic device, it can realize a number of functions leading to improved engine efficiency. Increase of the specific torque can be used for downsizing the engine. Reduction of gas transfer losses, an increase of the combustion speed, the disconnection of cylinders (another approach to downsizing), and variation of the compression ratio are other ways to improve the engine efficiency using variable valve lift systems. The advanced level is illustrated by the Valvetronic system put into serial production by BMW. This device facilitates continuous control of the intake valve lift level, which leads to a much more precise use of the engine settings and a relatively high fuel consumption decrease (10% reduction is claimed by BMW). Finally, most car manufacturers are studying the possibility of installing systems based on hydraulic or electromagnetic updating, which would be totally flexible. In this case the intake valve lift becomes independent from the camshaft and can occur at any time in the cycle. These solutions are unfortunately relatively complex and expensive but they do open up new and original methods of improvement and have the potential for further fuel consumption reduction.

However, the *downsizing approach* based on supercharging by a turbo-compressor and reduction of the engine capacity while maintaining the level of performance is probably the most promising from the point of view of potential to reduce CO_2 emissions. This approach can be refined and improved at various levels of performance according to the available technology. For example, the use of direct fuel injection is of major interest; indeed when fuel is introduced into the cylinder independently of air, it is possible to use the entering air flow to collect burned gases produced during the previous cycle and still remaining in the cylinder. These hot gases tend to generate knock, an abnormal form of combustion which can lead to the destruction of the engine, and are not compatible with proper combustion at high load. With turbocharged engines, we tend to struggle against the appearance of knock by decreasing the engine compression ratio, to the detriment of fuel efficiency. This is what is generally done in the case of indirect fuel injection, and explains why the hoped for fuel consumption improvements — a 10 to 15% reduction — are only achieved to a limited extent.

Direct fuel injection brings several advantages. First of all, when it is associated with a system of variable valve timing at the intake, it is possible to collect residual burned gases and thus increase the specific torque at low engine speed. By injecting the fuel directly into the chamber, there is the advantage of the cooling effect provoked by evaporation of the fuel. If the injection jet is correctly positioned in the chamber, cooling is directly applied to the fresh intake air and not to the cylinder walls. This cooling effect allows an increase in the knock resistance of the engine and the density of the air intake. These two effects have very positive consequences on the specific torque of the engine. The decrease in fuel consumption can reach 25% when the engine is also equipped with a variable valve timing device at the exhaust. A downsizing of the order of 50% — that is maintaining the performance of the engine in spite of a capacity divided by a factor of two — then becomes possible.

A study aimed at demonstrating the potential of this principle was launched in the framework of a collaboration between Renault and IFP. Figure 15 shows a comparison between two gasoline engines: a

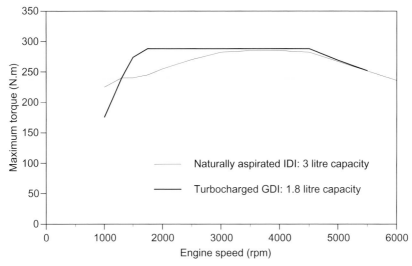

Source: IFP

Figure 15: Downsizing Approach: With about the same performances, the 1.8 l capacity gasoline engine allows, for urban driving, a fuel consumption reduction of about 21% with regard to the 3 l capacity engine (GDI : Gasoline Direct Injection, IDI: Gasoline Indirect Injection)

naturally aspirated one with a 3 l capacity and a turbocharged one with a 1.8 l capacity, derived from a modified Renault engine basis. It can be seen that the smaller capacity engine reaches a torque equivalent to that of the 3 l engine, except perhaps at very low engine speeds. However, the fuel consumption improvement measured on the standard homologation cycle is of the order of 21% to the advantage of the smallest engine.

A further refinement is possible that would see levels of fuel consumption reduced by about 30 percent. This can be achieved by combining the previous approach with a variable compression ratio system. This device, although relatively cumbersome and complex, enables fitting the compression ratio to the instantaneous engine request: a moderate compression ratio at full load to avoid knock and a high compression ratio at low load to maintain efficiency.

Figure 16 illustrates an arrangement allowing for the use of the variable compression ratio; this is an innovative system proposed by SAAB which causes the compression ratio to vary from 14:1 to 8:1 according to need. With this device the volume of the combustion chamber is modified by moving the cylinder head in relation to the piston by means of an articulation between the crankcase and the head.

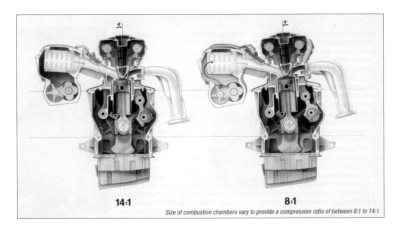

Source: SAAB

Figure 16: SAAB Variable Compression Ratio Engine

5.3   The Non-conventional Powertrains

By non-conventional powertrain, we mean either a deep modification of the architecture of the current systems, or engines dedicated to a new type of fuel such as natural gas, ethanol or hydrogen. Among the non-conventional systems that have reached an advanced stage of development, are already on the market or are being envisaged for mass production before 2020, are the following:

- battery-driven vehicles, the oldest non-conventional type
- vehicles with internal combustion engines dedicated to natural gas, already emerging in specific countries
- hybrid vehicles, the most promising technology
- vehicles powered with a fuel cell, the most challenging technology

Battery-driven vehicles have attracted a lot of interest in the past because of their intrinsic advantages. They were the object of significant governmental incentives (ZEV vehicles in California) and large-scale demonstrations were conducted. Nevertheless this type of vehicle has never met expectations and its market share has remained very low. Of course its advantages are very real: no emissions of pollutants, reduced noise emissions, a high torque for starting which makes the vehicle particularly pleasant to drive in the city. However the main problem remains its limited performance level and low mileage range, typically 100 to 200 km in real use. This situation is largely due to the inadequate performance of the batteries used for the electrical energy storage aboard the vehicle. The power batteries must be able to demonstrate a certain number of attributes to be suitable for vehicle application. The density of power expressed in kW/kg must be high enough to get sufficient acceleration and to recover the braking energy. The density of energy (expressed in W.h/kg) must provide sufficient mileage without excessive weight. Finally the number of cycles of charge-discharge that the battery can accept is an important parameter for its life expectancy in real use.

In spite of the application of other new technologies such as Nickel-cadmium, Nickel metal hydride or Lithium-ion batteries and in spite of the progress already achieved and to come, there is little hope that the density of energy of batteries will increase to any great extent. Figure 17 presents changes predicted between 2005 and 2020. The energy density of a high performance battery typically of 120 Wh/kg will remain much lower than that of a liquid fuel, approximately 12,500 Wh/kg.

The *hybrid vehicle* equipped with a thermal/electric mixed motorization system partially fills this gap. A typical schematic diagram illustrating the hybrid vehicle principle is presented in Figure 18. This one is equipped with two systems of energy storage, a fuel tank and a battery. It also possesses two types of motorization, an internal combustion engine and an electrical motor. In the most flexible configuration,

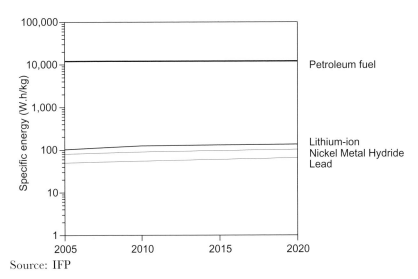

Source: IFP

Figure 17: Expected Evolution of the Specific Energy of Electrical Batteries Compared with a Petroleum Fuel

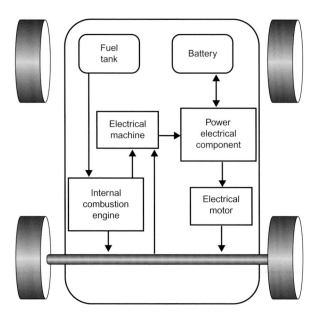

Source: IFP

Figure 18:   Schematic Diagram of a Hybrid Vehicle

many combinations are possible; the internal combustion engine can be used to recharge the batteries as well as to power the vehicle and the electrical motor can be used to move the vehicle as well as to recover the braking energy. The hybridization thus leads to numerous ways of optimizing the use of the energy aboard the vehicle. The hybrid vehicle allows emissions of pollutants to be reduced considerably (a full electric drive is even possible in a city for example) as well as the consumption (possible reduction from 40 to 50 percent). However, it is necessary to be aware that hybridization leads to the installation of two different systems of motorization, two systems of energy storage and more electrical components, which corresponds to a significant additional cost and an increase in the weight of the vehicle. But hybridization is also highly flexible; there are a whole range of possibilities from the light hybrid, which is moderate in cost and performance up to the full hybrid, which gives higher performance and costs more.

According to the level of technology that is implemented, the following strategies can be applied (Table 3, for a vehicle of approximately 1 300 kg):

If the strategy of full hybrid serie/parallel is applied, it is possible at

Table 3:   Possible Strategies for Motor Car Hybridization Depending on
On-board Electrical Power

Functions	Power of the Electrical Motor	Reduction of CO_2 Emissions
1. Stopping the internal combustion engine at idle	2 kW	8%
2. 1 + recovery of the braking energy	3 kW	13%
3. 1 + 2 + downsizing of the internal combustion engine and boost during acceleration	10 kW	30%
4. 1 + 2 + 3 + full hybrid serie/parallel	30 kW	45%

Source: IFP

any time to optimize the operating point of the engine according to the demand for power. The torque transmitted to the wheel can be totally uncoupled from the engine speed and the phases of battery recharge operated on the basis of a strategy of maximal efficiency.

The engine management system ensures that the operating conditions of the engine are always located on an optimal line (Figure 19).

Fuel cells are the subject of many developments linked to the possible occurrence of a 'hydrogen era'. In fact, hydrogen has a number of advantages. Firstly, it must be regarded as an energy carrier like electricity and not as an energy source, since there are no natural hydrogen fields. However, potentially it can be produced from almost any primary energy source – fossil, renewable or biomass. Compared to electricity, it can be stored much more easily for relatively long periods of time. Furthermore, when burned in a fuel cell which itself powers an electrical motor in a vehicle, it produces mechanical energy with emissions of only water vapour and no toxic pollutants. A vehicle equipped with a fuel cell does not emit any CO_2. However, some specific technologies have to be implemented in order to avoid CO_2 emissions during the hydrogen production stage. For example, hydrogen can be produced from fossil energy (coal, oil, natural gas) but CO_2 should then be captured during the process and stored in underground traps. Another method can be the electrolysis of water, provided that the production of electricity is not itself emitting CO_2. In other words, this electricity should be produced from nuclear or renewable energy.

These advantages explain why hydrogen has attracted so much interest during the past years. Nevertheless, it must be admitted that

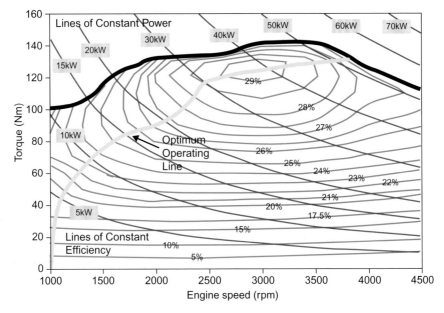

Source: Automotive Environment Analyst

Figure 19: Best Fuel Efficiency Line of the Powertrain System of a Hybrid Vehicle

the 'hydrogen era' is not going to happen in the near future because such a revolution would require significant technological breakthroughs and the mobilization of huge financial investments. And indeed the hydrogen chain from production to use in the vehicle is not yet ready to be implemented, even at the laboratory stage.

Taking into account the complexity of building up a hydrogen distribution network together with the technological issues to be addressed (such as production, transport, storage of hydrogen aboard the vehicle, and so on), and adding the currently excessive cost of all this, then most analysts consider that a significant production of vehicles equipped with fuel cells is not very likely before two decades.

In any case, many challenges need to be addressed before the fuel cell can be envisaged for large-scale production. In particular, it will be necessary to:

- reduce the cost of the components which today would be too high by at least one order of magnitude in the case of mass production. In particular, a significant reduction of the amount of platinum deposited on the membrane of the fuel cell must be achieved. Today the platinum used in a single PEM (Proton Exchange Membrane) fuel cell is by itself as expensive as a complete internal combustion engine of the equivalent power. When considering the price of all the electrical components as well as the fuel cell auxiliaries, a complete powertrain system based on a fuel cell is more than ten times the price of its internal combustion engine counterpart.

- improve the effective energy efficiency of (PEM) fuel cells which is far from the theoretical one. This is due in particular to the energy consumption of the auxiliaries (see Figure 20); furthermore the energy efficiency drops if an inboard reformer is used to transform a liquid fuel like methanol or gasoline into hydrogen.

- define safe and moderate cost equipment for hydrogen storage. The most common way to store hydrogen onboard a vehicle is to use a tank pressurized at 350 bars. However for the same stored energy, the tank volume is ten times larger than that of a diesel fuel tank. Higher service pressure (700 bars) would increase the tank weight and raise some safety issues. Another way is to use liquefied hydrogen stored in a cryogenic tank at a temperature of -253 °C. However if the vehicle is not used, about 1% of hydrogen will be vented every day in order to maintain the in-tank temperature. Moreover, 30% of the energy content will have to be consumed during the liquefying process.

- define and bring to fruition infrastructures for the distribution of

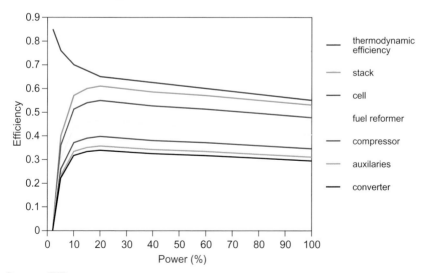

Source: IFP

Figure 20: Energy Efficiency of a Proton Exchange Membrane Fuel Cell

hydrogen; this requires huge financial investments which may not be undertaken by private investors as there is as yet no established market.

- define the procedures and safety regulations relative to the conception and use of vehicles and to the distribution networks of hydrogen.
- Define one or several approved industrial processes for the production of hydrogen from primary energy. The criteria of these processes are: production costs, availability of the primary source of energy, energy efficiency and a life cycle characterized by low net emissions of CO_2. For example, if we consider a process where nuclear energy is used to produce electricity, electrolysis to produce hydrogen which is then liquefied to be used in a fuel cell onboard a car, the overall energy efficiency of the process from 'well to wheel' is limited to only 8 percent. Several recent studies have also concluded that the 'well to wheel' CO_2 emissions of a hybrid vehicle powered by an internal combustion engine fuelled with natural gas was lower than those emitted when the vehicle was equipped with a fuel cell fuelled with hydrogen previously produced from natural gas.

If a hydrogen distribution network becomes available we can also envisage its use in *internal combustion engines dedicated to hydrogen*. The specific issues which then arise as far as fuel efficiency is concerned are close to those of the gasoline engines but with the specific drawback of a significant decline of the torque and of the specific power (40 to 50%) with regard to the gasoline engine. The performance must then be restored by the use of turbo-charging. This solution is often envisaged as a step towards the fuel cell because it will enable the implementation of the whole hydrogen chain to begin, from production to use in the vehicle, without the necessity of producing a large number of vehicles equipped with fuel cells.

The *engine dedicated to natural gas* is also considered to be a good candidate because of the specific qualities of this fuel. Reserves of natural gas are large and estimated at about 60 years of production in the current conditions. Natural gas fields are well distributed all over the planet and this limits the risk of a supply breakdown as a result of geopolitical problems. The emissions of pollutants are potentially lower than those from conventional engines, because of the properties of the gas, and their toxicity and chemical reactivity in the atmosphere are lower due to its composition. Besides, natural gas has a rather high octane number (of the order of 130) which can be used to increase the engine efficiency. Because of the low carbon to hydrogen ratio of the methane molecule, the emissions of CO_2 are heavily reduced compared to fuels derived from oil (23% less for the same energy). With an engine optimized for natural gas a reduction of CO_2 emissions of between 5% and 10% can be hoped for compared to its diesel engine counterpart. Finally, the use of natural gas in a hybrid vehicle is potentially one of the most successful solutions from the point of view of the CO_2 emissions once they are considered according to a life-cycle analysis.

On the technological side, natural gas engines are often derived from the conversion of already existing diesel or gasoline engines. In fact, the market for this type of powertrain is relatively small and does not encourage specific developments. When adapting an engine for the use of natural gas, two approaches are considered. First is downsizing associated to turbo-charging. In the even more successful next stage such an engine is integrated into a hybrid vehicle. The mass production of such advanced powertrains will certainly see developments for application to light vehicles and for buses and urban vehicles. This trend will probably be encouraged by public policies, through incentives and within the framework of the Kyoto Protocol. The European Commission has already published a directive aiming at a progressive replacement of conventional fuels by natural gas (2% in 2010, 5% in 2015, 10% in 2020). Moreover, natural gas vehicles have been promoted as a result of specific public policies and governmental incentives and have developed their market share during the last few decades. Argentina, Italy and Pakistan are good examples with a total of more than 1,300,000 natural gas vehicles on the road.

Ethanol has also been developed as a fuel for passenger cars in some countries like Brazil or the USA. Only minor modifications of the engine are required to make materials compatible with ethanol and these are now well known. Moreover an adaptation of the engine to take advantage of the favourable properties of ethanol would lead to further fuel efficiency improvements. The net well to wheel CO_2 emissions are reduced by 30–90% when compared to gasoline, varying according to the origin of ethanol (sugar cane, corn, beetroot, and so on). However, there are some drawbacks; the sensitivity of ethanol to water traces in the fuel distribution network needs to be controlled and there is a tendency to an increase of evaporation losses if the oil-derived fuel basis is not adapted to the introduction of ethanol.

6.   A Technological Roadmap

Most of the technologies described above require further development before they can be applied to mass production. Neither is it excluded that unforeseen difficulties would lead to the conclusion that their application is impossible. These difficulties may be economic, technical, or commercial.

The schedule of due dates given in Figure 21 concentrates on technologies that could bring about a significant reduction of pollutants and CO_2 emissions and which could be widely applied. As a rough guide, the terms of appearance in the market are defined as being the dates at which we consider that a particular technology could equip a significant part of the European vehicle share (typically 1 percent). We also adopt the hypothesis that the technology being considered should continue to expand from this point; this is notably the case for generic technologies which bring cumulative progress. The scale on the y axis represents the estimated risk of failure for the technology in question. Generally speaking, the more important are the developments needed, the further distant the term will be and the higher the risk of failure.

For the year 2005 we listed technologies that are already applied today and which we think possess enough potential for further development during the coming years. In the case of the fuel cell there is much uncertainty concerning mass production in 2020 as a large number of issues remain to be addressed.

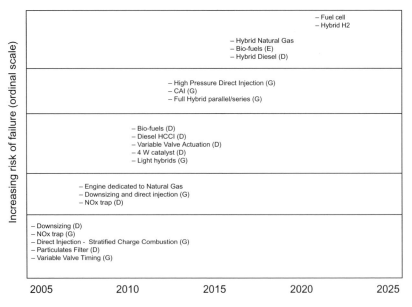

Source: IFP

Figure 21: Roadmap of Future Powertrain Technologies (G: Gasoline Engine. D: Diesel Engine)

Generally speaking, regulations and fiscal aspects will play a large part in the acceleration or slowing down of this schedule. In particular, the next steps of the agreement between the ACEA and the European Commission on passenger car CO_2 emissions after the year 2008 will be of fundamental importance. In the same way, incentives in favour of natural gas, biofuels or hydrogen will be very important in maintaining R&D efforts and bringing these various innovations to the market. Given the high risk of certain options, notably the latest ones, it is important to respect a balance between transitional and long-term solutions and to maintain the development of various rival or additional solutions.

It appears difficult to predict the worldwide situation of energy and transport in 2020. However, it can be expected that there will be more variety than today, both in terms of energy sources and passenger car powertrain systems. It is important to begin preparations now to manage this variety.

7. Conclusion

The motorcar dominates among individual means of transport and plays a very important role in modern economies. Despite major breakthroughs in finding alternative methods of transportation, these need a long time to become economically viable and suitable for mass production; hence the motorcar is likely to continue to grow in importance in the near future. Thus it becomes important to place the expansion of the motorcar and of road transport generally into a context of sustainable development and to take all necessary action to reduce or even remove some of the drawbacks.

First of all, the reduction in emissions of atmospheric pollutants due to road transport remains an important priority; it is thus likely that tightening of emission standard limits will continue during the next decade. In parallel, the contribution of road transport to greenhouse gas emissions must also be reduced in order to be compatible with continuous growth. This is an important issue in line with the commitment of the international community to drastically reduce the emissions of gases considered to be responsible for global warming. This trend towards control of the development of the motorcar is certain to include a significant role for technology, as has been the case in the previous decades.

Important technical innovations are in preparation for controlling emissions of pollutants from modern diesel engines, and these are very successful from the point of view of CO_2 emissions. NOx and soot will be reduced at source via the combustion control, which is made possible by the introduction of very advanced fuel injection systems actuated by a sophisticated electronic control. Advanced after-treatment systems such as particulate filters or NOx-traps will be combined in the exhaust to give the diesel engine ultra low pollutant emissions.

The gasoline engine shows about 25% over-consumption compared to its diesel counterpart and will also benefit from a large number of technological innovations. One of the most promising ways is the downsizing approach which involves reduction of engine capacity while maintaining performance. The combination of various technologies, in particular turbo-charging, direct injection of gasoline and variable valve timing, will allow consumption to be reduced and the gap to be filled with the diesel engine.

The introduction of internal combustion engine/electrical motor hybrid systems will also bring about a marked improvement in the fuel efficiency of the powertrain whether the engine is gasoline or diesel. This will be achieved at the expense of increased complexity and higher production costs. The importance of the hybrid system is that it can be implemented progressively, starting with relatively simple and quite low cost solutions and moving on to more complex and higher cost ones but with much higher fuel efficiency.

It is likely that the future will see a greater variety of sources of energy and of powertrain systems. In this context, engines dedicated to alternative fuels such as natural gas or biofuels should find a significant market share because of their excellent potential concerning net emissions of CO_2. In any case, increased passenger transport demand will induce such a growth in energy demand that different alternative fuels will probably be increasingly used. Synthetic fuels which can be produced from various sources are of particular interest and it is easy to predict that in the next decade liquid fuels produced from natural gas (Gas to Liquid), from coal (Coal to Liquid) and from biomass (Biomass to Liquid) will add their contribution to the pool of transport fuels. In all likelihood, however, the industrial production of these fuels will require technology to ensure the capture and sequestration of the extra CO_2 produced.

Fuel cells are also being considered but this is a rather long-term project as it will be necessary to address many issues arising from the use of a totally new energy chain. The CO_2-free production of hydrogen is one of the most important challenges, but moderate cost technologies for its storage and the fuel cell aboard the vehicle have yet to be developed and proven for real use application.

References

Communication from the Commission to the Council and the European Parliament, 'Implementing the Community Strategy to Reduce CO_2 Emissions from Cars: Fourth annual communication on the effectiveness of the strategy'. (Reporting year 2002) COM(2004) 78 final.

Communication from the Commission to the Council and the European Parliament, 'Implementing the Community Strategy to Reduce CO_2 Emissions from Cars: Fifth annual communication on the effectiveness of the strategy'. COM(2005) 269 final.

International Energy Agency (2004), *Biofuels for Transport. An International Perspective.*

IFP PANORAMA, (2005), February, 3 and 10. Proceedings.

Pinchon, P. 'Les Motorisations pour l'automobile du futur', *Revue de l'Energie*, 556, May 2004.

Mobility 2030, 'Meeting the Challenges to Sustainability. The Sustainable Mobility Project', Report WBCSD, 2004.

Abbreviations and Glossary

ACEA	Association of European Car Manufacturers
ADEME	French Environmental Agency
BTL	Biomass to Liquid (synthesis process)
CAI	Controlled Auto-Ignition (combustion process)
CO	Carbon Monoxide
CO_2	Carbon Dioxide
CRT	Continuously Regenerating Trap (Johnson Matthey)
DPNR	Diesel Particulates Matter and NOx Reduction system (Toyota)
ECU	Electronic Control Unit
EU	European Union
EURO III	European Emission Regulation (III: year 2000, IV: year 2005)
GDI	Gasoline Direct Injection
GDP	Gross Domestic Product
GHG	Greenhouse Gases
GTL	Gas to Liquid (synthesis process)
HC	Un-burnt Hydrocarbons
IEA	International Energy Agency
NADI	Narrow Angle Direct Injection (diesel combustion process from IFP)
NG	Natural Gas
NMVEG	New Motor Vehicle Exhaust Gas (European emissions standard driving cycle)
NOx	Nitrogen oxides
OECD	Organisation of Economic Co-operation and Development

PEM	Proton Exchange Membrane fuel cell
PM	Particulates Matter
R&D	Research and Development
SCR	Selective Catalyst Reduction
SMP	Sustainable Mobility Project, WBCSD
SULEV	Super Ultra Low Emissions Vehicles (California emissions standard)
VCR	Variable Compression Ratio
VGT	Variable Geometry Turbine (for turbo-charger)
VOC	Volatile Organic Compounds
VVL	Valves with Variable Lift
VVT	Valves with Variable Timing
WBCSD	World Business Council for Sustainable Development
WEO	World Energy Outlook, IEA publication
ZEV	Zero Emission Vehicles (California emissions standard)

CHAPTER 10

RENEWABLE ENERGY

Robert Mabro

1.   Introduction

Renewable sources of energy fall into two categories. Some are part of commercial energy and some belong to a set referred to as non-commercial. The first category includes solar, wind, tidal, wave and geothermal energy as well as hydroelectricity and biofuels; the second which virtually constitutes the whole group of non-commercial sources includes firewood, dung, and some agricultural waste.

The production of commercial energy involves capital investments and the output is sold and bought in markets. Non-commercial energy is collected by the rural population, and sometimes by the urban poor who live at the edge of towns. It is either a gift of nature or the by-product of animal husbandry and other agricultural activities. It is renewable in principle but sometimes not renewed because of either soil erosion or the devastating impacts of droughts.

The share of renewables in world total primary energy supplies (TPES) is very small. The most recent data, for year 2003, are as follows [IEA, 2005]:

Total primary energy supplies: 10579 Million tonnes oil equivalent
Of which: Hydro 2.2 percent
                Geothermal, solar, wind etc 0.5 percent
                Combustible renewables and waste 10.6 percent
Total primary commercial energy supplies: 9490 Million tonnes of oil equivalent
Of which: Hydro 2.3 percent
                Geothermal, solar, wind etc 0.56 percent
                Commercial combustible 0.32 percent

Interest in renewable commercial energy has been increasing in recent years for two major reasons. The first relates to environmental concerns and the second to security of supplies. This does not apply of course to non-commercial energy. On the contrary, their impact on the environment is often negative (soil erosion for example), and the poor are more immediately concerned with the availability of firewood and other biomass materials than by the security of oil and gas imports, items they can rarely afford. They also yearn for economic development which involves a shift from the use of non-commercial to commercial energy.

The environmental concern that underlies the interest in commercial alternatives to fossil fuels is essentially global warming. On that issue, one may ask whether nuclear energy provides a more efficient answer to the climate problem than the range of renewable energy, from wind to bio-fuels, which are capturing so much attention. But nuclear has its own problems because of the risks of accidents either in the plants or in the disposal of waste. These accidents release radiations with adverse effects on the health of those who become exposed to them. The fear of becoming a victim of leukaemia is more potent than that of suffering from the impact of climate change.

There are two elements in the supply security concerns that promote a drive towards renewable energy. The first element relates to supply disruptions. Oil is perceived as particularly vulnerable because a large share of world production, and an even larger share of traded oil, are or originate in parts of the world susceptible to political upheavals. Memories of 1973, the Iranian revolution of 1979, the Iraq-Iran war of the 1980s, the invasion of Kuwait by Iraq in 1990 remain vivid in some government departments of OECD countries and in Brussels. The fact that is often ignored is that in all these instances, and as recently as in 2003 when the US/UK intervention in Iraq coincided with supply problems in Nigeria and Venezuela, OPEC Members not affected by a particular crisis did their utmost, sometimes visibly, sometimes discreetly, to make up for the shortfall. The European Commission has now added natural gas to oil in the list of fuels that give rise to supply security concerns. This is a bit ironical. Similar concerns were expressed by the USA in the 1970s when a major pipeline was planned to supply Western Europe with gas from the Soviet Union. At the time these worries were dismissed by the European countries concerned. Today, despite the collapse of the Soviet Union, the emergence of a more friendly Russia and the end of the Cold War, it is Europe that is raising the issue once again. Finally, the energy security topic has been expanded to include the risks of electricity brown- and black-outs. However, these arise domestically because of inadequate investments in generation or regulatory failures. Replacing fossil fuels by renewable energy is irrelevant to this issue.

The second element of the security of supply concerns relates to the notion that conventional oil production is close to a peak. The pessimists think that we are within a short number of years – one to five perhaps − from this peak. Others believe that it may be reached in 2015 or thereabouts, which is soon enough. The peak oil theory begs more questions than it answers. The first question is about the oil concept used. Oil is one type of liquid fuel, and liquid fuels are those that matter. The second question is about the shape of the production curve. Many possibilities arise: the peak may take the shape of a plateau; the rate of decline following the peak may be very small; the production curve may take the form of a saddle with a second peak following the first as happened in the UK continental shelf in the 1980s. The third question is about when the peak will be reached. That oil production (unless discontinued) will reach a peak on one day or another is inevitable. The prediction is of the following type: all men are mortal; Socrates is a man; thus Socrates is mortal. As such it is of no great interest. I know that I am mortal. I know that an exhaustible resource is always on the way of being exhausted (unless displaced in use by another commodity). What we need to know is the date when I will die, and in this context the date when peak oil production will be reached. The fourth question is about the relationship of production to demand. By definition, production will begin to decline after the peak is reached. But this is of no significance by itself. The production decline rate must be compared with the behaviour of demand. If demand at that time happens to decline at the same, or at a faster, rate than production, worries about the adequacy of supplies will prove to have been in vain. And the final question is: assuming that a peak in oil production is correctly predicted to occur at a time of growing oil demand would the reliance on renewable energy provide the necessary solution? The adoption by various pro-renewable energy lobbies of the peak oil theory does not really enhance their case. Like other advocacies by lobbies which espouse any argument they can find, the recourse to the peak oil theory in this context only clouds the debate.

Commercial sources of renewable energy are developed in the first place when economic conditions favour an investment. Their development, however, may receive a boost from governments wishing to pursue objectives related to the environment and the security of supply as mentioned above. Governments may provide subsidies or indirect encouragement by taxing more heavily competing fuels. An interesting question is whether the pursuit of one of these objectives (say, the environment) necessarily yields benefits in terms of supply security, and vice versa. There is no doubt that the implementation of these two objectives produces overlapping sets of effects, but the two sets are not entirely identical.

Certain renewable energies – geothermal, solar, wind, wave hydro

– are used to generate electricity or heat. Their development displaces fossil fuels – coal, gas and oil – in electricity generation. Under *ceteris paribus* conditions this leads to a reduction in the emission of greenhouse gases. Displacing coal yields the greater benefit to the environment but does not serve very much the security objective. Coal supplies are generally considered safe because the greatest share of consumption consists of domestic production, and quantities that are traded internationally originate in countries perceived to be politically reliable. Furthermore, there are no worries about imminent exhaustion. Coal reserves at current production rates can last for at least 400 years.

Displacing oil from electricity generation serves the environmental and, to a very limited extent, the security objectives as conceived by policy makers in oil-importing countries. The limitation is due to the small proportion of world oil consumption that goes into electricity (and heat) generation. Only 7.3 percent of world oil consumption in 2002 was used to produce electricity or heat. Natural gas emits less CO_2 than either coal or oil, and is more benign as regards sulphur and particles. Methane is a potent greenhouse gas, yet its release in the atmosphere from leaking pipes or in production plants is not often talked about. Displacing gas is largely irrelevant in terms of supply security. It is odd that concerns about this issue are arising at a time when, as mentioned earlier, the Soviet ghost is not casting its shadow, and when international trade in gas is expanding and involving an increasing number of sources.

As we shall argue the problems that affect renewable energy, in many cases, are costs; for almost all of them there are problems over the specificity of location, and in critical instances over the fluctuating patterns of supply.

In this chapter, Section 1 is the introduction, Section 2 describes succinctly various sources of commercial renewable energy, and Section 3 compares their respective merits and drawbacks.

2.   Description[1]

Geothermal. The resource is heat generated at certain depths in the earth. The bulk of the heat flow is provided by radioactivity resulting from the disintegration of uranium, potassium or thorium. The bulk of usable heat is in the form of a stock accumulated over the centuries mainly in underground water formations. This raises the question of

---

1   These descriptions are based on Vernier (2005); Armstrong & Blundell (2005);

whether geothermal is a renewable source of energy. The answer is that, for all intents and purposes it is because the temperature of a particular heat reservoir is continually raised by hot earth. There is an important proviso however: the rate of depletion should not be higher than the replenishment. A resource that is in principle renewable does not necessarily imply that it will be renewed.

Geothermal temperatures are high in regions situated on the perimeter of tectonic plates. These regions are usually volcanic or prone to earthquakes. In other places temperatures are much lower, say, 30° Celsius at 1000 metres depth against several hundred degrees at that same depth in the former regions.

A good geothermal site combines high temperatures with underground water in a permeable rock. The water must be able to circulate. Yet, heat in dry rocks can also be exploited by injecting water on the formation. This approach is being tested raising hopes of a significant increase in the potential of geothermal energy.

Geothermal can be used to generate electricity if high temperature (greater than 150°) heat is captured. Wells are drilled to depths of 1500 to 2500 metres, even to greater depths. The temperature of the heat source in the ground that is used is usually in the 200°−280° range. Electricity is created by pumping water (sometimes oil) in the ground and using the hot gases resulting from evaporation to run turbines. In 2004, the capacity installed was 8912 MWe generating 56798 GWh/year. It is sobering to note that the total capacity of geothermal electricity in the world is less than 1.0 percent of the generating capacity extant in the USA.

Geothermal can also be used for heating either through deep earth pipes or heat pumps. When the water temperature is low (20°−40° Celsius) it is necessary to use a heat pump. The process is as follows. The warm water is used to bring to the boil a liquid that boils at low temperature. The gas obtained is then compressed to raise both its temperature and pressure. The gas now at a high temperature heats water which can be used for central heating for example. The use of geothermal in this system requires the injection of electricity for compressing the gas. There may still be a net gain in energy use.

Solar Power. The sun may not be the father of all things as Aristotle seems to have thought but is certainly the origin of most sources of

The Wikipedia, the free encyclopedia (*http://en.wikipedia.org/wiki/Renewable_energy*); US Energy Information Administration (*http://eia.doe.gov*); IEA (2005)

energy available to mankind. And this includes fossil fuels. It is expedient however to avoid an all-embracing concept of solar energy and focus in this section on the energy we receive *directly* from the sun. Others such as wind, tides and waves which are attributable indirectly to the sun will be discussed in subsequent sections.

The sun provides light and low temperature heat during daytime without any conversion or any other artificial process to harness its energy. These free benefits vary according to the time of the day, location, climate, weather conditions and seasons. We take all that for granted as we do for the air we breathe and the beauty of a landscape. The power thus received on the surface of the earth varies on average from 85 to 290W/m² per year. The variation is large but it is important to recall that no region is deprived of solar energy.

Architectural designs that maximize the exposure of buildings to the sun take good advantage of the 'passive benefits' of solar energy.

Solar energy is harnessed or converted to produce heat or electricity. Solar heating systems generally consist of thermal collectors, a fluid that transfers the heat from the collector to the point of use and a reservoir to stock the heat. Low temperature heat (30°−50°) is usually sufficient to heat water for household purposes, for space heating with radiators or preferably floor heating coils, or swimming pools. Solar installations for residences are of two types: compact and pumped systems. In cold climate, and often elsewhere, both types include an auxiliary energy source which is activated when the temperature of the water in the tank falls below a pre-set level, so that hot water becomes available when sunshine is scarce.

Compact systems consist of panels (usually flat thermal collectors), pipes and a tank for the heated water placed above the panels. Thanks to the thermo-siphon principle (a heated fluid becomes lighter and therefore rises) the heated fluid in the collector moves up into the tank and expels the cold water. There is no need therefore for a pump. Pumped systems are used to heat buildings like schools or hotels that are bigger than a normal residential house. In these the tank is placed inside the building and a pump is used to move the water. An input from another energy source is thus needed to operate the pump. This is in addition to the fuel requirements of the auxiliary heating system that complements the solar appliance for heat to be provided on a continuous basis.

Solar is also used for electricity production. Two different systems are involved. In the first the relationship is from solar energy to high temperature heat to electricity. This system produces electricity in solar power stations. To obtain the high temperatures (several hundred degrees Celsius) required the solar rays have to be concentrated towards a focal point. The principle is that well known to school pupils who amuse themselves with a lens in order to set alight a piece of paper. Solar power stations mainly differ by the manner in which the sun rays are concentrated. One type of plant uses cylinder-parabolic concentrators. These consist of long cylindrical mirrors that rotate to follow the sun. In this system the rays are concentrated along a line rather than on a point. The temperature that can be reached is 500°C at the maximum.

Other plants include heliostat mirror power plants, also known as power towers, and parabolic reflectors power plants. These can produce temperatures of up to 1000°C because they concentrate the solar rays in a focal point. Yet the difference between the first type of plant (cylinder-parabolic concentrators) which are in actual use and the power tower and the parabolic reflector types is that these two have been at the experimental stage until recently.

The second system for producing electricity with a solar energy input uses the photovoltaic phenomenon. The principle is simple and was in fact already known early in the nineteenth century (1839). Photons hitting some materials known as semi-conductors will release electrons, thus producing electricity. The semi-conductor most commonly used is silicon (Si), an extremely abundant resource which constitutes more than a quarter of the earth's crust. Photovoltaic cells are small. The power of a single cell is typically 1 Watt and the voltage 0.5V. In order to provide an electrical input to an appliance several cells have to be used by connecting them in series, and fitting a number of series in parallel. The cells have to be protected in some covered frame, be connected to the appliance and often to some storage system, and be installed in a way that keeps them correctly exposed to sunlight. All that can be cumbersome and expensive.

The main uses of photovoltaic systems are (a) for appliances in isolated locations not connected to electricity networks, as often is the case in developing countries, (b) for generating electricity to feed a distribution network.

Outside electricity generation, photovoltaic cells are used in a variety of small applications such as watches, calculators, small electronic gadgets and in space crafts.

Wind. Our distant ancestors knew how to harness the wind to produce mechanical energy in those windmills that inspired poets and painters and elicited so many fantasies in Don Quixote's mind. We have now moved from the use of mechanical energy for grinding wheat or pumping water to its transformation into electricity. This is achieved with the use of wind turbines. They consist of:

a) A tower which should be as high as possible because the wind speed is greater at heights than near the ground. And the power of a wind turbine is proportional to the cube of that speed.
b) A set of two, or more usually three, blades. The diameter of this set has been designed in the 40–70 metres range in recent years.
c) An electricity generator.

The maximum power that can be obtained is a function of the density of the air, the area swept by the blades, and as mentioned before the cube of the wind speed. This does not mean however that very high wind speeds can be accommodated. A French nursery song warned a sleeping miller that he should wake up because his mill is turning much too fast! The wind pushes the blades backward and then lifts them; and it is this lifting force that makes the blade turn. Brakes are installed to stop the blades turning wildly when the wind blows at too high a speed.

Wind energy, apart from its extensive use for pumping water, produces electricity either in isolated places not covered by distribution networks or where it can be connected to these networks.

Water Energy. This includes hydroelectricity and the generation of electricity using the movements of tides and waves. Hydroelectricity is considered a renewable source because it depends on rainfall which is a recurrent phenomenon in different seasons every year. The principle is simple. In sites where a waterfall exists or where it can be created with the construction of a barrage the potential energy of the falling water can be harnessed to generate electricity. The power that can be obtained is a function of the height of the fall and the water flow (volume per second). If the height H is measured in metres and the flow F in cubic metres per second, the power P in kilowatts will be given by:

$$P = 9.81FH$$

Losses have to be allowed for however, and these may reduce P to $8FH$ (Vernier, 2005, p.48).

A hydroelectricity system consists of:

a) A barrage always necessary to create or increase the height of the waterfall and which sometimes also creates a reservoir for the storage of water. Storage is necessary where rainfall is irregular.

b) Canals to feed the turbines and return to the river water leaving the turbines.
c) A power station with turbines.

Barrages are either big and heavy civil engineering structures which transmit through their weight the water pressure to the ground or vault-like structures which transmit the pressures to the vertical sides of the mountain. The former is expensive because of the volume of cement and earth fill material required, and the latter calls for great care in design and execution. The Aswan High Dam is of the former type.

Hydroelectricity is a renewable source of energy but the scope for growth is limited by the availability of suitable sites and the serious and complex environmental problems that affect many of them. There is little scope for growth in developed countries so that future expansion is most likely to take place in the developing world. But more on these points later.

Tides involve a difference in the water level between high and low tides. As for hydroelectricity the idea is to harness this potential energy in order to generate electricity. Near the coasts the difference in levels can be as high as 15 metres as is the case in the Bay of Fundy in Canada or close to 14 metres in the Severn in Great Britain. The system that turns tidal energy into electricity consists of a barrage with sluices and a power station. The sluices are opened when the tide rises and thus fills the reservoir created by the barrage. The sluices are closed when the tide begins to recede. At low tide there is a difference in levels between the reservoir and the sea, a water head which will move the turbines and generate electricity.

Waves are caused by wind hitting the surface of oceans and seas. The energy created and propagated by these shocks is small, but the area covered by the waves is immense. Much energy can thus be harvested near the coasts. There are several techniques to translate wave energy into electricity, either though the use of buoys, oscillating devices or narrowing channels that raise the height of the waves and create a water head used to move the turbines.

Wave energy varies from day to day as anybody on a beach holiday readily observes, and even more significantly from season to season. It shares this characteristic with other renewable sources of energy.

Biomass and Waste. A bio-fuel is a solid, liquid or gaseous fuel derived from biomass. The solids that can be used as fuels through combustion are wood, straw, dried plants, cattle dung and other animal droppings, husks and bagasse. The liquids are ethanol which is mainly produced from sugar cane in Brazil and corn in the USA, and biologically produced oils that can be used in diesel engines. The gaseous is methane produced by decaying garbage or manure in places where the methane can be collected.

Biomass is considered to be a renewable source of energy because plants and trees may grow again, and animals breed, and garbage is in continuous supply. Alas, there are many places where de-forestation and soil erosion prevent the renewal of plants, bushes and trees.

Plants grow through photo-synthesis which takes water plus CO_2 from the atmosphere and yields under the action of solar energy vegetal matters and oxygen which is released back to the atmosphere. At first sight this appears to be a magnificent system that removes carbon dioxide, releases oxygen and produces vegetation. From an energy perspective things are not as brilliant however. Biomass energy has a low density and its energy efficiency is very small.

There is not much to say about wood that is not familiar considering that it was used as a fuel throughout the centuries either directly or after its transformation into charcoal. It is still used but more in developing than in developed countries.

The 'modern' bio-fuels include ethanol which is an alcohol that can be produced from sugar plants, plants rich in starch, or from wood or straw as these contain cellulose. In Brazil, ethanol is used as an automotive fuel either on its own in specially designed car engines, or mixed with gasoline (approximately one part alcohol and three parts gasoline) in ordinary engines slightly adapted for the purpose. In the USA, the alcohol input in a car engine is limited to 10 percent of the mix, and elsewhere the percentage is even smaller. In these cases the alcohol is more of an additive that improves the octane index than a significant fuel.

Next to ethanol are bio-fuels used in diesel engines. These are biologically produced. They include ordinary vegetable oils from sunflower, soy beans, palm, rapeseed and so on, waste vegetable oils and bio-diesel obtained from the transesterification of animal fats and vegetable oils. Esters are superior to ordinary vegetable oils because they are less viscous and are better at self-combustion in the engine. The preferred bio-diesel today is the rapeseed oil ester.

3. An Assessment

Renewable energy sources differ in many of their important characteristics and should not be lumped together in a single category. Differences do matter. At the same time, there are certain similarities relevant to a comparative analysis.

Let us consider the similarities first. A number of renewable energies are site specific in many, if not all, their applications. Oil, natural gas and coal have a global reach as they can be, and indeed are, transported across the globe in international trade. By contrast, geothermal can only be used directly *in situ* because steam or hot water cannot be transported far away from their sources. Solar in applications other than electricity generation with links to a network is for local use only – heating water in a house or a communal building to give one example. Non-commercial biomass is not really a tradable commodity. However, many renewables advocates argue that an important value of renewables is precisely that they provide local solutions. From this perspective, noting the small share of renewables in global TPES is not a fair gauge of their success. Regarding geothermal, for example, a more valid evaluation is that in, say, the Philippines, more than 20 percent of the electricity generated comes from geothermal. To be sure, electricity generated by wind turbines, waves, tides, a solar or a hydroelectric power station will have a wider reach within a country or in neighbouring countries if the plants are connected to a national or regional grid. Electricity, however, does not travel as far as oil or coal, or as natural gas is increasingly able to do thanks to progress in LNG technology.

In many cases a renewable source supplies energy intermittently. This is clearly the case for wind, solar, waves and hydro. The wind 'bloweth where it listeth' as the Gospel of St John puts it. Exposure to the sun is subject to the alternation of days and nights, meteorological conditions which now cloud the sky and now keep it brilliantly clear, and to seasonal variations. Waves depend on the impact of winds on the surface of seas and oceans that is on a variable source. Hydro depends on rainfall.

The 'intermittency' can be managed as detailed in an IEA Working Paper by Timur Gül and Till Stenzel. Yet all solutions to the intermittency problem involve costs, sometimes small but often sufficiently significant to put a particular kind of renewable energy at a competitive disadvantage. Possible solutions include the recourse to a back-up source of energy, the geographical integration in a single system of several energy sources, the storage of either energy or in certain cases water, to mention but a few.

When water is the output of an operation using a renewable source of energy, intermittency is coped with by storing some of the water pumped up when the wind is blowing or the sun is shining. The water thus stored can then be used when the wind or the sun ceases to supply needed energy for a while. This solution is available to compensate for short-term (say, intra-day) fluctuations; not however for seasonal variations that may require huge volumes of storage. When the renewable source is water energy, water is stored at times of heavy rainfall, or through the diversion of a river, for use in dry seasons.

The complementarity of different energy sources can be taken advantage of when their natural cycles do not coincide. Winds, for example, sometimes blow when the sun shies away. As the actual patterns of these natural cycles are affected by uncertainties, solutions exclusively based on the use of two renewable energies are not often contemplated. Hybrid systems will usually involve a conventional energy source, for example wind turbines, photovoltaic sets *and* a diesel generator.

Storage of electricity (the output in these cases) is not an attractive option. '...there has been very little commercially available storage technology that operates in today's electricity grids' [IEA, 2005, p.19]. To smooth out very short-term fluctuations in electricity output one can use flywheels or batteries. These are micro solutions for variations over periods shorter than one minute and involve significant energy conversion losses.

Electricity systems usually involve different sets of plants for base load and peak load use for example. The latter will be used intermittently and this opens a window of opportunity for a renewable energy source with, say, a daily cycle conforming to the electricity demand cycle. There are other solutions that fit in the complexity of electricity supply systems [IEA, 2005].

A third common feature of commercial sources of renewable energy is their small shares in total primary energy supplies (TPES). Data for 2003 are presented at the beginning of this chapter. A further breakdown gives the following TPES shares:

Wind:	0.05 percent
Solar and tide:	0.04 percent
Geothermal:	0.41 percent
Total:	0.50 percent

Only the hydroelectric share, although small at 2.2 percent, cannot be considered to be negligible. These small initial shares do not by themselves tell us anything about growth potential. The proponents of this or that renewable energy source always emphasize the fact that the resource base is very large. Tony Batchelor, a speaker at a recent conference at St John's College, Oxford,[2] stated that available geothermal resources allow for a worldwide increase of geothermal use by a factor of 100. Similarly one can point to the fact that solar, wind and tide 'natural' resources are infinitely greater than the small amounts actually harnessed today. It is important however to sound a note of caution about what may be termed a 'physical fallacy'. The size of a natural resource is one thing; its transformation into usable energy is something else because economics, technology, physical constraints and environmental considerations then come into play. To give one example: the geothermal resources of Iceland may be very large but their use cannot exceed the volume of energy demand in that country. As trade is not an option, the surplus will remain unused.

Small initial shares of renewable types of energy in TPES simply tell us that high rates of growth sustained over many years will be necessary for these shares to be raised to significant levels.

To illustrate. Assume that TPES will grow annually at 1.8 percent (the recent historical trend) over the next 20 years, and the aggregate of wind/solar/tide/geothermal grows annually at the high rate of 15 percent over the same period. We thus find that the combined share of these four renewables would reach 5.73 percent in 2025. This is not a very high share despite our generous assumption of a 15 percent growth rate sustained for 20 years worldwide.

Differences between various sources of renewable energy mainly relate to costs and environmental impacts. Costs data are sparse and when available difficult to interpret partly because costs of the same type of renewable energy vary from location to location, different concepts are used by different sources, and the estimates are made at different times and in different currencies, sometimes US Dollars but more commonly Euros or Pounds.

Data in Vernier (2005) are as follows:

a) Photovoltaic: Photovoltaic electricity is expensive. The cost estimate is €0.4–0.7 per kWh for photovoltaic systems and €0.3–0.4 per kWh for solar power stations. These costs are a multiple of those of conventional electricity. (p.31)

b) Wind: €0.2–0.8 per kWh for autonomous systems partly because of equipment to transform the alternating current to continuous, store it in batteries and retransform it again to alternating for use in the relevant appliance. The costs are estimated at €0.04–0.08 per kWh for generation linked to a network. The lower limit of this range is close to the competitive threshold. (pp. 44–5)

---

2   'Energy…beyond oil', A Forum at St John's College, Oxford, 21 May 2005. See also Armstrong and Blundell (2005).

c) Hydro, tide and geothermal for electricity generation have a feature in common, namely very high capital and low operational costs. This is also a characteristic of nuclear energy. The implication is that assessments of the total costs per kWh are very sensitive to the rate of discount applied to capital costs. Should a hydroelectric structure be amortized over 20 or 50 years, to give but one example? Vernier assumes amortization over 15–30 years and then estimates that hydroelectricity costs, after the amortization period, fall to €0.02 per kWh. Unfortunately this does not tell us much about the huge capital costs involved. (p.65) Similarly, the investment costs of a tidal installation are very high and the operational costs almost negligibly low. A tidal barrage can survive century-long periods. Total costs per kWh may well range between €0.03 and €0.10. (p.68) Capital costs for geothermal energy are estimated at between €1200 and €2800 per kWh, a very wide range indeed. (p.123)

The St John's College panellists give the following cost estimates:

a) Photo voltaic: 'For the best system at well chosen sites the price per kWh is at present €0.25–0.65'. Michael Grätzel in Armstrong and Blundell (2005, p.14)
b) Wind: 'The price of onshore Wind Energy generated electricity is 3.4 p(ence)/kWh…It is variously estimated (the price for offshore Wind Power) to be 30%–100% greater than the price of onshore Wind Energy'. W.E Leithead in Armstrong and Blundell (2005, p.10)
c) Tidal: Electricity costs from one UK company are given as £0.18 per kWh in 2003, falling to £0.085 per kWh in 2005 and expected to fall to £0.06 per kWh in 2006 and £0.04 per kWh in 2010. The basis of the estimate computations is not given. (Dean Millar, speech at the St John's College Forum)
d) Geothermal: Estimates are $2000–4000 per kW installed for capital costs and $0.04–0.07 per kWh for production costs. (T. Batchelor, speech at the St John's College Forum)

Ethanol is the subject of a huge debate in the USA between those who believe that the costs are high and that the ratio of energy inputs in growing corn to the energy output in the ethanol produced is not favourable, and the powerful lobby of ethanol producers who deny these criticisms as outrageous. There is so much heat in this debate that it is impossible to find the truth. The fact that ethanol production in the USA benefits from excise tax exemptions and income tax breaks and that corn producers constitute a powerful political constituency must be taken into consideration in any assessment. I remain unconvinced that corn is the right crop for the production of ethanol. In France where ethanol is produced at 30 percent from wheat and 70 percent from beet, average costs (2004) are estimated by Vernier at €0.35–0.60 per litre against €0.35 per litre of automotive fuels (gasoline or diesel) assuming a crude oil price of $50 per barrel. If energy content is taken into consideration the cost comparison becomes even more unfavourable to ethanol.

The conclusions that may be derived from this information are that the case for solar is that photovoltaic offers the best economic solution in isolated places. Both wind and solar, optimistically, could perhaps become competitive in the future in some places thanks to expansion or technical progress. Wind, however, has a significant edge over solar in electricity generation in power stations. The economic advantages of hydro, tidal and geothermal electricity depend very much on how capital costs are treated. In these three cases operational costs are low and will emerge as competitive once the capital is amortized. Ethanol remains uncompetitive except perhaps in Brazil. The complication as regards ethanol is that it involves both external economies in agriculture and diseconomies on the environmental front.

Environmental impacts of different sources of renewable energy differ from one type to another. The commercial sources (solar, wind, tide, waves, geothermal and hydro) are said to have one feature in common in that they do not emit carbon dioxide. This is not absolutely true however. In some instances the auxiliary plants and appliances produce some emissions as is the case of back-up generators for example. For an accurate assessment it is important to consider each system as a whole, not only the core plant taken in isolation. It remains true that CO_2 emissions are minimal in these cases. Most non-commercial renewables emit carbon oxides, particles, nitrogen oxides, methane and other polluting materials through combustion. The growing of crops to produce ethanol yields nitrogen oxide emissions originating from the use of fertilizers.

The critics of wind energy complain about noise, the defilement of beautiful natural landscapes and the killing of birds. Building wind farms offshore solves the noise problem but many suitable locations happen to be beauty spots. The more serious impact on birds is in the paths of migration. Otherwise the number of birds killed by wind farms is minimal compared to devastation caused by other structures such as power lines, planes, tall buildings and glass surfaces, not to mention domestic cats. Solar has no significant side effects; geothermal energy involves some environmental problems when water or steam contains certain gases. There is also a risk of pollution of water sources from the disposal of geothermal waters containing salts and metals. The more serious environmental problems relate to hydroelectricity. Barrages can cause soil erosion downstream. They affect adversely both fauna and flora as they often deprive waters downstream of nutrients and silt. There are also considerable human problems when the hydroelectric project, as it sometime does, leads to the resettling of populations away from their traditional homes. Recall the plea of Nubians displaced by the Aswan High dam. Finally, the risks of accidents due to seismic shocks or a weakening of the natural rock supports in the foundation or the shoulders of a barrage are not negligible. The collapse of the Frejus barrage caused 423 deaths.

4.   Conclusion

The development of commercial renewable energy sources does not represent a serious threat for oil. The main output outside some micro applications is electricity and heat and the dominant fossil fuels in electricity/heat generation are coal and natural gas. The share of oil in these sectors is very small. The case in favour of commercial renewable energy sources is not that they are economically competitive. Broadly speaking, and there are exceptions of course, they are not. This becomes more evident if one were to treat capital costs in a hard nosed way. The case that supports their development is environmental, particularly as regards carbon emissions. Another case is the contribution that solar and wind can make in isolated locations where the opportunity costs of supplying electricity through long transmission lines are high. There is a case for geothermal where the resource is available with the right combination of heat and water, particularly where the output is low temperature heat.

The case for ethanol deserves dispassionate studies free from political advocacies and prejudices.

Those who argue in favour of these renewable energy sources for supply security reasons or in terms of the peak oil theory cannot convince. Current shares in total energy supplies are so small that a significant impact on supply security will have to wait decades and will require a very high rate of sustained growth over a long period of time. If oil production is to peak at a time of rising demand, renewables will not provide a solution to the problem. This is simply because the transport sector is captive to oil and renewables (except ethanol) are no substitute to petroleum products. The threat to oil can only come from a technological revolution in the transport sector.

Economic development is bound to have two opposite impacts on renewables. There will be an expansion of the use of commercial sources and a decline in the use of non-commercial ones. As the poor in developing countries become better off they will switch away from firewood and dung to kerosene. A displacement of oil on one front may be more than compensated for by an increase in the demand for oil in another area.

Policy makers concerned with climate change would be wise to allocate scarce funds to R&D for carbon sequestration rather than in subsidizing the development of some renewable energy. The impact on emissions of significant progress on the carbon sequestration front is likely to be greater than that of an expansion of subsidized wind farms and solar power stations. To emit less is a good thing; to take carbon out of the atmosphere may be a better option.

One can understand the attraction of water, wind and sun to our imagination. Yet, the attraction of 'poetic' energy should not distract from the need for rigorous reasoning particularly when policy making is involved.

References

Armstrong, F. and K. Blundell (eds) (2005), *Energy ... beyond Oil*, A Forum at St John's College, Oxford, 21 May 2005, pp.20.

Batchelor, T. (2005), Presentation on Geothermal Energy at the St. John's College, Oxford Forum on Energy ... beyond Oil.

Gül, T. and Stenzel, T. (2002), *Variability of Wind Power and other Renewables. Management Options and Strategies*, I.E.A. Working Paper.

I.E.A. (2005), *Renewables Information*, Paris: International Energy Agency.

Millar, D. (2005), Presentation on Wave and Tidal Energy Technologies at the St John's College, Oxford Forum on Energy beyond Oil.

Vernier, J., *Les Énergies Renouvelables, Paris:* Que sais-je? Presses Universitaires de France.

Wikipedia, the free encyclopedia, *http://en.wikipedia.prg/wiki/Renewable_energy*

INDEX